P9-DNT-428

THE EMOTIONAL BRAIN

The Mysterious Underpinnings of Emotional Life

JOSEPH LEDOUX

SIMON & SCHUSTER PAPERBACKS
New York London Toronto Sydney

SIMON & SCHUSTER PAPERBACKS
Rockefeller Center
1230 Avenue of the Americas
New York, New York 10020

For information about special discounts for bulk purchases,
please contact Simon & Schuster Special Sales:
1-800-456-6798 or business@simonandschuster.com.

Designed by Irving Perkins Associates

Manufactured in the United States of America

19 20

The Library of Congress has cataloged the hardcover edition as follows:

LeDoux, Joseph E.
The emotional brain: the mysterious underpinnings
of emotional life / Joseph LeDoux
p. cm.
Includes bibliographical references and index.
1. Emotions—Physiological aspects. 2. Neuropsychology.
QP401.L43 1996
152.4—dc20 96–24676
CIP

ISBN-13: 978-0-684-80382-1
ISBN-10: 0-684-80382-8
ISBN-13: 978-0-684-83659-1 (Pbk)
ISBN-10: 0-684-83659-9 (Pbk)

For the people who have had the greatest influence

on my emotional brain:

Nancy, Jacob, and Milo

and

Pris and Boo

CONTENTS

PREFACE

I FIRST STARTED WORKING on the brain mechanisms of emotion in the late 1970s. At that time very few brain scientists were interested in emotions. In the intervening years, and especially recently, the topic has begun to be fairly heavily investigated, and a good deal of progress has been made. I thought it was time to share some of this information with the general public.

The Emotional Brain provides an overview of my ideas about how emotions come from the brain. It is not meant as an all-encompassing survey of every aspect of how the brain produces emotions. It focuses on those issues that have interested me most, namely, issues about how the brain detects and responds to emotionally arousing stimuli, how emotional learning occurs and emotional memories are formed, and how our conscious emotional feelings emerge from unconscious processes.

I tried to write *The Emotional Brain* so that it would be accessible to readers not trained in science or versed in scientific jargon. But I also tried not to water down the science. I hope I've been successful in making the book readable and enjoyable for lay persons and scientists alike.

I'm extremely grateful to my family for tolerating me while I struggled to write this book. I owe much to my wife, Nancy Princenthal, for her tireless reading of endless drafts of my book proposal, and then of the chapters, and for her many useful suggestions. Our two boys, Jacob and Milo, kept my emotional brain in tip-top shape throughout.

Many students and postdoctoral researchers have helped greatly in my past and current research on emotions in the brain: Akira Sakaguchi, Jiro Iwata, Piera Chichetti, Liz Romanski, Andy Xagoraris, Christine Clugnet, Mike Thompson, Russ Phillips, Maria Morgan, Peter Sparks, Kevin LaBar, Liz Phelps, Keith Corodimas, Kate Melia, Xingfang Li, Michael Rogan, Jorge Armony, Greg Quirk, Chris Repa, Neot Doron, Gene Go, Gabriel Hui, Mian Hou, Beth Stutzmann, and Walter Woodson. I have also had some important collaborators, including Don Reis, David Ruggiero, Shawn Morrison, Costantino Iadecola, and Terry Milner at Cornell Medical School; David Servan-Schreiber and Jon Cohen at the University of Pittsburgh; Asla Pitkänen in Finland; and Chiye Aoki at NYU. And I will always be grateful to Claudia Farb for her many tangible and intangible contributions to my lab. Some of these people have had to do their work while I was writing the book. I apologize to them for being inaccessible, especially during those last days when it seemed that I might never finish. I also owe a great deal to Irina Kerzhnerman and Annette Olivero, who helped with many aspects of the final preparation of the book. Jorge Armony and Mian Hou assisted with the illustrations.

I also want to thank Mike Gazzaniga, my Ph.D. advisor, for showing me how to have fun while being a scientist, and for teaching me how to think about the mind. He encouraged me to write a book on emotions years before I actually got around to it. I'm also grateful to Don Reis, who took me into his lab as a postdoc, taught me neurobiology, and provided me with the resources I needed to start pursuing the brain mechanisms of emotion.

The Neuroscience Research Branch at the National Institute of Mental Health has generously funded my work. The research that this book is based on could not have been done without this support. New York University, especially the Faculty of Arts and Science Dean's Office, has also been very supportive. And I couldn't ask for better colleagues than those I have in the NYU Center for Neural Science.

Katinka Matson and John Brockman of Brockman, Inc., have been wonderful as literary agents. They were instrumental in helping me shape my proposal and in signing me up with Simon & Schuster, where I'm pleased to have had the opportunity to work with Bob Asahina, who made only good editorial suggestions. I wish him luck in his new job, which took him away just as the book went into production. Bob Bender, who took over, has been wonderful as well, and Johanna Li.

Some people go on sabbatical to write books. I'm now going on one to recover.

1
WHAT'S LOVE GOT TO DO WITH IT?

"Our civilization is still in a middle stage, scarcely beast, in that it is no longer guided by instinct, scarcely human in that it is not yet wholly guided by reason."

Theodore Dreiser, *Sister Carrie*[1]

MY FATHER WAS A butcher. I spent much of my childhood surrounded by beef. At an early age, I learned what the inside of a cow looks like. And the part that interested me the most was the slimy, wiggly, wrinkled brain. Now, many years later, I spend my days, and some nights, trying to figure out how brains work. And what I've wanted to know most about brains is how they make emotions.

You might think that this would be a crowded field of research. Emotions, after all, are the threads that hold mental life together. They define who we are in our own mind's eye as well as in the eyes of others. What could be more important to understand about the brain than the way it makes us happy, sad, afraid, disgusted, or delighted?

For quite some time now, though, emotion has not been a very popular topic in brain science.[2] Emotions, skeptics have said, are just too complex to track down in the brain. But some brain scientists, myself included, would rather learn a little about emotions than more about less interesting things. In this book, I'll tell you how far we've gotten. Skeptics be warned, we've gotten pretty far.

Of course, at some level, we know what emotions are and don't need scientists to tell us about them. We've all felt love and hate and fear and anger and joy. But what is it that ties mental states like these together into the bundle that we commonly call "emotions"? What makes this bundle so different from other mental packages, ones that we are less inclined to use the term "emotion" for? How do our emotions influence every other aspect of our mental life, shaping our perceptions, memories, thoughts, and dreams? Why do our emotions often seem impossible to understand? Do we have control over our emotions or do they control us? Are emotions cast in neural stone by our genes or taught to the brain by the environment? Do animals (other than human ones) have emotions, and if so do all species of animals have them? Can we have unconscious emotional reactions and unconscious emotional memories? Can the emotional slate ever be wiped clean, or are emotional memories permanent?

You may have opinions, and even strong ones, about the answers to some of these questions, but whether your opinions constitute scientifically correct answers can't be determined by intuitions alone. Occasionally, scientists turn everyday beliefs into facts, or explain the workings of intuitively obvious things with their experiments. But facts about the workings of the universe, including the one inside your head, are not necessarily intuitively obvious. Sometimes, intuitions are just wrong—the world seems flat but it is not—and science's role is to convert these commonsense notions into myths, changing truisms into "old wives' tales." Frequently, though, we simply have no prior intuitions about something that scientists discover—there is no reason why we should have deep-seated opinions about the existence of black holes in space, or the importance of sodium, potassium, and calcium in the inner workings of a brain cell. Things that are obvious are not necessarily true, and many things that are true are not at all obvious.

I view emotions as biological functions of the nervous system. I believe that figuring out how emotions are represented in the brain can help us understand them. This approach contrasts sharply with the more typical one in which emotions are studied as psychological states, independent of the underlying brain mechanisms. Psychological research has been extremely valuable, but an approach where emotions are studied as brain functions is far more powerful.

Science works by experimentation, which, by definition, involves the manipulation of some variables and the control of others. The brain is an enormously rich source of variables to manipulate. By studying emotion through the brain, we greatly expand opportunities for making new discoveries beyond what can be achieved with psychological experimentation alone. Additionally, studying the way emotion works in the brain can help us choose between alternative psychological hypotheses—there are many possible solutions to the puzzle of how emotions might work, but the only one we really care about is the one that evolution hit upon and put into the brain.

I got interested in how emotions come from brains one day in New England. It was the mid-1970s, and I was a graduate student doing my Ph.D. research at the State University of New York at Stony Brook. A decade earlier, my advisor, Mike Gazzaniga, had made a big splash with his thesis research involving the psychological consequences of split-brain surgery in humans, work that he had done at Cal Tech with the late Nobel Laureate Roger Sperry.[3]

Split-brain surgery is a procedure in which the nerve connections between the two sides or hemispheres of the brain are severed in an attempt to control very severe epilepsy.[4] A brand-new series of patients was being operated on at Dartmouth and the surgeon had asked Gazzaniga to study them.[5] We built a laboratory inside a camper-trailer attached to a pumpkin-colored Ford van, and frequently traveled from Long Island to see the patients at their homes in Vermont and New Hampshire.[6]

The earlier studies that Gazzaniga had done showed that when the brain is split, the two sides can no longer communicate with each other. And because language functions of the brain are usually in the left hemisphere, the person is only able to talk about things that the left hemisphere knows about. If stimuli are presented in such a way that only the right hemisphere sees them, the split-brain person is not able to verbally describe what the stimulus is. However, if you give the right hemisphere the opportunity to respond without having to talk, it becomes clear that the stimulus was registered. For example, if the left hand, which sends touch information to the right hemisphere, reaches into a bag of objects, it is able to sort through them and pull out the one that matches the picture seen by the right hemisphere. The right hemisphere can thus match the way the object feels with a memory of the way it looked a few moments earlier and pull out the correct one. The right hand can't do this because its touch information goes to the left hemisphere, which didn't see the object. In the split-brain patient, information put into one hemisphere remains trapped on that side of the brain, and is unavailable to the other side. Gazzaniga captured the essence of this remarkable situation in an early article on the topic called "One Brain—Two Minds."[7]

The split-brain experiment that set my scientific compass in the direction of emotion involved the presentation of stimuli with emotional connotations to the two half-brains of a special patient known as P.S.[8] He was special because unlike most previous patients of this type, he was able to read words in both hemispheres, although, as with the others, he could only speak through his left hemisphere. So when emotional stimuli were presented to the left hemisphere, P.S. could tell us what the stimulus was and how he felt about it—whether it signified something good or bad. When the same stimuli were presented to the right hemisphere, the speaking left hemisphere was unable to tell us what the stimulus was. However, the left hemisphere could accurately judge whether the stimulus seen by the right was good or bad. For example, when the right hemisphere saw the word "mom," the left hemisphere rated it as "good," and when the right side saw the word "devil," the left rated it as "bad."

The left hemisphere had no idea what the stimuli were. No matter how hard we pressed, the patient could not name the stimulus that had been presented to the right hemisphere. Nevertheless, the left hemisphere was consistently on the money with the emotional ratings. Somehow the emotional significance of the stimulus had leaked across the brain, even though the identity of the stimulus had not. The patient's conscious emotions, as experienced by his left hemisphere, were, in effect, being pushed this way and that by stimuli that he claimed to have never seen.

How did this occur? Most likely, the path taken by the stimulus through the right hemisphere forked. One branch brought the stimulus to parts of the right hemisphere that identify what the stimulus is. The split-brain surgery prevented the identification made by the right hemisphere from reaching the left. The other branch took the stimulus to parts of the right hemisphere that determine the emotional implications of the stimulus. The surgery did not prevent the transfer of this information over to the left side.

The left hemisphere, in other words, was making emotional judgments without knowing what was being judged. The left hemisphere knew the emotional outcome, but it did not have access to the processes that led up to that outcome. As far as the left hemisphere was concerned, the emotional processing had taken place outside of its realm of awareness (which is to say, had taken place unconsciously).

Split-brain surgery seemed to be revealing a fundamental psychological dichotomy—between thinking and feeling, between cognition and emotion. The right hemisphere was unable to share its thoughts about what the stimulus was with the left, but was able to transfer the emotional meaning of the stimulus over.

By the way, this work was not at all about the issue of possible hemisphere differences in emotion.[9] We were simply examining the kinds of information that could and could not flow between the hemispheres when the brain was split.

Freud of course told us long ago that the unconscious is the home of our emotions, which, he said, were often dissociated from normal thought processes. However, decades later, we still had little understanding of how this might take place, and whether it was true at all was often questioned. I set as my goal figuring how the brain processes the emotional meaning of stimuli, a goal that I have since pursued.

After completing my graduate work, I decided that the techniques available for studying the human brain were too limited and that I would never be able to understand the neural basis of emotion by studying humans. I therefore turned to studies of experimental animals, rats, for the purpose of trying to unlock the brain's emotional secrets. As important as the human split-brain observations were in getting me going on this topic, it has been the animal studies that have really shaped my view of the emotional brain.

This book will tell you what I've learned from my researching and thinking about brain mechanisms of emotions. It gives a scientific account of what emotions are, how they operate in the brain, and why they have such important influences on our lives.

Several themes about the nature of emotions will emerge and recur. Some of these will be consistent with your commonsense intuitions about emotions, whereas others will seem unlikely if not strange. But all of them, I believe, are well-grounded in facts about the brain, or at least in hypotheses that have been inspired by such facts, and I hope that you will hear them out.

- The first is that the proper level of analysis of a psychological function is the level at which that function is represented in the brain. This leads to a conclusion that clearly falls into the realm of the bizarre at first—that the word "emotion" does not refer to something that the mind or brain really has or does.[10] "Emotion" is only a label, a convenient way of talking about aspects of the brain and its mind. Psychology textbooks often carve the mind up into functional pieces, such as perception, memory, and emotion. These are useful for organizing information into general areas of research but do not refer to real functions. The brain, for example, does not have a system dedicated to perception. The word "perception" describes in a general way what goes on in a number of specific neural systems—we see, hear, and smell the world with our visual, auditory, and olfactory systems. Each system evolved to solve different problems that animals face. In a similar vein, the various classes of emotions are mediated by separate neural systems that have evolved for different reasons. The system we use to defend against danger is different from the one we use in procreation, and the feelings that result from activating these systems—fear and sexual pleasure—do not have a common origin. There is no such thing as the "emotion" faculty and there is no single brain system dedicated to this phantom function. If we are interested in understanding the various phenomena that we use the term "emotion" to refer to, we have to focus on specific classes of emotions. We shouldn't mix findings about different emotions all together independent of the emotion that they are findings about. Unfortunately, most work in psychology and brain science has done this.

- A second theme is that the brain systems that generate emotional behaviors are highly conserved through many levels of evolutionary history. All animals, including people, have to satisfy certain conditions to survive in the world and fulfill their biological imperative to pass their genes on to their offspring. At a minimum, they need to obtain food and shelter, protect themselves from bodily harm, and procreate. This is as true of insects and worms as it is of fish, frogs, rats, and people. Each of these diverse groups of animals has neural systems that accomplish these behavioral goals. And within the animal groups that have a backbone and a brain (fish, amphibians, reptiles, birds, and mammals, including humans), it seems that the neural organization of particular emotional behavioral systems—like the systems underlying fearful, sexual, or feeding behaviors—is pretty similar across species. This does not imply that all brains are the same. It instead means that our understanding of what it means to be human involves an appreciation of the ways in which we are like other animals as well as the ways in which we are different.
- A third theme is that when these systems function in an animal that also has the capacity for conscious awareness, then conscious emotional feelings occur. This clearly happens in humans, but no one knows for sure whether other animals have this capacity. I make no claims about which animals are conscious and which are not. I simply claim that when one of these evolutionarily old systems (like the system that produces defensive behaviors in the presence of danger) goes about its business in a conscious brain, emotional feelings (like being afraid) are the result. Otherwise, the brain accomplishes its behavioral goals in the absence of robust awareness. And absence of awareness is the rule of mental life, rather than the exception, throughout the animal kingdom. If we do not need conscious feelings to explain what we would call emotional behavior in some animals, then we do not need them to explain the same behavior in humans. Emotional responses are, for the most part, generated unconsciously. Freud was right on the mark when he described consciousness as the tip of the mental iceberg.

- The fourth theme follows from the third. The conscious feelings that we know and love (or hate) our emotions by are red herrings, detours, in the scientific study of emotions. This will surely be hard to swallow at first. After all, what is an emotion but a conscious feeling? Take away the subjective register of fear and there's not much left to a dangerous experience. But I will try to convince you that this idea is wrong—that there is much more than meets the mind's eye in an emotional experience. Feelings of fear, for example, occur as part of the overall reaction to danger and are no more or less central to the reaction than the behavioral and physiological responses that also occur, such as trembling, running away, sweating, and heart palpitations. What we need to elucidate is not so much the conscious state of fear or the accompanying responses, but the system that detects the danger in the first place. Fear feelings and pounding hearts are both effects caused by the activity of this system, which does its job unconsciously—literally, before we actually know we are in danger. The system that detects danger is the fundamental mechanism of fear, and the behavioral, physiological, and conscious manifestations are the surface responses it orchestrates. This is not meant to imply that feelings are unimportant. It means that if we want to understand feelings we have to dig deeper.
- Fifth, if, indeed, emotional feelings and emotional responses are effects caused by the activity of a common underlying system, we can then use the objectively measurable emotional responses to investigate the underlying mechanism, and, at the same time, illuminate the system that is primarily responsible for the generation of the conscious feelings. And since the brain system that generates emotional responses is similar in animals and people, studies of how the brain controls these responses in animals are a pivotal step toward understanding the mechanisms that generate emotional feelings in people. Studies of the neural basis of emotion in humans vary from difficult to impossible for both ethical and practical reasons. The study of experimental animals is, as a result, both a useful and a necessary enterprise if we are to understand emotions in the human brain. Understanding emotions in the human brain is

clearly an important quest, as most mental disorders are emotional disorders.

- Sixth, conscious feelings, like the feeling of being afraid or angry or happy or in love or disgusted, are in one sense no different from other states of consciousness, such as the awareness that the roundish, reddish object before you is an apple, that a sentence just heard was spoken in a particular foreign language, or that you've just solved a previously insoluble problem in mathematics. States of consciousness occur when the system responsible for awareness becomes privy to the activity occurring in unconscious processing systems. What differs between the state of being afraid and the state of perceiving red is not the system that represents the conscious content (fear or redness) but the systems that provide the inputs to the system of awareness. There is but one mechanism of consciousness and it can be occupied by mundane facts or highly charged emotions. Emotions easily bump mundane events out of awareness, but nonemotional events (like thoughts) do not so easily displace emotions from the mental spotlight—wishing that anxiety or depression would go away is usually not enough.
- Seventh, emotions are things that happen to us rather than things we will to occur. Although people set up situations to modulate their emotions all the time—going to movies and amusement parks, having a tasty meal, consuming alcohol and other recreational drugs—in these situations, external events are simply arranged so that the stimuli that automatically trigger emotions will be present. We have little direct control over our emotional reactions. Anyone who has tried to fake an emotion, or who has been the recipient of a faked one, knows all too well the futility of the attempt. While conscious control over emotions is weak, emotions can flood consciousness. This is so because the wiring of the brain at this point in our evolutionary history is such that connections from the emotional systems to the cognitive systems are stronger than connections from the cognitive systems to the emotional systems.
- Finally, once emotions occur they become powerful motivators of future behaviors. They chart the course of moment-to-

> moment action as well as set the sails toward long-term achievements. But our emotions can also get us into trouble. When fear becomes anxiety, desire gives way to greed, or annoyance turns to anger, anger to hatred, friendship to envy, love to obsession, or pleasure to addiction, our emotions start working against us. Mental health is maintained by emotional hygiene, and mental problems, to a large extent, reflect a breakdown of emotional order. Emotions can have both useful and pathological consequences.

As emotional beings, we think of emotions as conscious experiences. But when we begin probing emotion in the brain, we see conscious emotional experiences as but one part, and not necessarily the central function, of the systems that generate them. This does not make our conscious experiences of love or fear any less real or important. It just means that if we are going to understand where our emotional experiences come from we have to reorient our pursuit of them. From the point of view of the lover, the only thing important about love is the feeling. But from the point of view of trying to understand what a feeling is, why it occurs, where it comes from, and why some people give or receive it more easily than others, love, the feeling, may not have much to do with it at all.

Our journey into the emotional brain will take us down many different paths. We start with the curious fact that the study of emotion has long been ignored by the field of cognitive science, the major scientific enterprise concerned with the nature of the mind today (Chapter 2). Cognitive science treats minds like computers and has traditionally been more interested in how people and machines solve logical problems or play chess than in why we are sometimes happy and sometimes sad. We will then see that this shortcoming has been corrected in an unfortunate way—by redefining emotions as cold cognitive processes, stripping them of their passionate qualities (Chapter 3). At the same time though, cognitive science has been very successful, and has provided a framework that, when appropriately applied, provides an immensely valuable approach for pursuing the emotional as well as the cognitive mind. And one of the major conclusions about cognition and emotion that comes from this approach is that both seem to operate unconsciously, with only the outcome of cognitive or emotional processing entering awareness and occupying our conscious minds, and only in some instances.

The next stop along the way takes us into the brain, in search of the system that gives rise to our emotions (Chapter 4). We'll see that there is no single emotion system. Instead, there are lots of emotion systems, each of which evolved for a different functional purpose and each of which gives rise to different kinds of emotions (Chapter 5). These systems operate outside of consciousness and they constitute the emotional unconscious.

We then concentrate on one emotion system that has been extensively studied, the fear system of the brain, and see how it is organized (Chapter 6). The relation between unconscious emotional memory and conscious memories of emotional experiences is then discussed (Chapter 7). The breakdown of emotion systems, especially the fear system, is then considered (Chapter 8). We see how anxiety, phobias, panic attacks, and post-traumatic stress disorders emerge out of the depths of the unconscious workings of the fear system. Psychotherapy is interpreted as a process through which our neocortex learns to exercise control over evolutionarily old emotional systems. Finally, we explore the problem of emotional consciousness, and the relation of emotion to the rest of the mind (Chapter 9). I conclude with the hypothesis, based on trends in brain evolution, that the struggle between thought and emotion may ultimately be resolved, not simply by the dominance of neocortical cognitions over emotional systems, but by a more harmonious integration of reason and passion in the brain, a development that will allow future humans to better know their true feelings and to use them more effectively in daily life.

2
SOULS ON ICE

"Think, think, think."

Winnie the Pooh[1]

"Ahab never thinks, he just feels, feels, feels."

Herman Melville, *Moby-Dick*[2]

THE HUMAN BRAIN CONTAINS about 10 billion neurons that are wired together in enormously complex ways. Although the electrical sparks within and chemical exchanges between these cells accomplish some amazing and perplexing things, the creation of our emotions stands out as one of their most amazing and perplexing feats.

When we turn our mind's eye inward on our emotions, we find them at once obvious and mysterious. They are the states of our brain we know best and remember with the greatest clarity. Yet, sometimes we do not know where they come from. They can change slowly or suddenly, and their causes can be evident or opaque. We don't always understand what makes us wake up on the wrong side of the bed. We can be nice or nasty for reasons other than the ones we believe are guiding our actions. We can react to danger before we "know" we are in harm's way. We can be drawn toward the aesthetic beauty of a painting without consciously understanding what it is we like about it. Although our emotions are at the core of who we are, they also seem to have their own agenda, one often carried out without our willful participation.

It's hard to imagine life without emotions. We live for them, structuring circumstances to give us moments of pleasure and joy, and avoiding situations that will lead to disappointment, sadness, or pain. The rock critic Lester Bangs once said, "The only questions worth asking today are whether humans are going to have any emotions tomorrow, and what the quality of life will be if the answer is no."[3]

Scientists have had lots to say about what emotions are.[4] For some, emotions are bodily responses that evolved as part of the struggle to survive. For others, emotions are mental states that result when bodily responses are "sensed" by the brain. Another view is that the bodily responses are peripheral to an emotion, with the important stuff happening completely within the brain. Emotions have also been viewed as ways of acting or ways of talking. Unconscious impulses are at the core of an emotion in some theories, while others emphasize the importance of conscious decisions. A popular view today is that emotions are thoughts about situations in which people find themselves. Another notion is that emotions are social constructions, things that happen between rather than within individuals.

A scientific understanding of emotions would be wonderful. It would give us insight into how the most personal and occult aspects of the mind work, and at the same time would help us understand what may go wrong when this part of mental life breaks down. But, as the above comments indicate, scientists have not been able to agree about what an emotion is. The careers of many a scientist have been devoted to, if not devoured by, the task of explaining emotions. Unfortunately, one of the most significant things ever said about emotion may be that everyone knows what it is until they are asked to define it.[5]

This state of affairs might seem to pose a major stumbling block for our attempt to understand the emotional brain. If we can't say what emotion is, how can we hope to find out how the brain does it? But this book is not about mapping one area of knowledge (the psychology of emotion) onto another (brain function). It is instead about how studies of brain function allow us to understand emotion as a psychological process in new ways. I believe that we can get a unique and advantageous view of this puzzling part of the mental terrain by peering at it from inside the nervous system.

But I don't intend to ignore the psychology of emotion. Psychologists have had lots of insights. The problem is deciding which are correct and which are clever but wrong. Studies of the emotional brain can give us additional insights, but can also help us pick and choose from the psychological offerings. Aspects of the psychology of emotion are discussed in Chapter 3.

Our pursuit of the psychology of emotion, though, needs to be prefaced with an exploration of how emotion fits into a larger view of the mind—we need to delve into the nature of cognition, emotion's partner in the mind. The study of cognition, or just plain thinking, has advanced amazingly far in recent years. These advances provide a conceptual framework and methodology that is useful as an approach to all aspects of the mind, including emotion. The business of this chapter will therefore be to see what cognition is and how emotion and cognition relate.

Reason and Passion

Since the time of the ancient Greeks, humans have found it compelling to separate reason from passion, thinking from feeling, cognition from emotion. These contrasting aspects of the soul, as the Greeks liked to call the mind, have in fact often been viewed as waging an inner battle for the control of the human psyche. Plato, for example, said that passions and desires and fears make it impossible for us to think.[6] For him, emotions were like wild horses that have to be reined in by the intellect, which he thought of as a charioteer. Christian theology has long equated emotions with sins, temptations to resist by reason and willpower in order for the immortal soul to enter the kingdom of God. And our legal system treats "crimes of passion" differently from premeditated transgressions.

Given this long tradition of separation of passion and reason, it should not be too surprising that a field currently exists to study rationality, so-called cognition, on its own, independent of emotions. This field, known as cognitive science, tries to understand how we come to know our world and use our knowledge to live in it. It asks how we recognize a certain pattern of visual stimulation falling on the retina as a particular object, say an apple, or determine the apple's color, or judge which of two apples is bigger, or control our arm and hand in the act of catching an apple falling out of a tree, or remember where we were or who we were with when we last ate an apple, or imagine an apple in the absence of one, or tell or understand a story about an apple falling out of a tree, or conceive of a theory of why an apple falling out of a tree goes toward the earth instead of the sky.

Cognitive science emerged recently, around the middle of this century, and is often described as the "new science of mind."[7] However, in fact, cognitive science is really a science of only a part of the mind, the part having to do with thinking, reasoning, and intellect. It leaves emotions out. And minds without emotions are not really minds at all. They are souls on ice—cold, lifeless creatures devoid of any desires, fears, sorrows, pains, or pleasures.

Why would anyone want to conceive of minds without emotions? How could such a field focused on emotionless minds be so successful? How do we get emotion and cognition back together? To answer these questions we need to see where cognitive science came from and what it's all about.

Thinking Machines

Throughout much of the first half of this century, psychology was dominated by behaviorists, who believed that the subjective inner states of mind, like perceptions, memories, and emotions, are not appropriate topics for psychology.[8] In their view, psychology should not be the study of consciousness, as had been the case since Descartes said "Cogito, ergo sum,"[9] but instead should be the study of observable facts—objectively measurable behaviors. Being subjective and unobservable (except by introspection), consciousness could not, in the behaviorists' mind, be examined scientifically. Mental states came to be known pejoratively as "ghosts in the machine."[10] Behaviorists were known to ridicule those who dared to speak of mind and consciousness.

By mid-century, though, the behaviorist stronghold on psychology began to weaken.[11] Electronic computers had been developed,

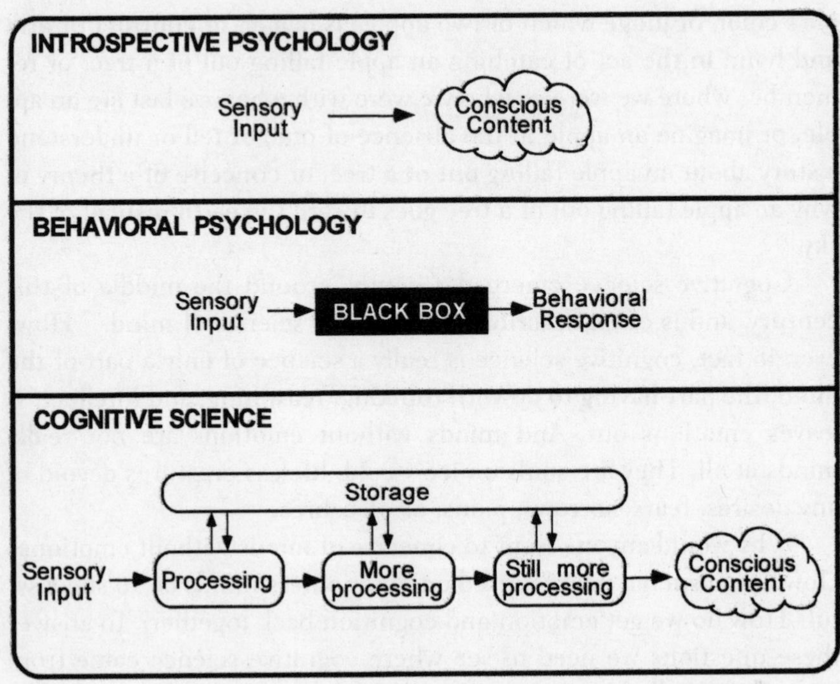

FIGURE 2-1

Three Approaches to the Science of Mind and Behavior.

Introspective psychology *is mainly concerned with the contents of immediate conscious experience.* Behaviorism *rejected consciousness as a legitimate subject matter for psychology and treated the events occurring between stimuli and responses as hidden in a black box.* Cognitive science *tries to understand the processes that occur inside the black box. These processes tend to occur unconsciously. In focusing on processes rather than conscious content, cognitive science did not exactly revive the view of the mind that the behaviorists rejected. More and more, however, cognitive scientists are beginning to try to understand the mechanisms of consciousness as well as the unconscious processes that sometimes do and sometimes do not give rise to conscious content.* (Bottom panel is based on figure 1 in U. Neisser [1976], *Cognition and Reality*. San Francisco: W.H. Freeman.)

and engineers, mathematicians, philosophers, and psychologists quickly saw similarities in the way computers process information and the way minds work. Computer operations became a metaphor for mental functions, and the field of artificial intelligence (AI), which seeks to model the human mind using computer simulations, was born. Pretty soon, anyone who bought into the notion of the mind as an information-processing device came to be known as a cognitive scientist. Cognitive science caused a revolution in psychology, dethroning the behaviorists and bringing the mind back home. But the impact of cognitive science reached far beyond psychology. Today, cognitive scientists can be found in linguistics, philosophy, computer science, physics, mathematics, anthropology, sociology, and brain science, as well as psychology.

One of the most important conceptual developments in the establishment of cognitive science was a philosophical position known as functionalism, which holds that intelligent functions carried out by different machines reflect the same underlying process.[12] For example, a computer and a person can both add 2 + 5 and come up with 7. The fact that both achieve the same answer cannot be explained by the use of similar hardware—brains are made of biological stuff and computers of electronic parts. The similar outcome must be due to a similar process that occurs at a functional level. In spite of the fact that the hardware in the machines is vastly different, the software or program that each executes may be the same. Functionalism thus holds that the mind is to the brain as a computer program is to the computer hardware.

Cognitive scientists, carrying the functionalist banner, have been allowed to pursue the functional organization of the mind without reference to the hardware that generates the functional states. According to functionalist doctrine, cognitive science stands on its own as a discipline—it does not require that we know anything about the brain. This logic was a shot in the arm to the field, giving it a strong sense of independence. Regardless of whether they do experiments on humans or use computer simulations of the human mind, many cognitive scientists today are functionalists.

It would be natural to presume that the cognitive revolution resulted in the return of consciousness as the number one topic of psychology. But this was not the case. The cognitive movement brought the mind back to psychology, but not exactly the all-knowing conscious mind that Descartes had popularized. For Descartes, if it wasn't conscious it wasn't mental: mind and consciousness became synonymous after him.[13] In contrast, as we'll soon see, cognitive scientists tend to think of the mind in terms of unconscious *processes* rather

FIGURE 2-2
Functionalism.

This is a philosophical position which proposes that mental functions (thinking, reasoning, planning, feeling) are functional rather than physical states. When a person and a computer add 2 to 5 and come up with 7, the similar outcome cannot be based on similar physical makeup, but instead must be due to a functional equivalence of the processes involved. As a result, it is possible to study mental processes using computer simulations. Minds might in principle even exist without bodies. (Based on J.A. Fodor, The Mind-Body Problem. *Scientific American* [January 1981], Vol. 244, p. 118.)

than conscious *contents.* And in leaving out consciousness, cognitive science left behind those conscious states called emotions. Later, we'll see why this occurred. For now, we want to explore the unconscious nature of cognitive processes.

The Cognitive Unconscious

Rooted in the idea of mind as an information processing device, cognitive science has been geared toward understanding the functional organization and processes that underlie and give rise to mental events, and much less toward understanding the nature of consciousness and its subjective contents. In order for you to consciously perceive an apple in front of you in space, the apple must be represented in your brain and that representation must be made available to the conscious part of your mind. But the mental representation of the apple that you consciously perceive is created by the unconscious turnings of mental gears. As Karl Lashley long ago pointed out, conscious content comes from processing, and we are never consciously aware of the processing itself but only of the outcome.[14] These mental processes are the bread and butter of cognitive science. Cognitive scientists sometimes speak of consciousness as the end result of processing, but are usually far more interested in the underlying processes than in the contents of consciousness that occur during and as a result of the processing. This emphasis on unconscious *processes* as opposed to conscious *content* underlies much work in cognitive science.[15] And for adherents of strong versions of functionalism these processes can be studied equally well in any device that can solve the functional problem at hand, regardless of whether the device is made of neurons, electrical components, mechanical parts, or sticks and stones.[16]

The psychologist John Kihlstrom coined the term "cognitive unconscious" to describe the subterranean processes that have been the main preoccupation of cognitive science.[17] These processes span many levels of mental complexity, all the way from the routine analysis of the physical features of stimuli by our sensory systems, to remembrance of past events, to speaking grammatically, to imagining things that are not present, to decision making, and beyond.

Like Freud before them, cognitive scientists reject the view handed down from Descartes that mind and consciousness are the same. However, the cognitive unconscious is not the same as the Freudian or dynamic unconscious.[18] The term cognitive unconscious merely implies that a lot of what the mind does goes on outside of consciousness, whereas the dynamic unconscious is a darker, more malevolent place where emotionally charged memories are shipped to do mental dirty work. To some extent, the dynamic unconscious can be conceived in terms of cognitive processes,[19] but the term cognitive unconscious does not imply these dynamic operations. We are going to also discuss the dynamic unconscious in some detail in later chapters. But for now we are focused on the tamer cognitive unconscious, which consists of processes that take care of the mind's routine business without consciousness having to be bothered. Let's consider some examples.

The first level of analysis of any external stimulus by the nervous system involves the physical properties of the stimulus. These low-level processes occur without awareness.[20] The brain has, for example, mechanisms for computing the shape, color, location, and movement of objects we see, and the loudness, pitch, and location of sounds we hear. If we are asked to say which of two objects is closer or which of two sounds is louder, we can do so, but we cannot explain what operations the brain performed to allow us to reach these conclusions. We have conscious access to the outcome of the computation but not to the computation itself. The processing of physical stimulus features makes possible all other aspects of perception, including our conscious awareness of perceiving something. It is just as well that we are unaware of these processes, as we would be so busy doing the computations that we would never get around to actually perceiving anything if we had to do it all with deliberate concentration.

On the basis of its analysis of physical features of stimuli, the brain begins to construct meaning. In order to know that the object you are looking at is an apple, the physical features of the stimulus have to find their way into your long-term memory banks. Once there, the stimulus information is matched up with stored information about similar objects and is classified as an apple, allowing you to "know" that you are looking at an apple and perhaps even leading you to remember past experiences you've had that involved apples. The end result is the creation of conscious memories (conscious contents) but through processes that you have little conscious access to. Presumably you can remember what you had for dinner last night, but you are not likely to be able to explain the machinations your brain went through to pull that information out.

Even that most ghostly of cognitions, the mental image, is the product of unconscious processes. For example, the cognitive psychologist Stephen Kosslyn asked subjects to draw an imaginary island that contained certain objects (tree, hut, rock, etc.).[21] The subjects were then asked to imagine the map and focus on one of the objects. A test word was then given and the subjects had to press a button to indicate whether the word named one of the objects on the map. The amount of time taken to press the button was directly related to the distance between the object named by the test word and the object being imagined. This suggested to Kosslyn that the brain actually computes geometric distances in mental images. But the subjects did not deliberately perform these calculations. They just gave the answers by pressing a button. All the work was done by the brain operating unconsciously.

Just because your brain can do something does not mean that "you" know how it did it. If it seems odd that the brain can unconsciously solve geometric problems, imagine the kinds of automatic calculations that go on in the brain when we turn the steering wheel to navigate a curve at 60 mph, or better yet, the kinds of processes that go on in the nervous system of homing pigeons or honeybees as they fly out into the world in search of food and then effortlessly find their way home using an internal compass.

Speech, consciousness' favorite behavioral tool, is also the product of unconscious processes.[22] We do not consciously plan the grammatical structure of the sentences we utter. There simply isn't enough time. We aren't all great orators, but we usually say things that make sense linguistically. Speaking roughly grammatically is one of the many things that the cognitive unconscious takes care of for us.

The cognitive unconscious also extends to complex judgments about the mental origins of beliefs and actions. In 1977, Richard Nisbett and Timothy Wilson published an extremely interesting paper, "Telling More Than We Can Know: Verbal Reports on Mental Processes."[23] They created a number of carefully structured experimental situations in which people were required to do things and then say why they did what they did. In one study, they lined up several pairs of stockings on a table. Female subjects were then allowed to examine the stockings and to choose which one they liked best.

When the women were questioned, they had all sorts of wonderful answers about the texture and sheerness of the stockings that justified their choices. But unbeknownst to them, the stockings were identical. The subjects believed that they had decided on the basis of their internal judgments about the quality of the stockings. In this and a host of other studies, Nisbett and Wilson showed that people are often mistaken about the internal causes of their actions and feelings. Although the subjects always gave reasons, the reasons came not from privileged access to the processes that underlay their decisions, but from social conventions, or ideas about the way things normally work in such situations, or just plain guesses. Accurate introspective reports, Nisbett and Wilson say, often occur in life because the stimuli involved in causing the behavior or the belief are salient and plausible causes of these. But when salient and plausible stimuli are not available, people make up reasons and believe in them. In other words, the inner workings of important aspects of the mind, including our own understanding of why we do what we do, are not necessarily knowable to the conscious self.[24] We have to be very careful when we use verbal reports based on introspective analyses of one's own mind as scientific data.

Around the same time that Nisbett and Wilson were performing their studies, Michael Gazzaniga and I were engaged in studies of split-brain patients that led us to a similar conclusion.[25] It was well-known from earlier work by Gazzaniga and others that information presented exclusively to one hemisphere of a split-brain patient is unavailable to the other.[26] We capitalized on this as a model of how consciousness deals with information generated by an unconscious mental system. In other words, we secretly instructed the right hemisphere to perform some response. The left hemisphere observed the response but did not know why the response was performed. We then asked the patient why he did what he did. Since only the left hemisphere could talk, the verbal output reflected that hemisphere's understanding of the situation. Time after time, the left hemisphere made up explanations as if it knew why the response was performed. For example, if we instructed the right hemisphere to wave, the patient would wave. When we asked him why he was waving, he said he thought he saw someone he knew. When we instructed the right hemisphere to laugh, he told us that we were funny guys. The spoken explanations were based on the response produced rather than knowledge of why the responses were produced. Like Nisbett and Wilson's subjects, the patient was attributing explanations to situations as if he had introspective insight into the cause of the behavior when in fact he did not. We concluded people normally do all sorts of things for reasons they are not consciously aware of (because the behavior is produced by brain systems that operate unconsciously) and that one of the main jobs of consciousness is to keep our life tied together into a coherent story, a self-concept. It does this by generating explanations of behavior on the basis of our self-image, memories of the past, expectations of the future, the present social situation, and the physical environment in which the behavior is produced.[27]

Although a good deal remains uncertain about the cognitive unconscious,[28] it seems clear that much of mental life occurs outside of conscious awareness. We can have introspective access to the outcome of processing (in the form of conscious content), but not all processing gives rise to conscious content. Stimulus processing that does not reach awareness in the form of conscious content can nevertheless be stored implicitly or unconsciously (see chapter 7) and have important influences on thought and behavior at some later time.[29] Further, it is worth emphasizing that information can be simultaneously processed separately by systems that do and do not give rise to conscious content, leading to the conscious representation in some and the unconscious representation in other systems. We may sometimes be able to introspect and verbally describe the workings of the systems that create and use conscious representations, but introspection is not going to be very useful as a window into the workings of the vast unconscious facets of the mind. This will be an especially important point when we consider the emotional unconscious in the next chapter.

The field of cognitive science has been incredibly successful in its stated mission of understanding information processing, which turns out to mean the unconscious processing of information. We now have excellent models of how we perceive the world in an orderly fashion, remember events from the past, imagine stimuli that are not present, focus our attention on one stimulus while ignoring many others, solve logical problems, make decisions on the basis of incomplete information, make judgments about our beliefs, attitudes, and behaviors, and many other aspects of mental functioning.[30] That much of the processing involved in these functions occurs unconsciously has allowed cognitive science a luxury that earlier forms of mentalism did not have—the field could get on with the business of studying the mind without having to first solve the problem of consciousness.[31] This does not mean that consciousness is irrelevant or unimportant. It is so important that when it has come up in the past it has completely dominated the scientific pursuit of the mind. This time around, though, scientists have figured out that the unconscious aspects of the mind are also important. In fact, it is probably not too far off the mark to say that consciousness will only be understood by studying the unconscious processes that make it possible. In this regard, cognitive science seems right on track. We'll return to the topic of consciousness, and especially emotional consciousness, in Chapter 9.

The Mental Health of Machines

The *cognitive mind* (the mind being studied by cognitive scientists) can do some very interesting and complicated things. For example, it can play chess so well that real grand masters can be given a run for their money.[32] But the cognitive mind, when playing chess, does not feel driven to win. It doesn't enjoy putting its partner in checkmate, or feel saddened or annoyed if it loses a match. It is not distracted by the presence of an audience at a big game, by sudden anxiety over the realization that a mortgage payment is late, or by the need to go to the little chip's room. The cognitive mind can even be programmed to cheat at chess, but not to feel guilty when it does.

As one thumbs through some of the attempts to define cognitive science it is striking how often this field is characterized by saying that it is not about emotion. For example, in *The Mind's New Science: A History of the Cognitive Revolution,* Howard Gardner lists the deemphasis of affective or emotional factors as one of five defining features of cognitive science.[33] In his seminal 1968 textbook, *Cognitive Psychology,* Ulric Neisser states that the field is not about the dynamic factors (like emotions) that motivate behavior.[34] Jerry Fodor, in *The Language of Thought,* a groundbreaking book in the philosophy of cognitive science, describes emotions as mental states that fall outside the domain of cognitive explanation.[35] And Barbara von Eckardt, in a book titled *What Is Cognitive Science?* says that most cognitive scientists do not consider the study of emotions to be part of the field.[36] These cognitive scientists each pointed out that emotional factors are important aspects of the mind, but also emphasized that emotions are just not part of the cognitive approach to the mind.

What is it about emotion that has compelled cognitive scientists to separate it out from attention, perception, memory, and other bona fide cognitive processes? Why was emotion banned from the rehabilitation of the mind that took place in psychology's cognitive revolution?

For one thing, as we have seen, philosophers and psychologists have for millennia found it useful to distinguish thinking and feeling, cognition and emotion, as separate facets of mind. And following the work of philosophers like Bertrand Russell[37] in the early twentieth century, thinking came to be viewed as a kind of logic, now known, thanks to Fodor, as the language of thought.[38] When the computer metaphor came along, it was seen as more applicable to logical reasoning processes than to so-called *illogical* emotions. But, as we will see, cognition is not as logical as it was once thought and emotions are not always so illogical.

AI researchers realized early on that knowledge was needed in problem-solving machines—problem solvers with impeccable logic but without facts didn't get very far.[39] However, knowledge was a crutch to logic in these models. It is now believed that thinking does not normally involve the pure reasoned rules of logic.[40] This has been demonstrated in research by Philip Johnson-Laird.[41] He examined people's ability to draw logical conclusions from statements like: all artists are beekeepers, all beekeepers are chemists. He found that quite often people draw logically invalid conclusions, suggesting that if the human mind is a formal logic machine, it is a pretty poor one. People are rational, according to Johnson-Laird, they just don't achieve their rationality by following formal laws of logic. We use what Johnson-Laird calls mental models, hypothetical examples drawn from our past experiences in real life or from imagined situations. Other studies by Amos Tversky and Daniel Kahneman led to a similar view, but from a different angle.[42] They showed that people use their implicit understanding of the way the world works, often relying on educated guesswork rather than formal principles of logic, to solve the problems that they face in their daily lives. Economist Robert Frank, however, goes further.[43] He argues that decision making is often not rational at all: "Many actions, purposely taken with full knowledge of their consequences, *are* irrational. If people did not perform them, they would be better off and they know it." He cites examples such as battling endless red tape to get a small refund on a defective product or weathering a snowstorm to cast a ballot that will on its own have little impact in a race. Jorge Luis Borges' description of the British and Argentine battle over the Falkland Islands, quoted by Frank, says it all: "two bald men fighting over a comb." If cognition is not just logic, and is sometimes illogical, then emotion might not be as far afield from cognition as it was initially thought.

Many emotions are products of evolutionary wisdom, which probably has more intelligence than all human minds together. The evolutionary psychologists John Tooby and Leda Cosmides say that the species' past goes a long way toward explaining the individual's present emotional state.[44] What is irrational about responding to danger with evolutionarily perfected reactions? Daniel Goleman gives lots of examples of emotional intelligence in his recent book.[45] Success in life, according to Goleman, depends on a high EQ (emotional quotient) as much or more than a high IQ. True, derailed emotions can lead to irrational and even pathological consequences, but emotions themselves are not necessarily irrational. Aristotle, for example, saw anger as a reasonable response to an insult, and a number of philosophers have taken this view.[46] Antonio Damasio, a neurologist, also stresses the rationality of emotion in his book *Descartes' Error.*[47] He emphasizes the importance of *gut feelings* in making decisions. And while early AI programs were most successful at modeling logical processes, more recent models have gone far beyond this truly artificial approach and some even try to model aspects of emotions. Some programs use emotional *scripts* or *schemas* (built-in information that suggests what is likely to happen in certain situations: for example, in baseball games, classrooms, business meetings) as aids to decision making and action, others try to simulate the processes through which people evaluate or appraise the emotional meanings of stimuli, and still others attempt to make use of our understanding of the emotional brain in order to model how emotions are processed.[48] The logical/illogical or rational/irrational distinction is not a very sharp one when it comes to separating emotion and cognition, and is certainly not a clean way of defining what a science of mind should be about.

The second reason why emotion was not rehabilitated in the cognitive revolution may have been because emotions have traditionally been viewed as subjective states of consciousness. To be afraid, angry, or happy is to be aware that you are having a particular kind of experience, to be conscious of that experience. Computers process information rather than have experiences (at least by most people's way of thinking). To the extent that cognitive science was the science of information processing, rather than a science of conscious content, then emotion, being an aspect of consciousness, did not necessarily fit comfortably in the program. Recently, though, as we'll see in Chapter 9, consciousness has come to be more and more a part of cognitive science. Consequently, the excuse that emotions are subjective states loses much of its appeal. But the subjective argument should have never carried much weight. There is really nothing more or less subjective about the experience of an emotion than about the experience of the redness of an apple or the memory of eating one. The study of visual perception or memory has not been held back simply because these brain functions have subjective correlates, and neither should the study of emotion.

As we will see in the next chapter, subjective emotional states, like all other states of consciousness, are best viewed as the end result of information processing occurring unconsciously. Just as we can study how the brain processes information unconsciously in perceiving visual stimuli and using visual information to guide behavior, we can study how the brain processes the emotional significance of stimuli unconsciously and uses this information to control behaviors appropriate to the emotional meaning of the stimuli. And just as we hope that studying how the brain processes visual stimuli will help us understand how it creates the accompanying subjective perceptual experiences, we hope that studying how the brain processes emotional information will help us understand how it creates emotional experiences. This does not mean that we will program computers to have these experiences. Instead, it means we can use information-processing ideas as the conceptual apparatus for understanding conscious experiences, including subjective emotional feelings, even if such experiences are themselves not computational states of computers.[49] More about this when we get to consciousness in Chapter 9.

So, emotion could have fit into the cognitive framework. The question is whether it should have been included in cognitive science, or, more to the point, whether the boundaries of cognitive science should now be expanded to include emotion, placing all of the mind under one big conceptual tent.

All along some cognitive scientists have recognized that emotion is important. AI pioneer Herbert Simon,[50] for example, argued in the early 1960s that cognitive models needed to account for emotions in order to approximate real minds, and around the same time social psychologist Robert Abelson[51] suggested that the field of cognitive psychology needed to turn toward "hot cognitions," as opposed to the "cold" logical processes that it had been focusing on. Philip Johnson-Laird and George Miller, two leading cognitive psychologists, made a similar point in the 1970s.[52] And recently, Alan Newell, another AI pioneer, writing about emotions, noted, "no satisfactory integration yet exists of these phenomena into cognitive science. But the mammalian system is clearly constructed as an emotional system."[53] These suggestions by leading cognitive scientists have finally begun to have an impact—more and more cognitive scientists are getting interested in emotions. The problem is, instead of heating up cognition, this effort has turned emotion cold—in cognitive models, emotions, filled with and explained by thoughts, have been stripped of passion (we're going to go into the cognitive theory of emotion and its unfortunate consequences in great detail in the next chapter).

In the final analysis, then, the processes that underlie emotion and cognition can be studied using the same concepts and experimental tools. Both involve unconscious information processing and the generation of conscious content (sometimes) on the basis of this processing. At the same time, though, it does not quite seem right that emotion should be subsumed under cognitive science. The experimental study of the mind should be done in a framework that conceives of the mind in its full glory. The artificial separation of cognition from the rest of the mind was very useful in the early days of cognitive science and helped establish a new approach to the mind. But now it is time to put cognition back into its mental context—to reunite cognition and emotion in the mind. Minds have thoughts as well as emotions and the study of either without the other will never be fully satisfying. Ernest Hilgard, an eminent psychologist, makes the point nicely when he says that sibling rivalry is as important a concept to child development as is the maturation of thought processes.[54] "Mind science" is the natural heir to the united kingdom of cognition and emotion. To call the study of cognition and emotion cognitive science is to do it a disservice.

Minds, Bodies, Emotions

The idea of what the mind is has changed a number of times since the early Greeks, many of whom were preoccupied with rationality, but tended to view the mind as having both knowable and unknowable facets. Descartes redefined the mind to include only what we are aware of, making mind and consciousness the same thing. Since consciousness was viewed as a unique human gift, other animals were treated as mindless creatures. Freud, in formalizing the unconscious as the home of primitive instincts and emotions, helped reestablish a mental link between animals and humans, and began to dethrone consciousness as the sole occupant of the mind. The behaviorists dismissed the whole idea of mind, and took a step that really put animals and people on the same continuum, but one involving behavioral rather than mental functions. Cognitive science resurrected the Greek idea of mind, mind as reason and logic. And because the kind of mental states that were being suggested in the earlier days were based on the rules of logic, which is closely tied up with the human capacity for language, cognitive science was, for some time, not very friendly to the idea of animal minds. The idea of the human mind as a carefully engineered machine seemed more appealing than the idea of the mind as a biological organ with an evolutionary history.

The emergence of ideas about unconscious processing, and the re-realization that mind is more than cognition, again puts major parts of the mental life of humans and other animals on a continuum and encourages cognitive scientists to study mental functions in the context of the machine in which the functions are housed rather than as complete abstractions. Reacting to the functionalist credo that the mind can be modeled independent of knowledge of how the brain works, philosopher Patricia Churchland and computational neuroscientist Terrence Sejnowski have argued, "Nature is more ingenious than we are. And we stand to miss all that power and ingenuity unless we attend to neurobiological plausibility. The point is, *evolution has already done it,* so why not learn how that stupendous machine, our brain, actually works?"[55]

The functionalist conception of mind as a program that can run on any machine (mechanical, electronic, biological) has been fairly easy to accept, or at least tolerate, in the area of cognition. The biological machine of relevance to cognition, of course, is the brain. And the idea that the brain is a cognitive computer is now commonplace. However, in emotions, unlike in cognitions, the brain does not usually function independently of the body. Many if not most emotions involve bodily responses.[56] But no such relation exists between cognitions and actions. In the case of cognitively driven responses, the response is arbitrarily linked to cognition. This is partly why cognition is so powerful—cognitions allow us to be flexibile, to choose how we will respond in a certain situation. Such responses are used by but are not essential to the cognition. The capacity to understand language, one of man's highest forms of cognition, and the form of cognition most closely tied to a specific set of expressive responses, works just fine in people who live their lives without being able to express this capacity in speech. In the case of emotion, though, the response of the body is an integral part of the overall emotion process. As William James, the father of American psychology, once noted, it is difficult to imagine emotions in the absence of their bodily expressions.[57]

We know our emotions by their intrusions (welcome or otherwise) into our conscious minds. But emotions did not evolve as conscious feelings. They evolved as behavioral and physiological specializations, bodily responses controlled by the brain, that allowed ancestral organisms to survive in hostile environments and procreate. If the biological machine of emotion, but not cognition, crucially includes the body, then the kind of machine that is needed to run emotion is different from the kind needed to run cognition. Even if the functionalist argument (that the hardware is irrelevant) could be accepted for mind as cognition (and it is not clear that it can), it would not seem to work for the emotional aspects of the mind (since the hardware does seem to make a difference when it comes to emotion).

Programming a computer to be conscious would be an essential first step toward programming it to have a full-blown emotional experience, since the feelings through which we know our emotions occur when we become conscious of the unconscious workings of emotional systems in the brain. However, even if a computer could be programmed to be conscious, it could not be programmed to have an emotion, as a computer does not have the right kind of composition, which comes not from the clever assembly of human artifacts but from eons of biological evolution.

3
BLOOD, SWEAT, AND TEARS

"My love was so hot as mighty nigh to burst my boilers."

Davy Crockett, *A Narrative of the Life of David Crockett.*[1]

IN SPITE OF THE benign neglect of the topic of emotion by cognitive science, scientists who study emotion have by no means ignored cognition. In fact, psychologists interested in emotion, seduced by the intellectual excitement and appeal of cognitive science, have for some time been preoccupied with attempts to explain emotions in terms of cognitive processes. By this way of thinking, an emotion is not different from a cognition—emotions are just thoughts about situations we find ourselves in. Although this approach has had its share of successes, these have come at a high price. In trading in the passion of an emotion for thoughts about it, cognitive theories have turned emotions into cold, lifeless states of mind. Lacking sound and fury, emotions as cognitions signify nothing, or at least nothing very emotional. Our emotions are full of blood, sweat, and tears, but you wouldn't know this from examining modern cognitive research on emotion. Emotion research wasn't always this way, so let's see how and why the transformation occurred.

Body Heat

Why do we run away if we notice that we are in danger? Because we are afraid of what will happen if we don't. This obvious (and incorrect) answer to a seemingly trivial question has been the central concern of a century-old debate about the nature of our emotions.

It all began in 1884 when William James published an article titled "What Is an Emotion?"[2] The article appeared in a philosophy journal called *Mind*, as there were no psychology journals yet. It was important, not because it definitively answered the question it raised, but because of the way in which James phrased his response. He conceived of an emotion in terms of a sequence of events that starts with the occurrence of an arousing stimulus and ends with a passionate feeling, a conscious emotional experience. A major goal of emotion research is still to elucidate this stimulus-to-feeling sequence—to figure out what processes come between the stimulus and the feeling.

James set out to answer his question by asking another: do we run from a bear because we are afraid or are we afraid because we run? He proposed that the obvious answer, that we run because we are afraid, was wrong, and instead argued that we are afraid because we run:

> Our natural way of thinking about . . . emotions is that the mental perception of some fact excites the mental affection called emotion, and that this latter state of mind gives rise to the bodily expression. My thesis on the contrary is that the bodily changes follow directly the PERCEPTION of the exciting fact, and that our feeling of the same changes as they occur IS the emotion.[3]

FIGURE 3-1
The Stimulus-to-Feeling Sequence.

Identification of the processes that intervene between the occurrence of an emotion-arousing stimulus and the conscious emotions (feelings) it elicits has been one of the major goals of emotion research. Unfortunately, this goal has often been pursued to the exclusion of some other equally important goals.

The essence of James' proposal was simple. It was premised on the fact that emotions are often accompanied by bodily responses (racing heart, tight stomach, sweaty palms, tense muscles, and so on) and that we can sense what is going on inside our body much the same as we can sense what is going on in the outside world. According to James, emotions feel different from other states of mind because they have these bodily responses that give rise to internal sensations, and different emotions feel different from one another because they are accompanied by different bodily responses and sensations. For example, when we see James' bear, we run away. During this act of escape, the body goes through a physiological upheaval: blood pressure rises, heart rate increases, pupils dilate, palms sweat, muscles contract in certain ways. Other kinds of emotional situations will result in different bodily upheavals. In each case, the physiological responses return to the brain in the form of bodily sensations, and the unique pattern of sensory feedback gives each emotion its unique quality. Fear feels different from anger or love because it has a different physiological signature. The mental aspect of emotion, the feeling, is a slave to its physiology, not vice versa: we do not tremble

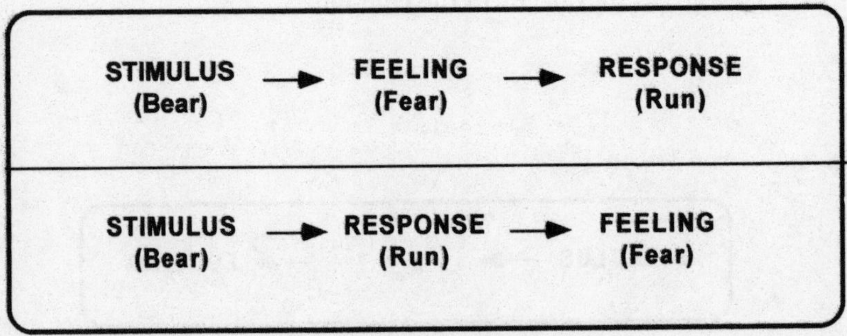

FIGURE 3-2
William James' Two Chains of Emotion.

The modern era in emotion research began when James asked whether feelings cause emotional responses or responses cause feelings. In answering that responses cause feelings, he started a century-old debate about where feelings come from. The question of what causes responses in the first place has, unfortunately, often been ignored.

FIGURE 3-3
James' Feedback Theory.

James' solution to the stimulus-to-feeling sequence problem was that feedback from responses determine feelings. Since different emotions have different responses, the feedback to the brain will be different and will, according to James, account for how we feel in such situations.

because we are afraid or cry because we feel sad; we are afraid because we tremble and sad because we cry.

Fight or Flight

James' theory dominated the psychology of emotion until it was called into question in the 1920s by Walter Cannon, a prominent physiologist who had been researching the bodily responses that occur in states of hunger and intense emotion.[4] Cannon's research led him to propose the concept of an "emergency reaction," a specific physiological response of the body that accompanies any state in which physical energy must be exerted. According to Cannon's hypothesis, the flow of blood is redistributed to the body areas that will be active during an emergency situation so that energy supplies, which are carried in the blood, will reach the critical muscles and organs. In fighting, for example, the muscles will need energy more than the internal organs (the energy used for digestion can be sacrificed for the sake of muscle energy during a fight). The emergency reaction, or "fight-or-flight response," is thus an adaptive response that occurs in anticipation of, and in the service of, energy expenditure, as is often the case in emotional states.

The bodily responses that make up the emergency reaction were believed by Cannon to be mediated by the sympathetic nervous system, a division of the autonomic nervous system (ANS). The ANS is a web of neural cells and fibers located in the body that controls the activity of the internal organs and glands, the so-called internal milieu, in response to commands from the brain. The characteristic bodily signs of emotional arousal—like pounding hearts and sweaty palms—were known in Cannon's day to be the result of the activation of the sympathetic division of the ANS, which was believed to act in a uniform way, regardless of how or why it was activated. Given this supposed singularity of the sympathetic response mechanism, Cannon proposed that the physiological responses accompanying different emotions should be the same regardless of the particular emotional state that is experienced. As a result, James could not be right about why different emotions feel different, since all emotions, according to Cannon, have the same ANS signature.[5] Cannon also noted that ANS responses are too slow to account for feelings—we're already feeling the emotion by the time these responses occur. So even if different emotions had different bodily signatures, these would be too slow to account for whether we feel love, hate, fear, joy, anger, or disgust in a particular situation. The answer to the riddle of emotion, according to Cannon, is found completely within the brain, and does not require that the brain "read" the bodily response, as James had said.[6] We'll discuss the neural views espoused by James and Cannon in the next chapter, and we'll return to the issue of bodily feedback contributions to emotional experience in Chapter 9.

Cannon felt that while bodily feedback could not account for differences between emotions, it nevertheless played an important role, giving emotions their characteristic sense of urgency and intensity. Although James and Cannon disagreed about what distinguishes different emotions, they would seem to have agreed that emotions feel different from other (nonemotional) states of mind because of their bodily responses.

Passions as Reasons

During the behaviorists' reign in psychology, emotions, like other mental processes, were treated as ways of acting in certain situations.[7] There was little or no effort to explain what gives rise to conscious emotional experiences, as these were not recognized as legitimate phenomena for scientific investigation. The stimulus-to-feeling sequence was simply not an issue. In fact, the concept of emotion as a subjective state was often singled out by behaviorists as a prime example of the kind of fuzzy idea that needed to be dispensed with in a scientific psychology. It was one of the prime mental fictions, ghosts in the machine, created by psychologists to overcome their ineptness at explaining behavior.[8]

In the early 1960s, though, all this began to change. Stanley Schachter and Jerome Singer, social psychologists at Columbia University, revived the issue of where our feelings come from and proposed a new solution to the James-Cannon debate.[9] Like James, Schachter and Singer suggested that bodily arousal or feedback was indeed crucial in the genesis of an emotional experience, but not quite in the way that James had proposed. And, like Cannon, they believed that physiological feedback lacked specificity. Riding the tide of the cognitive revolution, which had begun to penetrate deep into the heart and soul of most areas of psychology by this time, they argued that cognitions (thoughts) fill the gap between the nonspecificity of feedback and the specificity of felt experiences.

Schachter and Singer started with the assumption that physiological responses in emotion (sweaty palms, rapid heart beat, muscle tension) inform our brain that a state of heightened arousal exists. However, since these responses are similar in many different emotions they do not identify what kind of aroused state we are in. Schachter and Singer suggested that, on the basis of information about the physical and social context in which we find ourselves, as well as knowledge about what kinds of emotions occur in these particular kinds of situations, we label the aroused state as fear or love or sadness or anger or joy. According to Schachter and Singer, this labeling of the aroused state is what gives rise to and accounts for the specificity of felt emotion. In other words, emotional feelings result when we explain emotionally ambiguous bodily states to ourselves on the basis of cognitive interpretations (so-called attributions) about what the external and internal causes of the bodily states might be.

The major prediction from the Schachter-Singer theory was that if ambiguous physiological arousal was induced in human subjects it should be possible to bias the kind of emotion experienced by arranging the social context in which the arousal occurs. Schachter and Singer tested this hypothesis by giving subjects injections of adrenaline, a drug that induces physiological arousal by artificially activating the sympathetic division of the ANS. The subjects were then exposed to either a pleasant, unpleasant, or emotionally neutral situation. As predicted, mood varied in accord with the context for the subjects given adrenaline but not for the control group that received placebo injections: adrenaline-treated subjects exposed to a joyful situation came out feeling happy, those exposed to an unpleasant situation came out feeling sad, and the neutral ones felt nothing in particular. Specific emotions were produced by the combination of artificial arousal and social cues. By inference, then, when emotionally ambiguous physiological arousal occurs naturally in the presence of real emotional stimuli, the aroused feeling is labeled on the basis of social cues. Emotions, in short, result from the cognitive interpretation of situations.

Stuart Valins, another social psychologist, performed a series of experiments to try to elucidate the nature of the cognition-arousal-emotion interaction.[10] Subjects were given inaccurate information about how their body was responding to some situation. For example, Valins showed male subjects pictures of partially nude women. The subjects were at the same time listening to a sound that was supposed to be indicative of the rate at which their heart was beating. Valins manipulated the sounds independent of true heart beat so that some

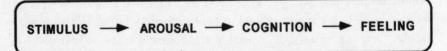

FIGURE 3-4
The Schachter-Singer Cognitive Arousal Theory.

Schachter and Singer, like Cannon, accepted that feedback is not specific enough to determine what emotion we feel in a given situation, but, like James they felt it was still important. Their idea was that feedback from bodily arousal is a good indicator that something significant is going on, even though it is not able to signal exactly what is happening. Once we detect bodily arousal (through feedback) we are then motivated to examine our circumstances. On the basis of our cognitive assessment of the situation, we then label the arousal. The labeling of arousal is what determines the emotion we feel. Cognitions thus fill the gap between the nonspecificity of bodily feedback and feelings for Schachter and Singer.

pictures were associated with high false heart rates and others with low rates. Later, the subjects judged as more attractive the pictures that had been associated with the high heart rate sounds, even though their actual heart rate was not high during exposure to these pictures. Valins concluded that in order for physiological activity to contribute to an emotional experience, the activity has to be represented cognitively. He argued that it is the cognitive representation of the physiological arousal, not the arousal itself, that interacts with thoughts about the situation in the generation of feelings.

The Schachter-Singer theory and the research that followed were criticized on many points.[11] The real impact of this work, though, was not so much that it explained where our emotions come from but instead that it revitalized an old notion, one that was implicit in the philosophical writings of Aristotle, Descartes, and Spinoza—that emotions might be cognitive interpretations of situations.[12] Schachter and Singer put the idea into a package that fit nicely into the cognitive pandemonium that was everywhere in psychology. The success of their efforts is exemplified by the fact that the psychology of emotion, to this day, is mostly about the role of cognition in emotion.

The Big Chill

Something was missing in the cognitive theory espoused by Schachter and Singer. They tried to explain how we deal with emotional responses once they occur (when you notice your heart beating and your forehead sweating as you begin to run away from a bear in the woods, you label the experience fear) but did not give an account of what generates the responses in the first place. Obviously, the brain has to figure out that the bear is a source of danger and has to arrange for the responses that are appropriate to danger to occur. The brain's emotional business is thus well underway by the time Schachter and Singer's mechanism kicks in. So what happens first? What makes us run from danger? What comes between the stimulus and the response? Cognitive evaluations, according to appraisal theorists, fill this gap.

The concept of appraisal was crystallized by Magda Arnold in an influential book on emotion published at about the same time that

Schachter and Singer were doing their experiments.[13] She defined appraisal as the mental assessment of the potential harm or benefit of a situation and argued that emotion is the "felt tendency" toward anything appraised as good or away from anything appraised as bad. Although the appraisal process itself occurs unconsciously, its effects are registered in consciousness as an emotional feeling.

Arnold's interpretation of James' bear-in-the-woods story would go like this: we perceive the bear and appraise it unconsciously, and our conscious experience of fear results from the tendency to run. In contrast to James, for Arnold the response does not need to occur in order to have the feeling—a feeling only requires an action tendency rather than an actual action. Emotions thus differ from nonemotional states of mind by the presence of appraisals in their causal sequence, and different emotions are distinguished from one another because different appraisals elicit different action tendencies, which give rise to different feelings.

In Arnold's view, once the appraisal outcome is registered in consciousness as a feeling, it becomes possible to reflect back on the experience and describe what went on during the appraisal process. This is possible because, according to Arnold, people have introspective access to (conscious awareness of) the inner workings of their mental life, and in particular access to the causes of their emotions. Arnold's approach assumes that we can, after an emotional experience, gain access to the unconscious processes that gave rise to the emotion. As we will see, this assumption is open to challenge.

The appraisal concept was adopted by other researchers in the 1960s. One of these was Richard Lazarus, a clinical psychologist who used the concept to understand the way people react to and cope with stressful situations.[14] Studies by Lazarus clearly showed that interpretations of situations strongly influence the emotion experienced. For example, in a classic experiment, subjects watched a gruesome film clip of a circumcision ritual involving teenage members of an aboriginal Australian tribe. For some subjects, the soundtrack verbally played up the gory details, whereas for others the episode was either minimized or intellectualized by the voice overlay. The group that had the first soundtrack, in which the gruesome details were emphasized, had stronger ANS responses and their self-reports suggested that they felt worse afterward than the other two

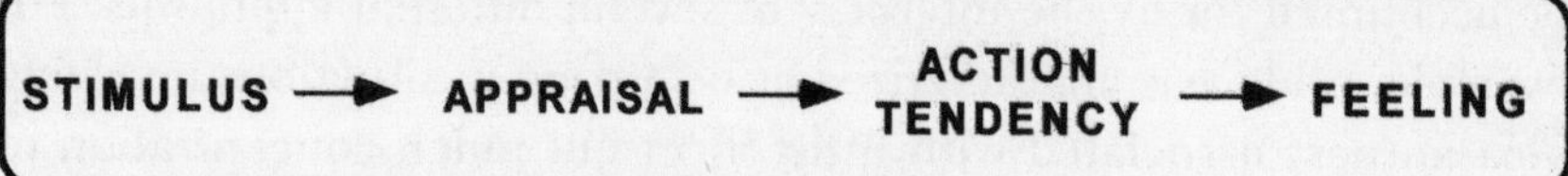

FIGURE 3-5
Arnold's Appraisal Theory.

Arnold argued that in order for a stimulus to produce an emotional response or an emotional feeling, the brain must first appraise the significance of the stimulus. Appraisals then lead to action tendencies. The felt tendency to move toward desirable objects and situations and away from undesirable ones is what accounts for conscious feelings in this model. Although appraisals can be either conscious or unconscious, we have conscious access to the appraisal processes after the fact.

groups, in spite of the fact that the arousing parts of the film were the same for all. Lazarus suggested that the different soundtracks caused the subjects to appraise the films in different ways and this led to different feelings about the situation. Lazarus argued that emotions can be initiated automatically (unconsciously) or consciously, but he emphasized the role of higher thought processes and consciousness, especially in coping with emotional reactions once they exist. Summarizing his position, he recently noted that "cognition is both a necessary and sufficient condition of emotion."[15]

Appraisal remains the cornerstone of contemporary cognitive approaches to emotion.[16] In the tradition started by Arnold, most work in this field has proceeded under the assumption that the best way to find out about appraisals is the old-fashioned way—to ask the subjects to introspect and figure out what went through their minds when they had some past emotional experience. For example, in a seminal study of these *emotion-antecedent appraisal processes* by Craig Smith and Phoebe Ellsworth, people were asked to recall a past experience implied by emotion words (pride, anger, fear, disgust, happiness, etc.) and to rate the recalled experiences on different dimensions (pleasantness, effort involved, self-other involvement, attentional activity, controllability, etc.).[17] They found that remembered experiences triggered by thoughts about emotion words could be accounted for by the interplay of several different appraisals. For example, pride was characterized as occurring in situations involving pleasantness associated with little effort but much concentration of attention and personal responsibility, whereas anger involved unpleasantness associated with much effort, lack of control, and someone else being responsible. Smith and Ellsworth concluded that people's emotions are intimately related to their cognitive appraisals of their circumstances and that it is possible to gain insights into them by asking people to reflect back on what different emotions are like. These and other researchers assume that the kind of information that subjects use when they reflect back on an emotional experience is the same kind of information that the brain uses in creating the emotional experiences.[18]

To my mind, appraisal theories came very close to getting things right: the evaluation of a stimulus is clearly the first step in the initiation of an emotional episode; appraisals occur unconsciously; emotion involves action tendencies and bodily responses, as well as conscious experiences. But appraisal theories took two wrong turns on the road to understanding the emotional mind. First, they based their understanding of appraisal processes largely on self-reports—introspective verbal reflections. As we saw in the last chapter, introspection is often a blurry window into the workings of the mind. And if there is one thing about emotions that we know well from introspection, it is that we are often in the dark about why we feel the way we do. Second, appraisal theories overemphasized the contribution of cognitive processes in emotion, thereby diminishing the distinc-

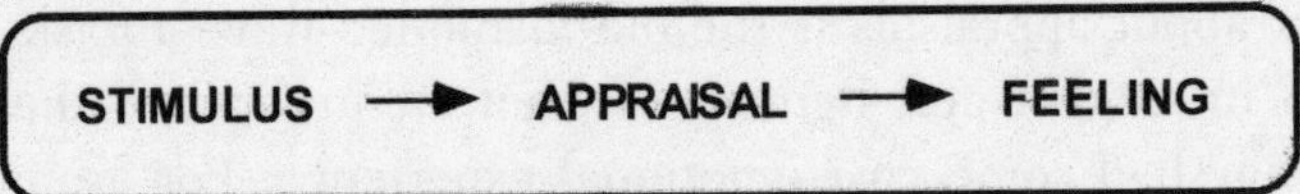

FIGURE 3-6
General Purpose Appraisal Model.

Following Arnold, many psychologists today recognize the importance of appraisal processes in emotional phenomena, but they do not necessarily accept Arnold's equation of emotional feelings with action tendencies. The general-purpose appraisal model shown here thus simply suggests that appraisals fill the stimulus-to-feeling gap.

tion between emotion and cognition. Given that a major failing of cognitive science as a science of mind is its lack of concern with emotion (see Chapter 2), it is not too surprising that the cognitive approach to emotion suffers from the same problem—in emphasizing cognition as the explanation of emotion, the unique aspects of emotion that have traditionally distinguished it from cognition are left behind.

The Psychologist Who Came in from the Cold

By 1980, the cognitive approach to emotion was just about the only approach. But this began to change with the publication of a paper by social psychologist Robert Zajonc (pronounced, zy-unce).[19] The paper was called "Feeling and Thinking: Preferences Need No Inferences." It very persuasively argued, on the basis of logic and clever experiments, that preferences (which are simple emotional reactions) could be formed without any conscious registration of the stimuli. This, he said, showed that emotion has primacy over (can exist before) and is independent of (can exist without) cognition. The net effect was a stall, rather than the demise, of the cognitive approach to emotion, as much appraisal research has occurred in the years following Zajonc's paper. Nevertheless, Zajonc had a major impact on the field, keeping alive the idea that an emotion is not just a cognition.

Zajonc summarized several experiments that he and his colleagues had performed using a psychological phenomenon, called the mere exposure effect, that he had discovered earlier. If subjects are exposed to some novel visual patterns (like Chinese ideograms) and then asked to choose whether they prefer the previously exposed or new patterns, they reliably tend to prefer the preexposed ones. Mere exposure to stimuli is enough to create preferences.

The twist to the new experiment was to present the stimuli subliminally—so briefly that the subjects were unable in subsequent tests to accurately state whether or not they had seen the stimulus before. Nevertheless, the mere exposure effect was there. The subjects judged the previously exposed items as preferable over the new (previously unseen) ones, in spite of the fact that they had little ability to consciously identify and distinguish the patterns that they had seen from those that they had not. As Zajonc put it, these results go against common sense and against the widespread assumption in psychology that we must know what something is before we can determine whether we like it or not. If in some situations emotion could be present without recognition of the stimulus, then recognition could not be viewed as a necessary precursor to emotion.

The subliminal mere exposure effect has been confirmed by many different laboratories and the idea that preferences can be formed for stimuli that do not enter consciousness seems rock solid.[20] However, Zajonc's interpretation was controversial. He argued that the absence of conscious recognition meant that preferences (emotions) were forming without the aid of cognition—that emotion and cognition are separate functions of the mind. As we saw in Chapter 2, many information-processing functions that are considered prototypical examples of cognition also occur without conscious awareness. The absence of conscious recognition is not, strictly speaking, a useful basis for exclusion of cognition from emotional processing. At the same time, although Zajonc's studies did not prove that emotion and cognition are separable aspects of the mind, this does not mean that the opposite is correct, a point that we will return to at the end of the chapter.

Regardless of the relevance of Zajonc's subliminal mere exposure studies for understanding whether emotion depends on cognition, the experiments provided incontrovertible evidence that affective re-

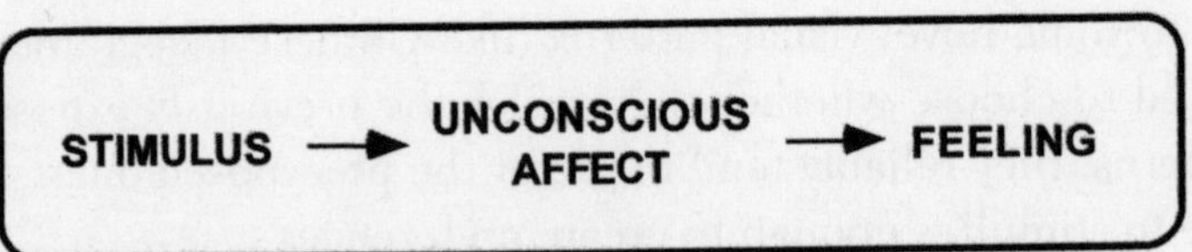

FIGURE 3-7
Zajonc's Affective Primacy Theory.

Contrary to much work in psychology, Zajonc has argued that affect precedes and occurs independent of cognition. This controversial hypothesis has been heatedly debated. What seems clear now is that emotional processing can occur in the absence of conscious awareness, but that this is a different issue from whether emotion and cognition are independent.

actions could take place in the absence of conscious awareness of the stimuli. Although some appraisal theories accept that appraisal is, or can be, unconscious, they have tended to also suggest that the individual has conscious access to the processes underlying appraisals (thus justifying the use of verbal reports to identify emotion-antecedent appraisal processes). If unconscious occurrences such as those found by Zajonc were commonplace, rather than esoteric outcomes of a clever experimental design, the conscious introspections that make up the database of appraisal theory would not be a very solid foundation for an understanding of the emotional mind.

The Emotional Unconscious

Zajonc was certainly not the first experimental psychologist to be interested in the emotional unconscious. There was a time, back around mid-century, when the emotional unconscious was quite the rage in psychology. It all began with the New Look movement,[21] which challenged the stimulus-response view of perception espoused by the behaviorists. The New Look argued that perceptions are constructions that integrate sensory information about physical stimuli with internal factors, such as needs, goals, attitudes, and emotions. When New Look psychologists started doing experiments showing that subjects could have ANS responses to emotionally charged stimuli in the absence of conscious awareness of the stimuli (see below), it seemed as though the gap between two strange (if not estranged) bedfellows, psychology and psychoanalysis, might be closing.[22] After all, the unconscious, and especially the emotional unconscious, is the linchpin of psychoanalytic theory.

After a brief period of enthusiastic reception, the unconscious perception studies of the New Look were extensively criticized and they were essentially dismissed. Unconscious perception just did not make sense to many psychologists, as there was no adequate framework for thinking about perception without awareness of the perceived stimulus. The cognitive movement and its emphasis on unconscious processing was knocking on the door, but psychology was still strongly behavioristic, and verbal responses were the primary behaviors of interest in research on humans. As one commentator,

Matthew Erdelyi of Brooklyn College, noted, there is a certain irony in this history.[23] Studies of unconscious processing were being buried at just the time that cognitive science was beginning to discredit the behaviorist preconceptions that made non-verbalizable perceptions seem impossible. But there is another irony here—that behaviorists, whose field was created to rid psychology of ghostly concepts like consciousness, should have befriended conscious introspections (verbal reports) as a method for validating psychological ideas.[24] Below, we will take a look at some early unconscious perception studies and the criticisms of them, and then turn to the new wave of research on this topic.

One of the major areas of research on unconscious processing to come out of the New Look involved *perceptual defense,* the demonstration that "dirty" words have a higher threshold for stimulus recognition than comparable words that lack sexual, scatological, or other taboo connotations. In a typical experiment, subjects were shown words on a screen. By varying the amount of time that the words were shown, it was possible to determine the amount of time a particular subject needed to recognize a given word. It was discovered that the exposure time required for "taboo" words (e.g., bitch, fuck, Kotex, cancer) was longer than for words lacking taboo connotations.[25] The results were interpreted in terms of Freudian defense mechanisms, particularly repression: the taboo words were perceived subconsciously and censured (prevented from entering consciousness) because their appearance in consciousness would have elicited anxiety.

A related line of work involved *subliminal perception.* One of the classic studies was performed by Richard Lazarus, before his appraisal theory days.[26] In that experiment, patterns of letters were briefly flashed on a screen using exposure durations that were determined to be too short to allow verbal identification. Some of the patterns had been previously paired with an electric shock in order to transform the meaningless letters into emotionally charged stimuli capable of eliciting ANS responses. When these conditioned emotional stimuli were presented subconsciously, but not when emotionally neutral stimuli occurred, the ANS responded, indicating that the emotional meaning of the conditioned stimuli had been registered, in spite of the fact that the subjects reported no awareness of the stimuli (ANS responses have been a favorite in this kind of work since they do not depend on verbal processes and can thus be used to track emotions that occur in the absence of the ability to verbally describe the stimulus).

Marketing experts seized upon the implications of subliminal perception research, hoping to surreptitiously influence consumers to buy products. A theater in New Jersey, for example, gave audiences quick flashes of the phrase "Drink Coke" or "Eat popcorn" in order to promote visits to the concession stand.[27] Whether the tactic actually worked or not is not clear, but the public was outraged over this unethical act of manipulation and the invasion of privacy.[28] In point of fact, though, the advertising industry uses emotional cues (implicitly and explicitly) to persuade consumers to buy products all the time. Persuasion is, after all, their business, as Vance Packard noted in a famous book, *The Hidden Persuaders*.[29] Persuasion always works better when the persuadee is not aware that he or she is being influenced.[30] Implicit messages are the bread and butter of many advertising campaigns.

In spite of a great deal of initial interest in the theoretical implications of the perceptual defense and subliminal perception experiments, the interpretation of the results in terms of unconscious perception of emotional meaning was called into question by Charles Eriksen in the early 1960s.[31] Eriksen believed that unconscious perception was a logical impossibility[32] and he challenged this interpretation of the findings. He argued that the failure of the subjects in perceptual defense studies to verbally identify the taboo stimuli was due, not to the failure of the stimulus to enter consciousness, but to an unwillingness of the subjects to say these embarrassing words in public. And the inability of subjects in the subliminal perception experiments to verbally identify the secret stimuli was due, not to a failure to consciously perceive the stimuli, but to imperfections of verbal processes when it comes to accurately characterizing perceptual experiences.

Widespread acceptance of Eriksen's critique sealed research on the emotional unconscious into what seemed to be a coffin, but turned out to be a time capsule. After somewhat of a hiatus in the 1960s and 1970s, a new surge of interest in unconscious emotional processes emerged, spurred on by Zajonc's studies and by Matthew Erdelyi's reinterpretation of the perceptual defense and subliminal perception work in terms of the principles of cognitive science.[33] Nevertheless, within the psychology of emotion, especially amongst the cognitively minded appraisal theorists, the emphasis has remained on the conscious, verbally accessible aspects of emotion. The evidence for the existence of unconscious aspects of emotion is often ignored or denied, or when accepted is given second billing to the conscious aspects. As several researchers who work on unconscious processes have stated, they are so busy trying to prove that unconscious processing exists that there is little time to actually explore how it works.[34] But due to the creation of new and improved techniques for studying unconscious processing,[35] the existence proofs now seem clear. Below I will review some of the evidence showing that emotional processing can take place outside of conscious awareness. Some of the work involves subliminal stimulation, whereas other studies utilize stimuli that are consciously perceived but their emotional implications are implicit and not noticed at the time the stimulus is seen or heard.

Zajonc's subliminal mere exposure studies were some of first to use the new techniques that made unconscious processing seem undeniable. In the wake of this research, many similar experiments were performed. In one particularly interesting variation by Robert Bornstein, subjects were brought into the laboratory and given very brief exposures to pictures of faces.[36] As expected, they were unable to identify which ones they had seen before, but when asked to rate how much they liked the pictures, the preexposed ones received more positive ratings. Mere exposure works for faces. In a second part of the study, the subjects were given brief (subliminal/unconscious) exposures to pictures of person A or of person B. Then, the subject, together with persons A and B, was asked to try to decide on the gender of the author of several poems. By a prearrangement unknown to the subject, A and B disagreed and the subject had to break the tie. As predicted by the mere exposure hypothesis, the subjects tended to side with the person whose face they had been unconsciously exposed to. Familiarity does not necessarily breed contempt. Bornstein later performed what is called a "meta-analysis" of subliminal mere exposure research, which means he analyzed the published data from many different studies.[37] This led him to conclude that the mere exposure effect is much stronger when the stimuli are subliminally presented than when the stimuli are freely available for conscious inspection. This turns out to be a common finding in a number of different kinds of studies of unconscious emotional processing, and it emphasizes a point that we will see time and again—our emotions are more easily influenced when we are not aware that the influence is occurring.

The emotional unconscious has also been studied with a procedure called subliminal emotional priming that has been used extensively by Zajonc and several of his associates in recent years.[38] In this kind of experiment, a priming stimulus with some emotional connotation, such as a picture of a frowning or smiling face, is presented very briefly (5 milliseconds, or 1/200th of a second) and is immediately followed by a masking stimulus, which eliminates the subject's ability to consciously recall the prime—the mask displaces the prime from consciousness, essentially blanking it out. Following a delay, a target stimulus pattern is presented. It remains on a comfortable amount of time (seconds) and is consciously perceived. After seeing many patterns in this way, the subject is asked to rate how much they liked the target stimuli. Zajonc found that whether the subjects liked or disliked a stimulus (for example, a Chinese ideogram) was related to whether the stimulus was primed by an unconscious smile or frown. The target stimulus acquired emotional significance by virtue of its relationship with an emotional meaning activated subliminally by the unconsciously processed smile or frown. And, as in the mere exposure work, the emotional priming was much more effective for subliminal (masked, thus unconscious) presentations than for presentations that were not masked and where conscious awareness of the stimulus was possible.

And then there is the Pöetzl effect.[39] Otto Pöetzl, a Viennese psychiatrist, performed studies in 1917 in which a complex visual picture, like a landscape, was shown to subjects subliminally. He then asked the subjects to draw as much of the picture as possible. Afterward, the subjects were instructed to go home and have a dream that night, and then come back the next day. When they returned to the laboratory, they were asked to report on the dream and draw pictures related to the dream. Poetzl claimed that features of the original picture that were not included in the first drawing emerged in the drawing of the dream.

Matthew Erdelyi has profitably exploited the Pöetzl effect to explore the nature of unconscious processes.[40] In one study Erdelyi presented subjects with a complex visual scene for 500 milliseconds. This is not a subliminal presentation, as there is plenty time for parts of the stimulus to enter into awareness. The purpose of using this duration was to allow some but not all of the scene to be consciously perceived. In fact, though, you can get the same result by allowing the subject to just look at the picture freely since in any complex scene there will always be stimulus elements that are noticed and others that are not,[41] and of those that are noticed some will be recalled and some not. In Erdelyi's study, after the flash, the subjects were then asked to draw as much of the scene as possible. Some then engaged in a period of free association and fantasy while others played a game of darts. They then made drawings of the picture again. Erdelyi found that the second drawings often reflected previously unremembered aspects of the stimulus for the subjects allowed to fantasize and free associate but not for the dart game group. Erdelyi calls this effect "hypermnesia," by which he means an improvement in memory—the recovery of a previously inaccessible memory. Hypermnesia has been shown by Erdelyi using his modified Pöetzl procedure and several other kinds of techniques, and he believes that the recovery of memory, by dreaming, and by fantasy and free association while awake, represents the release of memories from suppression by other factors.

Through therapeutic sessions with patients, psychoanalyst Howard Shevrin identified words related to their conscious experience of a symptom or to the unconscious conflict underlying the symptom.[42] For example, a patient may come to the analyst complaining of being extremely uncomfortable in social situations. The patient is thus fully conscious of this social phobia, but does not consciously know the cause of the problem. After the analytic sessions, Shevrin came up

FIGURE 3-8
Hypermnesia Stimulus:

Complex visual scene used by Erdelyi to study the effects of fantasy and dreaming on memory recall. Subjects examined the picture briefly. The next day they were asked to recall as much as they could about the picture. See text for details. (From *Psychoanalysis: Freud's Cognitive Psychology* by Erdelyi. Copyright © 1985 by Mathew Hugh Erdelyi. Used with permission of W.H. Freeman and Company.)

with a set of individually tailored words that he felt captured aspects of either the unconscious conflict or the conscious symptoms. He then presented them subliminally or openly to the patients while "brain waves" were recorded from their scalps. For the words related to the unconscious conflict (the underlying cause of the social phobia), the brain waves were more strongly elicited by subliminal presentations, whereas for the words related to the conscious symptom (fear of social situations), the brain waves were more strongly elicited when the stimuli were consciously perceived. Again, the emotional mind seems to be particularly susceptible to stimuli that its conscious counterpart does not have access to.

Finally, social psychologist John Bargh has performed many experiments showing that emotions, attitudes, goals, and intentions can be activated without awareness, and that these can influence the way people think about and act in social situations.[43] For example, physical features (like skin color or hair length) are enough to activate racial or gender stereotypes, regardless of whether the person possessing the feature expresses any of the behavioral characteristics of the stereotype. This kind of automatic activation of attitudes occurs in a variety of different situations and appears to constitute our first reaction to a person. And once activated, these attitudes can influence the way we then treat the person, and can even have influences over our behavior in other situations. In one dramatic example, Bargh had subjects participate in what they thought was a language test. They were given words on cards and had to make sentences out of them. For some subjects, the sentences were about elderly people, whereas other subjects received sentences about other topics. After completing the task, the subjects left the room. Unbeknownst to them, the amount of time taken to walk down the hall to a designated location was timed by the experimenters. Remarkably, the subjects that had unscrambled sentences about elderly people took longer to cover the distance than the other subjects. The sentences did not include any specific statements about old people being slow or weak, but simply thinking (and pretty indirect thinking at that) about old people was enough to activate this stereotype and influence their behavior. In other studies subjects unscrambled sentences having to do with either "assertiveness" or "politeness." They were then told to walk down the hall and find the experimenter, who, by prearrangement, was involved in a conversation with someone. The amount of time the subjects waited before interrupting was recorded. Those primed with assertiveness interrupted sooner than those primed with politeness. Bargh notes that this automatic activation of unconscious processes has an upside and a downside. If we are nice to someone they may indeed be nice to us in return. On the other hand, if seeing someone from another racial group activates a negative attitude (e.g., that persons of that group are hostile and aggressive), we may act negatively toward them, prompting them in return to act negatively toward us, creating a vicious circle that further perpetuates the stereotype.

In the two studies described above, the priming stimuli were consciously available but their meanings were implicit. Nevertheless, other studies show similar effects when the social primes are presented subliminally. Bargh argues that whether the subjects are aware of the priming stimulus is less critical than whether they are aware of the ways in which the stimuli are implicitly (without awareness) categorized and interpreted. The fact that emotions, attitudes, goals, and the like are activated automatically (without any conscious effort) means that their presence in the mind and their influence on thoughts and behavior are not questioned. They are trusted the way we would trust any other kind of perception. In other words, the perception in oneself of an attitude (disguised as fact) about a racial group can seem to be as valid as the perception of the color of their skin. When one is aware of biases and possesses values against having these, he or she can exercise control over them. However, the ability to do this depends on being aware of the unconscious influences, which is quite another matter. As cognitive psychologist Larry Jacoby asks and answers: "When are unconscious influences expected to have their largest effects? . . . When you least expect them."[44] According to Bargh, a goal of social psychology should be to make people aware of these nonintuitive, scientifically discovered unconscious factors that affect thought and behavior. But he admits that this is an uphill battle: "Inasmuch as people check such a proposition against their own phenomenal experience to test its validity, we will never be persuasive, because by definition one can never have any phenomenal experience of perception without awareness."[45]

As we look back on almost a half a century of research on unconscious processing, it is fair to say that some of the early studies may indeed not have used techniques that allow one to completely rule out the possibility of some awareness of the stimuli. But science is progressive, and the mistakes of the past are part of the wisdom of the present. We have learned a great deal about how subliminal perception research should be conducted and interpreted, and research today has higher standards of what counts as an unconscious perception.[46] When we apply the new, clever, and strict ways of evaluating whether information processing takes place unconsciously, we still reach the conclusion that emotional meanings can be processed at subconscious levels. Just because the research methods of the past may have lacked perfection does not mean that the results were wrong. It now seems undeniable that the emotional meanings of stimuli can be processed unconsciously. The emotional unconscious is where much of the emotional action is in the brain.[47]

A Reappraisal

From James onward an important gap was left in the causal chain leading to emotional responses and emotional experiences, and something like appraisal was needed. The gap occurs between the arrival of the emotion-provoking stimulus and the resulting physiological responses and/or feelings. In James' theory, the perception of the stimulus automatically (without conscious participation) produces the responses that provide the feedback that defines the feeling. But not all stimuli that are perceived do this. Something else has to happen. The physical features of the stimulus have to be evaluated—appraised; their significance to the individual has to be determined. It is this computed significance that starts the emotion ball rolling. This is the case for all of the theories that have been described. The brain has to evaluate a stimulus and decide whether that stimulus should be ignored or should lead to some reaction. Appraisal, in other words, fills the gap between stimuli and responses and between stimuli and feelings. But, in my view, appraisal theories did not quite get it right, as they required that the appraisal mechanism get all involved in introspectively accessible levels of higher cognition from the start.

The inadequacy of any approach to emotion based solely or mainly on introspectively accessible aspects of the mind is apparent from the experimental studies described above showing that much of emotional processing occurs (or can occur) unconsciously, as well as by the fact that people often find their emotions puzzling. Consciously accessible appraisal processes cannot be the way, or at least not the only way, the emotional brain works. Even when we are conscious of the outcome of some emotional appraisal (for example, knowing that you dislike someone), this does not mean that you consciously understand the basis of the appraisal (knowing why you dislike the person). The conscious outcome might be based on nonverbalizable intuitions, so-called gut feelings,[48] rather than on some verbalizable set of propositions.

Proponents of folk psychology (a kind of introspective psychology), argue that people know what is in their minds and they use this information all the time.[49] They point out that people are very good at accounting for their mental life and behavior on the basis of self-knowledge and their general understanding of the way other people's minds work. For example, if I say I will pick up my son at school at a certain time, chances are I will do it. If I see you arguing with your spouse, chances are I'll be correct if I assume you are mad. The folk psychologist says that examples like this add proof to the idea that age-old wisdom constitutes a scientifically correct theory of mind that we all have in our heads. But even though I am consciously aware of my decision to pick my son up and may even consciously remember to carry out the plan, this does not mean that I know how I remembered to do it. And even though I may be correct when I decide you are angry, that does not mean that I know how I made my decision or that I know what it is in your brain that accounts for your anger. The biologist Stephen J. Gould makes a good point: "Science is not 'organized common sense'; at its most exciting, it reformulates our view of the world by imposing powerful theories against the ancient, anthropocentric prejudices that we call intuition."[50] When I say I'm angry, I may be, but I might also be wrong. I might really be afraid or jealous or some combination of all of these. Donald Hebb pointed out long ago that outside observers are far more accurate at judging a person's true emotional state than is the person himself.[51] I'm not denying that people are consciously aware of certain things and that they can consciously do things. All I'm saying is that some, perhaps many, of the things we do, including the appraisal of the emotional significance of events in our lives and the expression of emotional behaviors in response to those appraisals, do not depend on consciousness, or even on processes that we necessarily have conscious access to.

Noting that emotions can sometimes be puzzling, the philosopher Amelie Rorty makes a distinction between the apparent cause of an emotion (the stimuli immediately available and consciously perceived) and the actual cause.[52] The real cause of an emotion is not necessarily some immediately present stimuli, but instead may involve the interaction of these with a causal history stored in memory. As we have seen, unnoticed events can activate memories, including emotional memories, implicitly (without awareness), and implicit and undetected meanings of consciously perceived stimuli can do the same. A father who yells at his children may rationalize his outburst by saying that the children were misbehaving. But the outburst may also be due in part to the fact that he had a bad day at the office, or even to the way his parents treated him as a child, and at the time he may not be consciously aware of these influences at all. The cause of an emotion can, in other words, be very different from the reasons we use to explain the emotion to ourselves or others after the fact. Appraisal theories have dealt with reasons rather than causes.

Two of the leading appraisal theorists, Nico Frijda and Klaus Scherer, have both recently acknowledged significant limitations of the research foundation of much of cognitive appraisal. Frijda says: "investigating the relationships between appraisals and emotion labels is research into emotion word meanings or into the structures of experience, as distinct from research that qualifies as investigation of emotion antecedents. . . . Emotions may well result from appraisal processes, but these need not be those suggested by the self reports."[53] Along similar lines, Scherer says that the emphasis of appraisal research on mapping emotion words onto emotion experiences has left the field concentrating on the content of experiences and the way experiences are verbally labeled to the exclusion of the true processes that give rise to appraisals.[54] And in an insightful discussion of unconscious processes, Kenneth Bowers makes the interesting point that if our understanding of the causation of thought and action were directly available to introspection, we would not need the field of psychology.[55] Indeed, it was the inadequacy of introspection that led to behaviorism, and the success of cognitive science as an alternative to behaviorism is due in large measure to its ability to investigate the mind without relying exclusively or mainly on introspection.

Some appraisals lead to conscious awareness of the appraisal outcome, whereas others do not. Introspections are often going to be a poor window into how processing that gives rise to conscious content works and are no window at all into processing that does not give rise to immediate conscious content. Although cognitive appraisal theorist Richard Lazarus has emphasized conscious appraisal processes in emotions, he has always accepted that unconscious appraisals occur and recently he argued: "although it is a daunting task, I believe we must . . . find effective ways of exploring what lies below the surface, how it relates to what is in awareness, and how it influences the entire emotion process."[56] Similarly, Klaus Scherer recently challenged his colleagues who study human appraisal processes to rely more on techniques that do not depend on verbal reports. Scherer also suggests that appraisal researchers turn to brain science to try to validate mechanisms that psychologists uncover.[57] I go one step further and argue that we might turn to brain research to find novel mechanisms that psychologists have not thought of or to find novel interpretations of existing mechanisms.

Introspective understanding of the causes of emotion states can be weak, especially when people are asked to reflect back on an episode after it is over.[58] And even if they are asked right away they may still not know the actual cause. There is much more to explain about an emotion than what one can get at from retrospective consciously accessible thoughts about the situation. But this does not mean that introspection is useless. There are some kinds of mental events that we have introspective access to and others that we do not. The trick, obviously, is to figure out where the dividing line is. However, the line is both thin and fuzzy—it may not be in the same place in different people and in a given individual it may move from moment to moment.[59] There is much to be learned about conscious experience by studying introspections. But if emotions reflect processes that also occur unconsciously, as it seems they do, then we need to take these into consideration as well.

Emotion and Cognition: Two Sides of the Same Coin, or Different Currency?

So far I have tried to present a strong case for the argument that much emotional processing occurs unconsciously and therefore that there is more to an emotion than what we can glean from our introspections about it. But the same argument was made about cognition in the previous chapter—that not all aspects of thinking, reasoning, problem solving, and intelligence can be understood on the basis of introspections. Given that emotional and cognitive processing both largely occur unconsciously, it is possible that emotional and cognitive processing are the same, or, as it is usually said, that emotion is just a kind of cognition.

There is a benign and a not-so-benign version of the idea that emotion is a kind of cognition. In both versions, the terms "cognitive" and "mental" are used interchangeably. This is clearly a departure from the approach of early cognitive scientists, who saw cognition as the part of the mind having to do with thinking and reasoning, but not with emotion and some other mental processes like motivation and personality (see Chapter 2).

In the benign version, the boundaries of cognition are moved so that, in addition to including thinking, reasoning, and intelligence, it also includes emotion. In this scheme, nothing fundamental changes in the way emotion is conceived—cognition and emotion are given equal billing in a field that studies both. This is simply a semantic issue about what the mind, and its science, should be called. I prefer the term "mind science" over "cognitive science" for this all-encompassing approach to the mind. Although this is somewhat a matter of preference, it is not an idle one. It is an attempt to prevent one from sliding from the benign to the not-so-benign version, in which emotion is perversely reconceived as thinking and reasoning.

In the not-so-benign version, then, "cognitive" and "mental" are equated by squeezing emotion into the traditional view of cognition—cognition as thinking and reasoning. This, as we have seen, is the unfortunate way the study of emotion has gone since the early 1960s—the essence of an emotion has been altered in order that emotions could be conceived as reasoned thoughts about situations. It is this trend that Zajonc was reacting to when he proposed that emotion and cognition should be kept separate. But the heated debate over the relation of emotion to cognition got caught up in a variety of technical issues and this broader concern was lost.[60]

My desire to protect emotion from being consumed by the cognitive monster comes from my understanding of how emotion is organized in the brain. Although the brain organization of emotion is the subject of other chapters, I will summarize several key points that justify my belief that emotion and cognition are best thought of as separate but interacting mental functions mediated by separate but interacting brain systems.

- When a certain region of the brain is damaged, animals or humans lose the capacity to appraise the emotional significance of certain stimuli without any loss in the capacity to perceive the same stimuli as objects. The perceptual representation of an object and the evaluation of the significance of an object are separately processed by the brain.
- The emotional meaning of a stimulus can begin to be appraised by the brain before the perceptual systems have fully processed the stimulus. It is, indeed, possible for your brain to know that something is good or bad before it knows exactly what it is.
- The brain mechanisms through which memories of the emotional significance of stimuli are registered, stored, and retrieved are different from the mechanisms through which cognitive memories of the same stimuli are processed. Damage to the former mechanisms prevents a stimulus with a learned emotional meaning from eliciting emotional reactions in us, whereas damage to the latter mechanism interferes with our ability to remember where we saw the stimulus, why we were there, and who we were with at the time.
- The systems that perform emotional appraisals are directly connected with systems involved in the control of emotional responses. Once an appraisal is made by these systems, responses occur automatically. In contrast, systems involved in cognitive processing are not so tightly coupled with response control systems. The hallmark of cognitive processing is flexibility of responses on the basis of processing. Cognition gives us choices. In contrast, activation of appraisal mechanisms narrows the response options available to a few choices that evolution has had the wisdom to connect up with the particu-

lar appraisal mechanism. This linkage between appraisal process and response mechanisms constitutes the fundamental mechanism of specific emotions.

- The linkage of appraisal mechanisms with response control systems means that when the appraisal mechanism detects a significant event, the programming and often the execution of a set of appropriate responses will occur. The net result is that bodily sensations often accompany appraisals and when they do they are a part of the conscious experience of emotions. Because cognitive processing is not linked up with responses in this obligatory way, intense bodily sensations are less likely to occur in association with mere thoughts.

The conversion of emotions into thoughts has allowed emotion to be studied using the tools and conceptual foundations of cognitive science. There are now numerous computer simulations of appraisal and other emotional processes[61] and some proponents of this AI approach to emotion believe that emotions can be programmed in computers.[62] The following limerick spun by an AI researcher summarizes the beliefs and hopes of the field:

A computer so stolid and stern
Can simulate man to a turn.
Though it lacks flesh and bones
And erogenous zones,
It can teach—but, oh can man learn?[63]

Simulations do indeed have much to offer as an approach to modeling aspects of the mind. However, as the next limerick (though tasteless) reminds us, minds feel as well as think, and feelings involve more than thinking.

There was once an ardent young suitor
Who programmed a female computer,
But he failed to connect
The affective effect,
So there wasn't a thing he could do to 'er.[64]

And, finally, we are also reminded by a limerick that there may be some things that a computer just can't do. This limerick needs to be prefaced with a reminder that in the old days computers were fed information on cards with holes punched out of them that were read by special sensing devices, and that some aspects of computer memory were stored on endless spools of magnetic tape.

There was once a passionate dame
Who wanted some things made plain,
So she punched up the cards,
Filled tape by the yards,
*But—somehow—*it just wasn't the same![65]

Where Do We Go from Here?

I have tried to make a clear statement about what emotion is not. It is not merely a collection of thoughts about situations. It is not simply reasoning. It cannot be understood by just asking people what went on in their minds when they had an emotion.

Emotions are notoriously difficult to verbalize. They operate in some psychic and neural space that is not readily accessed from consciousness. Psychiatrists' and psychologists' offices are kept packed for this very reason. Yet, much of our understanding of the way the emotional mind works has been based on studies that have used verbal stimuli as the gateway to emotions or verbal reports to measure emotions.

Consciousness and its sidekick, natural language, are new kids on the evolutionary block—unconscious processing is the rule rather than the exception throughout evolution. And the coin of the evolutionarily old unconscious mental realm is nonverbal processing. Given that so much work on unconscious processing (cognitive and emotional) has focused on verbal processes, we probably have a highly inaccurate picture of the level of sophistication of unconscious processes in humans. And we will not likely begin to fully understand the workings of human unconscious processes until we turn away from the use of verbal stimuli and verbal reports.

It is a testament to human vanity and linguistic chauvinism that the ancestral functions of the brain are characterized as the negation of newly evolved ones. Animals were unconscious and nonverbal long before they were conscious and verbal. Fortunately, ancestral functions, like certain emotional processing functions, are preserved in the human brain, and we can turn to studies of animals to discover how these work in humans as well.

Obviously, we cannot explain everything about human emotions by studying animals. But, as I hope to show you, we have been able to come to a very good understanding of some basic emotional mechanisms that are common to humans and other animals. With this information in hand, we are in a much better position to understand how newly evolved functions, like language and consciousness, contribute to emotions, and particularly how language and consciousness interact with the underlying nonverbal and unconscious systems that make up the heart and soul of the emotional machine.

4
THE HOLY GRAIL

The brain is my second favorite organ.

Woody Allen[1]

A MAJOR GOAL OF modern brain science is to figure out in as much detail as possible where different functions live in the brain. Knowing "where" a function is located is the first step toward understanding "how" it works. Not surprisingly, emotions are functions that scientists have traditionally been interested in localizing in the brain.

For more than a century, crusades have been made through the cerebral promised land in search of the emotional Holy Grail, the brain region or network that will explain where guilt and shame and fear and love come from. Around mid-century it seemed that the prize was finally in hand when the limbic system theory of emotion was proposed.[2] This remarkable conception gave an explanation of emotional life in terms of a brain network that evolved to subserve functions necessary for the survival of the individual and species. The limbic system theory claimed nothing short of having found the physical basis of the Freudian id.

But by the early 1980s, very little research on the brain mechanisms of emotion was being conducted. No doubt, the extension of the cognitive revolution (which excluded emotion as a research topic) into brain science contributed to this, but so too did the apparent thoroughness of the limbic system theory as an account of emotion. The emotional brain seemed, at least in general terms, to be understood.

It would be hard to overestimate the impact of the limbic system concept. It had a tremendous influence not only on the way we think of emotional functions but also on the way we think of the structural organization of the brain. Each year, legions of neuroscience students are taught where the limbic system is and what it does. Unfortunately, though, there is a problem. The limbic system theory is wrong as an explanation of the emotional brain and some scientists even say that the limbic system does not exist. So before I go further and give you my view of the emotional brain, I want to tell you where the limbic system idea came from and explain why it is inadequate as an account of emotional life.

Bumps on the Head

In many ways, we owe the notion that functions are localized to specific parts of the brain to a strange movement in the nineteenth century called phrenology.[3] Phrenologists were scientists, or some would say pseudoscientists, who analyzed personality traits and mental disorders by feeling the surface geography of the human skull.

Phrenology originated through the work of a respected scientist, Franz Josef Gall. Like many before and after him, Gall felt that the mind is composed of a variety of specific faculties (like sensing, feeling, speech, memory, intelligence). Gall went further and made the interesting suggestion that each faculty has its own "organ" in the brain. This was the birth of modern ideas about functional localization. So far so good.

Unfortunately, Gall, and especially his followers, took another step,[4] arguing that more developed faculties had larger brain organs, and that the skull above these protruded more than above less developed organs. As a result, it was possible to characterize personality traits and intellectual abilities by feeling the bumps on one's head, and to detect disorders of thinking and mood by finding deviations from the norm. Phrenologists cast caution to the wind, mapping the cranial location not only of regular old mental faculties (sensation, feeling, memory, and language) but also such exotic traits as veneration, benevolence, friendship, sublimity, suavity, and even philoprocentiveness (whatever that might be).

With so little known about the brain from a scientific point of view, phrenology captured the imagination of inquiring Victorian

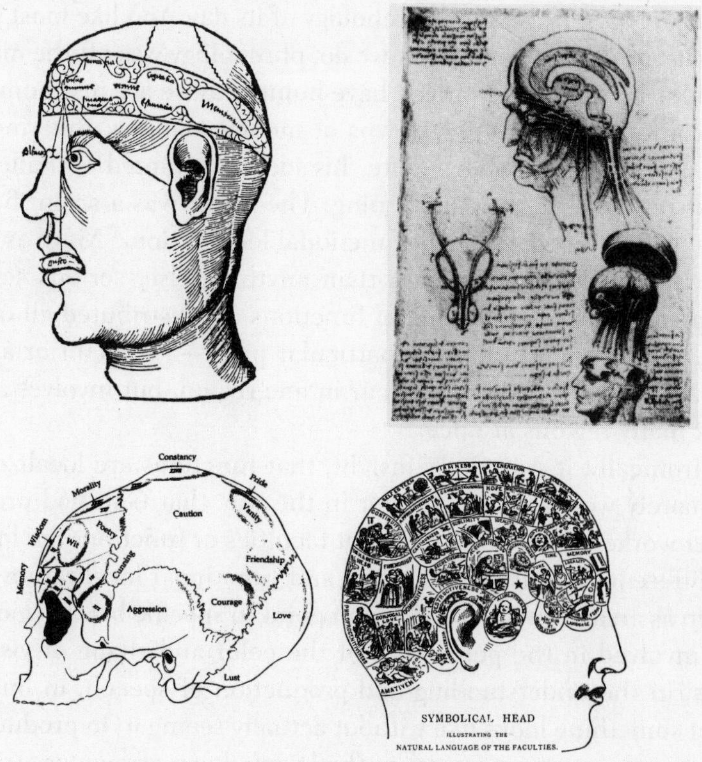

FIGURE 4-1

Localization of Brain Function Through the Ages.

Around A.D. *500, St. Augustine proposed that higher mental functions come from the brain's ventricles, the caverns that contain cerebrospinal fluid. This view persisted for centuries. A version of this idea from around 1500 is depicted in a drawing from Gregor Reisch's book (upper left). Around the same time, Leonardo sketched his view of brain function (upper right). With the emergence of phrenology in the late nineteenth century, localization of function came to be associated with specific areas of the brain, especially of the neocortex. By feeling the shape of one's skull, phrenologists proposed that they could tell the extent to which different parts of the underlying brain were developed (bottom left). Extremists amongst the phrenologists identified a great variety of skull locations with different psychological functions (bottom right).* (Left side illustrations reprinted with permission from M. Jacobson [1993], *Foundations of Neuroscience.* New York: Plenum [upper left reprinted from figure 1.7 and lower left from figure 1.11]. Right side illustrations reprinted with permission from F. Plum and B.T. Volpe [1987], Neuroscience and higher brain function: From myth to public responsibility. In F. Plum, ed., *Handbook of Physiology, Section 1: The nervous system,* vol. V. Bethesda: American Physiological Society.)

minds. It became the pop psychology of its day. And like most pop explanations of why we do what we do, phrenology was off the mark. Although the skull does indeed have bumps, these are just bumps, not indicators of the ups and downs of mental abilities. But since Gall was a scientist of some repute, his ideas warranted a challenge by other respectable men of learning. The result was a scientific backlash against the very idea of functional localization.[5] More as a reaction to phrenology's excesses than anything else, serious scientists adopted the belief that mental functions are distributed all over the brain rather than located in particular parts—a thought or an emotion, in this view, does not occur in one region, but involves all or at least many regions at once.

Ironically, it was Gall's insight, that functions are localized, that ultimately won out, though not in the way that Gall had proposed. Later workers found that different faculties or functions are localized in different regions of the brain, and functional localization is now taken as an accepted fact.[6] We can point to specific brain regions that are involved in the perception of the color and shape of visual objects, in the understanding and production of speech, in imagining what something looks like without actually seeing it, in producing accurate movements in space, in the laying down of memory traces, in smelling the difference between a rose and lilac, in detecting danger, finding food and shelter, selecting mates, and on and on.

Nevertheless, mental processes are not, strictly speaking, functions of brain areas. Each area functions by way of the system of which it is a part. For example, the visual cortex, a region in the rear of the cerebral cortex (the wrinkled outer layer of the brain), is crucially involved in our ability to see. If this region is damaged, you will, for all intents and purposes, be blind.[7] This does not mean that vision is localized to the visual cortex. It means that the visual cortex is a necessary part of the system that makes seeing possible. This system includes the visual cortex as well as a variety of other areas that transfer information from the eyes into the brain and ultimately to the visual cortex. And the visual cortex itself is a complicated structure, being composed of many subregions and subsystems that each contribute in unique ways to the act of seeing.[8] Damage anywhere along the path from the eye through the final stages of processing in the visual cortex can disrupt vision, just as removal of any link breaks a chain.

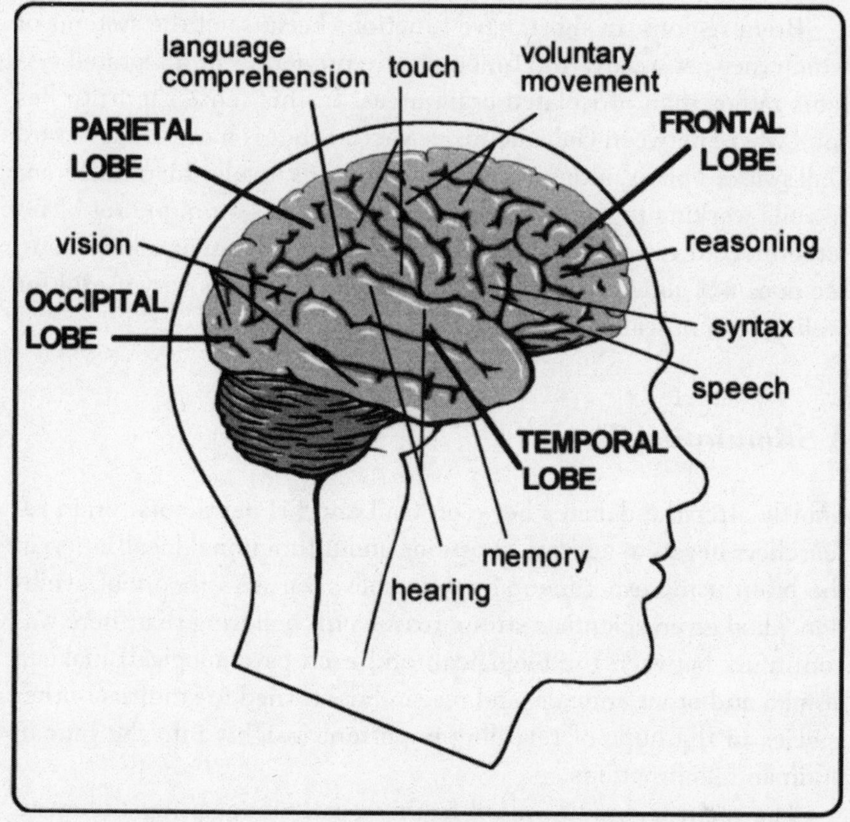

FIGURE 4-2
A Contemporary Map of Some Cortical Functions.

Contemporary understanding of cortical function is based on studies that show the effects of damage to specific regions on the ability to perform behavioral or mental tasks; that reveal behavioral or mental consequences of stimulating brain areas; or that record neural activity or image neural activity in different locations during the performance of behavioral or mental tasks. However, the identification of brain regions associated with specific functions should not be taken too literally. Functions are mediated by interconnected systems of brain regions working together rather than by individual areas working in isolation. In general, studies of experimental animals allow much more precision in identifying the functions to which specific brain regions contribute. In fact, without the animal research, it would be difficult to interpret some of the less precise findings from humans. Nevertheless, studies of the human brain make unique contributions, especially to our understanding of functions that are primarily present in the human brain. (Modified clipart from Canvass and Corel Draw.)

Brain regions, in short, have functions because of the systems of which they are a part. And functions are properties of integrated systems rather than of isolated brain areas. In this sense the truth lies somewhere between Gall and his detractors, but is more tilted toward Gall's view. That is, although mental functions involve many different regions working together, each function requires a unique set of interconnected regions, its own system. The system that allows us to see does not allow us to hear, and neither of these is very useful for feeling pain or walking.[9]

A Stimulating Time

Shortly after the debates between Gall and his detractors, brain researchers began to address questions about functional localization in the brain using experimental approaches. Darwin's theory of evolution[10] had given scientists strong reasons for believing that there was continuity between the biological (and even psychological) makeup of man and other animals, and researchers turned to studies of other species in the hope of revealing important insights into the human brain and its functions.

The main techniques used in these early pioneering studies involved stimulating cortical areas or ablating (removing) them surgically. Brain stimulation involves passing small amounts of electric current through an electrode, a tiny wire that is inserted in the brain. Since the brain operates on the basis of electrical signals transmitted from neurons (brain cells) in one area to neurons in another, electrical stimulation artificially reproduces the effects of natural information flow in the brain. Ablation studies reveal the functions of a brain region by the behavioral or mental capacities that are lost as a result of the damage, whereas with stimulations functions are suggested by responses that are produced. These were the yin and yang of early brain science methodology, and remain key techniques even today.

One of the first discoveries from experimental studies of the brain was that electrical stimulation of certain areas of the cortex elicited movements of specific bodily parts and that surgical lesions of the same regions produced deficits in performing movements of these same parts.[11] The areas in question were in the front part of the cortex, in a region now referred to as the motor cortex, which is known to be crucially involved in the control of voluntary movement. This region has connections with neurons in the spinal cord, which in turn send messages out that control the movement of limbs and other body parts. Stimulation of areas in the rear of the cortex produced no movements, but damage to these regions interfered with the normal perception of information from the eyes, ears, or skin, rendering the animals blind, deaf, or insensitive to touch, all depending on where the lesion was located. These areas are now known as the visual, auditory, and somatosensory regions of the cerebral cortex.

Early neurologists were making very similar discoveries on the basis of observations in brain-injured humans suffering from strokes or tumors.[12] The correspondence between the clinical observations in humans and the more exacting findings from animal experimentation provided strong evidence in support of a Darwinian continuity of brain organization across species.[13]

Quite the Rage

In the early explorations of the functions of the cerebral cortex it was noted that animals with massive ablations showed strikingly normal patterns of emotional reactivity.[14] For example, following removal of the entire cerebral cortex, cats still exhibited characteristic signs of emotional arousal. When provoked, they crouched down, arched their backs, retracted their ears, unsheathed their claws, growled, hissed, showed their teeth, and bit any object around.[15] In addition, they exhibited strong signs of autonomic arousal, including piloerection (puffing of the hair), pupil dilation, and elevations of blood pressure and heart rate. This was surprising given the belief at the time that the complex behaviors, including emotional behaviors, were controlled by the sensory and motor cortex. For example, in the neural version of his feedback theory, William James had proposed that emotions are mediated by sensory and motor areas of the cortex—the motor areas were needed for producing responses and the sensory areas for detecting the stimulus in the first place and then for "feeling" the feedback from the responses (see Figure 4-3). James got this one wrong, since cortical damage had no effect on emotional responses.

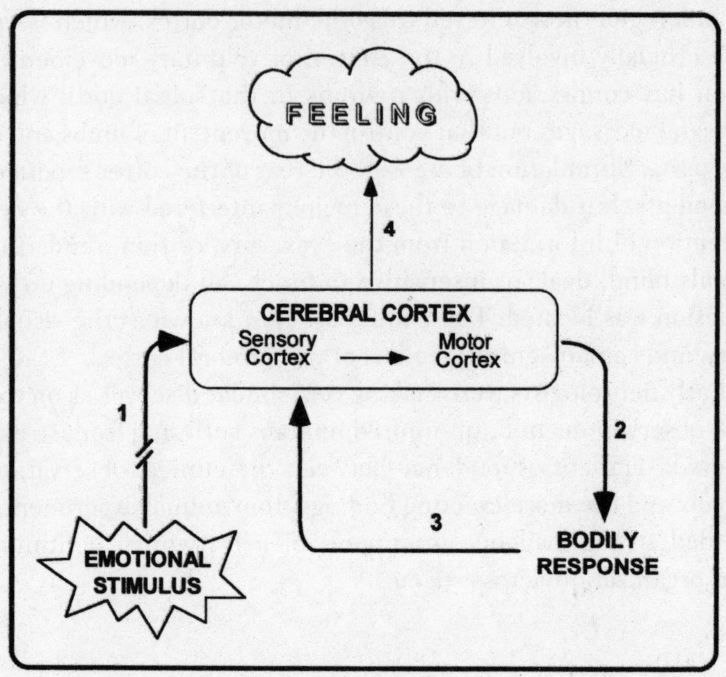

FIGURE 4-3
William James' Brain Pathways of Emotion.

An external stimulus, such as the sight of a bear, is perceived by the sensory areas of the cerebral cortex. Through the motor cortex, responses, such as running away, are controlled. Sensations produced by the responses are fed back to the cerebral cortex, where they are perceived. The perception of bodily sensations associated with the emotional responses is what gives the peculiar quality to the emotion in James' theory.

Yet, the emotional behavior of decorticate animals (animals in whom the cerebral cortex was removed) was not completely normal. These creatures were very easily provoked into emotional reactions by the slightest events. They seemed to be lacking any regulation of their rage, which suggested that cortical areas (like Plato's charioteer) normally rein in these wild emotional reactions and prevent their expression in inappropriate situations.[16]

Walter Cannon was famous not only for his attack on William James' theory, described in Chapter 3, but also for his own neural theory of emotion, which was based on research conducted in his laboratory by Philip Bard. Bard carried out a systematic series of lesion studies aimed at finding just what parts of the brain are required for the expression of rage.[17] He made larger and larger lesions, starting with the cortex and working his way down, until he found a pattern of destruction that eliminated rage responses. The critical lesion was one that encroached upon an area called the hypothalamus. In the absence of the hypothalamus, only fragments of emotional reactivity, rather than fully integrated reactions, could be mustered, and only in response to very intense, painful stimuli. The animals might crouch, snarl, hiss, unsheathe their claws, retract their ears, bite, and/or exhibit some autonomic reactivity, but these did not all occur together in coordinated fashion as they did when the hypothalamus was intact, and only very intense stimulations evoked the responses. Such findings suggested to Bard and Cannon that the hypothalamus is the centerpiece of the emotional brain (see Figure 4-4).

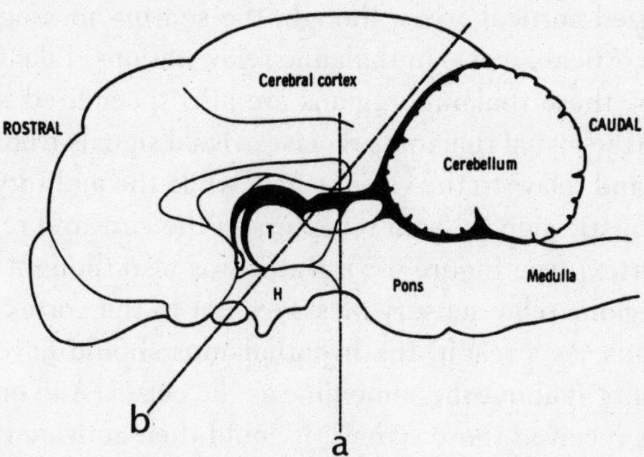

FIGURE 4-4

Lesion Strategy Used by Bard to Isolate the Hypothalamus as an Emotional Center.

First, Bard removed all brain areas above (to the left of) line "b." Included were the cerebral cortex and all other parts of the forebrain except the hypothalamus (H) and some portions of the thalamus. These lesions did not prevent emotional reactions. However, when the lesion was extended to include the areas between lines "b" and "a" as well as the areas rostral to "b," emotional reactions were essentially eliminated. Rostral = Front; Caudal = Back. (Based on J.E. LeDoux [1987], Emotion. In F. Plum, ed., *Handbook of Physiology, Section 1: The nervous system,* vol. V. Bethesda: American Physiological Society.)

The brain can be divided into three divisions along the vertical axis, the hindbrain, midbrain, and forebrain. As we ascend from hindbrain to forebrain, the functions represented go from psychologically primitive to psychologically elaborate. The hypothalamus, about the size of a peanut in the human brain, sits at the base of the forebrain and forms the interface between the psychologically sophisticated forebrain and the more primitive lower areas. In Cannon and Bard's day, the hypothalamus was known to be involved in the regulation of the autonomic nervous system,[18] and it made sense to them that the hypothalamus might be the place where bodily reactions occurring in strong emotions might be controlled by the forebrain.

The Cannon-Bard theory built upon the well-known fact that the sensory systems that take in information from the outside world send the information they receive to specialized regions of the cerebral cortex—information from the eyes goes to the visual cortex and information from the ears to the auditory cortex. In their travels toward the specialized cortical areas, though, the sensory messages make a stop in subcortical areas—in thalamic relay stations. Like their cortical partners, these thalamic regions are also specialized for sensory processing (the visual thalamus receives visual signals from receptors in the eyes and relays to the visual cortex while the auditory thalamus receives acoustic signals from receptors in the ears and relays to the auditory cortex) (see Figure 4-5). But it was also thought that some thalamic regions relay sensory messages not to the cortex but to the hypothalamus. As a result, the hypothalamus should have access to sensory inputs at about the same time as the cortex. And once the hypothalamus received these signals, it could then activate the body to produce the autonomic and behavioral responses characteristic of emotional reactions (see Figure 4-6). This explained to Cannon and Bard why decortication failed to prevent the expression of emotion and why James' cortical theory was therefore wrong (emotional responses are controlled by the hypothalamus, not the motor cortex, and sensations can activate the hypothalamus directly, without passing through the sensory cortex).

Although Cannon and Bard eliminated the cortex from the chain of events leading to emotional responses, they did not completely rule out a role for the cortex in emotion. In fact, Cannon and Bard felt that the conscious experiences of emotions, the feelings, depend

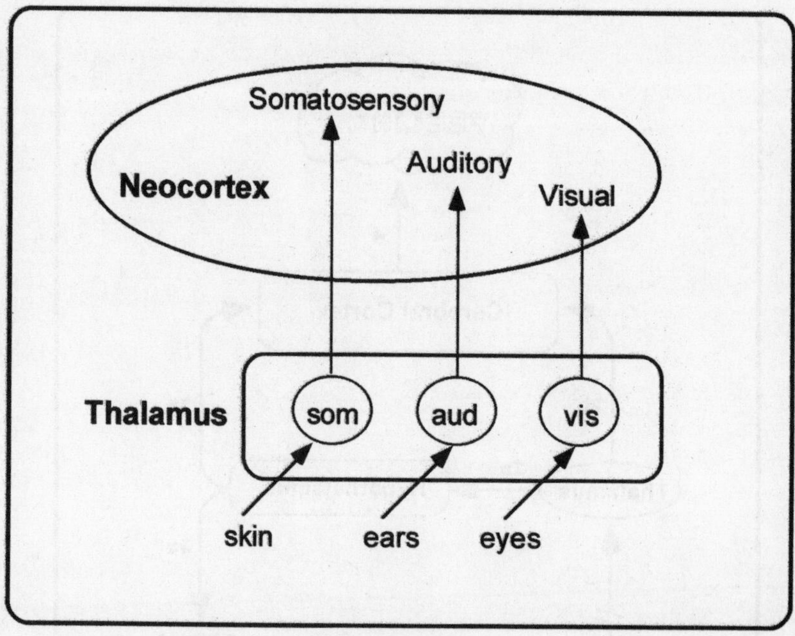

FIGURE 4-5
Relation of Sensory Thalamus and Sensory Cortex.

Sensory messages are transmitted from external receptors (e.g., in the eyes, ears, and skin) to specific areas of the thalamus that then process the signals and relay the results to specialized areas of the neocortex. Abbreviations of thalamic areas: som, somatosensory thalamus; aud, auditory thalamus; vis, visual thalamus.

upon the activation of the cortex by nerve fibers ascending from the hypothalamus. So in the absence of the cortex angry behavior is produced but is unaccompanied by the conscious feeling of rage. For this reason, Cannon used the term "sham rage" to describe the emotional outbursts of decorticate animals.[19]

For James, the peculiar quality of an emotional experience was determined by feedback to the brain from the bodily responses—responses thus occur before feelings. But for Cannon emotions are defined by processes completely contained within the brain and centered around the hypothalamus. The hypothalamus discharged to the body to produce emotional responses and to the cortex to pro-

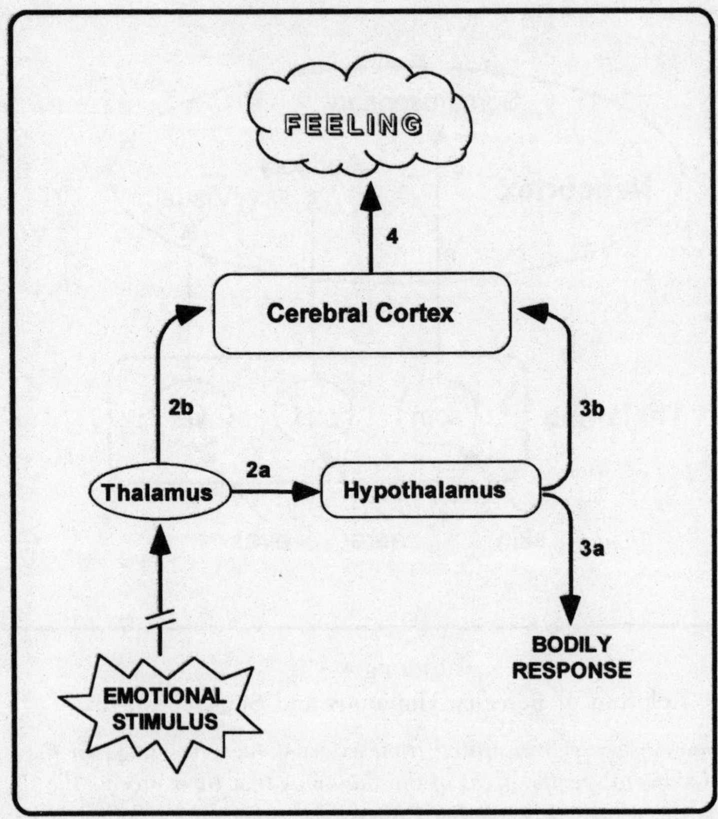

FIGURE 4-6
Cannon-Bard Theory of the Emotional Brain.

Cannon and Bard believed that external stimuli processed by the thalamus were routed to the cerebral cortex (path 2b) and to the hypothalamus (path 2a). The hypothalamus, in turn, sent messages to both bodily muscles and organs (path 3a) and the cortex (path 3b). The interaction of messages in the cortex about what the stimulus is (path 2b) and about its emotional significance (path 3b) results in the conscious experience of emotion (feelings). Emotional responses and feelings occur in parallel in this theory.

duce emotional experiences. And since the fibers descending to the bodily response systems and the fibers ascending to the cortex are activated simultaneously by the hypothalamus, emotional feelings and emotional responses occur in parallel, rather than in sequence.

Although Cannon disagreed with James on what causes emotional experiences, on another very important point, and one that is not often recognized, Cannon seems to have agreed with James—emotional responses (running from the bear) are not caused by conscious emotional experiences (being afraid). For James, emotional responses precede and determine conscious experiences, whereas for Cannon responses and experiences occur simultaneously. It thus seems likely that James and Cannon would both go along with the notion, developed in the last chapter, that conscious emotional experiences are a consequence of prior emotional processes (evaluations or appraisals) that occur outside of conscious awareness, which is to say, unconsciously.

The Stream of Feeling

James Papez, a Cornell University anatomist, never specifically did research on emotion, but in 1937 he proposed one of the most influential theories of the emotional brain.[20] Rumor has it that Papez found out that an American benefactor had donated a large sum of money to a British laboratory for the purpose of figuring out how emotions work.[21] In a stroke of national pride, he dashed off his famous article in a few days to show that Americans also had some ideas about emotions. However, had he known that his theory would sink into the back pages of science until rediscovered and revived around mid century, Papez might have developed his theory at a more leisurely pace.

Papez was strongly influenced by the work of C. Judson Herrick, an anatomist who specialized in brain evolution. Herrick had earlier pointed out a distinction between two parts of the cortex, the lateral and medial parts.[22] Imagine the brain as a hot dog bun, with the two halves of the bun being the two cerebral hemispheres. The brownish, toasted part of the bun that we can see from the outside is like the lateral part of the cerebral cortex. This is the part of the cortex that has all the sensory and motor functions that we've talked about and is generally believed to be the place where all our highest thought processes occur. Now, imagine prying the bun apart down the seam in the middle, pulling the two hemibuns away from each other. The white, untoasted part of hemibuns down the middle is like the medial part of the cortex. This part, according to Herrick, is evolutionarily older and is involved in more primitive functions than the newer cortex, known as the neocortex, a designation that reflects its supposedly more recent origin in evolution (see Figure 4-7).

Herrick's medial cortex had earlier been called *le grand lobe limbique* by the great French anatomist Paul Pierre Broca.[23] Broca noted that the medial cortical areas have an oval shape, almost like the rim of a tennis racquet. In fact, *limbique* is the French version of the Latin word *limbus,* which means rim. Broca had a structural description of the medial cortex in mind when he labeled the area, but somewhat later the limbic lobe was renamed the rhinencephalon, which

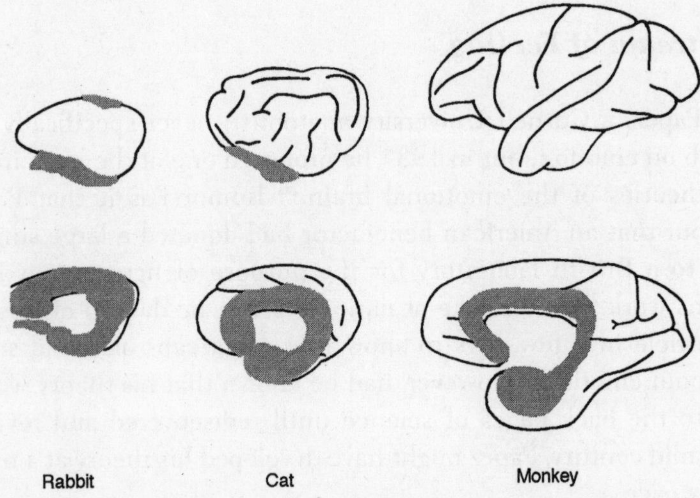

FIGURE 4-7
Lateral and Medial Cortex in Mammalian Evolution.

The cerebrum of a rabbit (left), cat (center), and monkey (right), is shown in lateral (top row) and medial (bottom row) views. The so-called limbic lobe, supposedly made up of evolutionarily old cortex, is shown in gray, and the evolutionarily new cortex, the neocortex, is in white. The limbic lobe is mostly located in the medial wall of the cerebrum. In the rabbit, the limbic lobe accounts for most of the medial cortex. In cats and primates, the limbic lobe accounts for progressively less of the medial cortex, and thus for less of the cortical mass. The changes across mammals reflect the expansion of the neocortex. Cortical expansion reaches its greatest extent (so far) in humans. (Based on P.D. MacLean [1954], Studies on limbic system ["visceral brain] and their bearing on psychosomatic problems. In E. Wittkower and R. Cleghorn, eds., *Recent Developments in Psychosomatic Medicine*. London: Pitman.)

means "smell brain," to account for its apparent involvement in the perception of odors and in controlling behaviors guided by smell.

Herrick noted that in primitive animals smell plays an important role in feeding, sexual, and defensive behaviors. He proposed that the higher intellectual functions mediated by the lateral neocortex evolved from the sense of smell and that the lateral cortex itself is an evolutionary outgrowth of the smell brain. According to Herrick, the basic sensory and motor functions controlled by the medial cortex in primitive animals were transferred over to the newly evolved lateral cortex, allowing room for the elaboration of sensation into higher thought processes and the expansion of primitive motor functions of early vertebrates into complex human behaviors.

Papez was a great synthesizer and put together Herrick's idea about the evolutionary distinction between the medial and lateral cortex with two other kinds of findings—observations about the consequences of brain damage in the medial cortex in humans and research on the role of the hypothalamus in the control of emotional reactions in animals. The outcome was a theory that explained the subjective experience of emotion in terms of the flow of information through a circle of anatomical connections from the hypothalamus to the medial cortex and back to the hypothalamus. This is now known as the Papez circuit.

Papez, in the tradition of Cannon, emphasized the importance of the hypothalamus in the reception of direct sensory inputs about emotional stimuli from the thalamus, in the control of bodily responses during emotion, and in the regulation of emotional experience by ascending fibers to the cortex. However, Papez was particularly concerned with further illuminating how emotional experience might come out of the brain and proposed a more elaborate and detailed emotional network than Cannon.

The Papez hypothesis begins with the idea that sensory inputs transmitted into the brain are split at the level of the way stations in the thalamus into the *stream of thought* and the *stream of feeling*. The stream of thought is the channel through which sensory inputs are transmitted, by way of paths through the thalamus, to the lateral areas of the neocortex. Through this stream, sensations are turned into perceptions, thoughts, and memories. The stream of feeling also involved sensory transmission to the thalamus, but at that point the information was relayed, as Cannon proposed, directly to the hypothalamus, allowing the generation of emotions.

Cannon treated the hypothalamus as a homogeneous structure. But Papez pinpointed the hypothalamic mammillary bodies, so named because of their breastlike protrusion from the bottom of the brain, as the place that receives the thalamic sensory inputs and then relays the messages toward the cortex. And he was also quite specific about which part of the cortex was involved, proposing that the cingulate cortex (a part of the older, medial cortex) is the cortical region for the perception of emotion, just as the visual cortex is the region for visual perception. And pursuing the analogy with sensory systems, he proposed that the anterior thalamic nucleus, which connects the mammillary bodies to the cingulate cortex, is a thalamic relay in the emotional system. But the circuit did not stop there. The cingulate cortex then sends its outputs to the hippocampus, another old, medial cortical area, the output of which was directed back to the hypothalamus, thus completing the circle of emotion (see Figure 4-8).

In Papez's time, the actual connections of the brain were very poorly understood, as the methods for tracing connections between regions were crude. Papez therefore based his circuit partly on the known connections, but also on the clinical effects of damage to the various brain regions and speculation about what connections might exist. The hippocampus was included in the circuit because it was known to be a major site of brain damage associated with rabies, a disease characterized by "intense emotional, convulsive, and paralytic symptoms," in which the patient presents the "appearance of intense fright and of mingled terror and rage."[24] The cingulate was given a central role because damage there resulted in apathy, drowsiness, delirium, depression, loss of emotional spontaneity, disorientation in time and space, and sometimes coma.

Descartes had put the soul in the pineal gland, the only part of the brain not divided in two halves.[25] Papez was more sympathetic with the suggestion of a later Frenchman, La Peyronie, the professor of surgery at Montpellier, who located the seat of the soul near the cingulate region. But Papez had a more modest aim than La Peyronie, referring to the cingulate cortex as the place where "environmental events are endowed with an emotional consciousness."[26] If

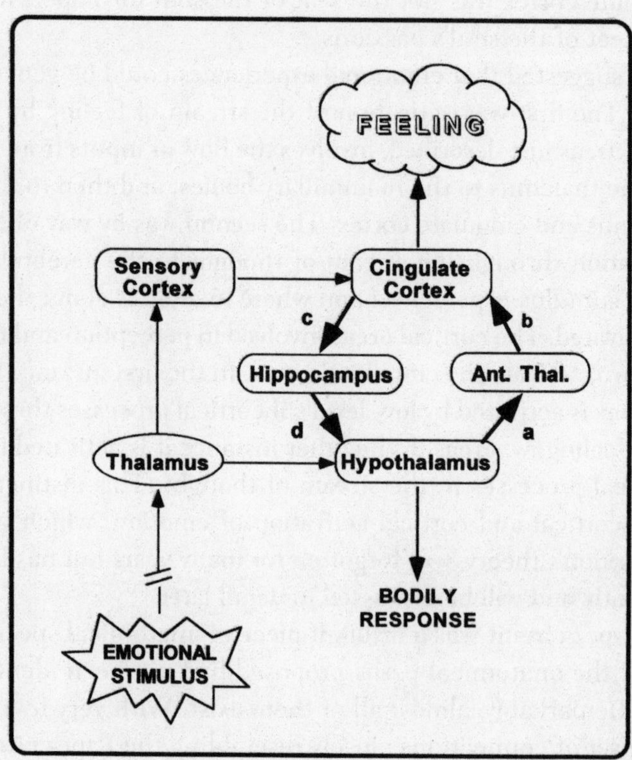

FIGURE 4-8
The Papez Circuit Theory.

Like Cannon and Bard, Papez believed that sensory messages reaching the thalamus are directed to both the cerebral cortex and the hypothalamus; the outputs of the hypothalamus to the body control emotional responses; and outputs to the cortex give rise to emotional feelings. The paths to the cortex were called the "stream of thinking" and the ones to the hypothalamus the "stream of feeling." Papez was considerably more specific than Cannon and Bard about how the hypothalamus communicates with the cortex and about which cortical areas are involved. He proposed a series of connections from the hypothalamus to the anterior thalamus, to the cingulate cortex (part of the evolutionarily old, medial cortex). Emotional experiences occur when the cingulate cortex integrates signals from sensory cortex (part of the evolutionarily new, lateral cortex) and the hypothalamus. Outputs from the cingulate cortex to the hippocampus and then the hypothalamus allow thoughts occurring in the cerebral cortex to control emotional responses.

the cingulate cortex was not the seat of the soul for Papez, it was at least the seat of the soul's passions.

Papez suggested that emotional experiences could be generated in two ways. The first was activation of the stream of feeling by sensory objects. This, as just described, involves the flow of inputs from sensory areas of the thalamus to the mammillary bodies, and then to the anterior thalamus and cingulate cortex. The second was by way of the flow of information through the stream of thought to the cerebral cortex, where the stimulus is perceived and where memories about the stimulus are activated. The cortical areas involved in perception and memory then, in turn, activate the cingulate cortex. In the first instance the cingulate cortex is activated by low-level subcortical processes through the stream of feeling, whereas in the other instance it is activated by high-level cortical processes in the stream of thought. This distinction between subcortical and cortical activation of emotion, which was also part of Cannon's theory, was forgotten for many years but has been revived recently and will be discussed in detail later.

The Papez circuit was a brilliant piece of anatomical speculation, as most of the anatomical paths proposed had not been identified at the time. Remarkably, almost all of them exist. With very few modifications, a set of connections closely resembling the Papez circuit has been found. Unfortunately, at least for the Papez theory of emotion, this circuit appears to have little involvement in emotion. Nevertheless, the Papez circuit theory is a crucial part of the history of the emotional brain, as it was the takeoff point for the limbic system theory, which will be considered shortly.

Psychic Blindness

Nineteen thirty-seven was a banner year for the emotional brain. Not only was the Papez theory published then, but so too was the first in a series of reports by Heinrich Klüver and Paul Bucy.[27] These investigators were involved in a study of the brain regions that mediate drug-induced visual hallucinations when they stumbled upon a remarkable set of observations on the effects of damage to the temporal lobes of monkeys.

The lateral cerebral cortex can be divided into four subparts, known as lobes. The occipital lobe is in the back and is where the visual cortex is located. The frontal lobe, obviously, is in the front, and rests just above the eyes on each side. Between the occipital and frontal lobes are the parietal and temporal lobes. The parietal lobe sits on top, and the temporal lobe is just below it, right behind and slightly above your ears (see Figure 4-2).

In their first report on one monkey, a case study of the effects of temporal lobe removal, Klüver and Bucy noted:

> the animal does not exhibit the reactions generally associated with anger and fear. It approaches humans and animals, animate as well as inanimate objects without hesitation and although there are no motor defects tends to examine them by mouth rather than by the use of the hands. . . . Various tests do not show any impairment in visual acuity or in the ability to localize visually the position of objects in space. However, the monkey seems to be unable to recognize objects by the sense of sight. The hungry animal, if confronted with a variety of objects, will, for example, indiscriminately pick up a comb, a bakelite knob, a sunflower seed, a screw, a stick, a piece of apple, a live snake, a piece of banana, and a live rat. Each object is transferred to the mouth and then discarded if not edible.[28]

They referred to this collection of symptoms as "psychic blindness," by which they meant that the animals had perfectly good visual acuity but were blind to the psychological significance of stimuli.

Later studies confirmed the major facets of what has come to be known as the Klüver-Bucy syndrome.[29] Animals with such lesions are "tame" in the presence of previously feared objects (people and snakes); they will put almost anything in their mouths, being unable to identify by sight alone whether something is edible; and they become hypersexual, attempting to copulate with other monkeys of the same sex or with members of other species (sexual activities seldom if ever practiced by "normal" monkeys).

In the wake of Klüver and Bucy's publications, a plethora of research was conducted to try to further understand the nature of the syndrome. This work had a significant impact on several areas of brain research, including the search for the brain mechanisms of visual perception, long-term memory, and emotion. But the impact of most relevance to us at the moment was its influence on Paul MacLean and his limbic system theory of the emotional brain.

The Emotional Keyboard

Research on the neural basis of emotion was interrupted by World War II. But things began to pick up steam again in 1949 when Paul MacLean revived and expanded the Papez theory, integrating it with the Klüver-Bucy syndrome and with Freudian psychology.[30] The Papez theory, in fact, might have faded quietly into the past had it not been a major inspiration for MacLean's treatise.

MacLean sought to construct an all-encompassing theory of the emotional brain. Drawing on the work of Cannon and Papez, as well as that of Klüver and Bucy, MacLean noted the importance of the hypothalamus in emotional expression and the importance of the cerebral cortex in emotional experience. He sought to identify some way in which these regions might communicate and thereby allow the affective qualities of experience to act on the autonomic and behavioral control systems in the generation of emotional responses and in the establishment and maintenance of psychosomatic diseases, such as hypertension, asthma, and peptic ulcers.

MacLean, like many others before him and since, believed that the capacity to appreciate all of the various affective or emotional qualities of experience and to differentiate them into feeling states (like fear, anger, love, and hate) required the cerebral cortex. At the same time, the newly evolved part of the cortex, the neocortex, was known to lack significant connections with the hypothalamus, and therefore could not act on the autonomic centers to produce visceral responses. However, as Papez and Herrick had argued, the evolutionarily older areas of the medial cortex, the so-called rhinencephalon, are intimately connected with the hypothalamus. That the rhinencephalon was not just a smell brain in higher mammals was evident from the fact that dolphins and porpoises, which have no sense of smell, have highly elaborate rhinencephalic regions, and that in humans the olfactory sense is of comparatively less importance but certain areas of the rhinencephalon (especially the hippocampus and cingulate regions) reach their greatest development.

Clinical evidence was added to the list supplied by Papez to support the role of the rhinencephalon in emotion. For example, MacLean noted that in temporal lobe epilepsy, which often involves pathology in the hippocampal region, the patient is sometimes in a "dreamy state" and can have a feeling of fear or even terror right before having a seizure. Further, patients with epilepsy may be afflicted with severe emotional and psychological disturbances (nervousness, obsessive thinking, depression) between seizures. The role of the mammillary bodies in emotion was further supported by the fact that damage to this region, often as a consequence of vitamin deficiencies resulting from poor dietary habits in prolonged and severe alcoholism, can result in psychotic behavior. Further, stimulation of the brain in the region of the mammillary bodies was known to produce blood pressure elevations, which MacLean took to mean that this region is involved in psychosomatic diseases, such as high blood pressure due to excess stress. And, as further support of importance of the cingulate cortex in emotion, he suggested that the changes in respiration elicited by stimulation of this region might be related to psychosomatic forms of asthma. And he also mentioned the case of a fifty-five-year-old woman with a tumor in the vicinity of the cingulate region. The main symptoms exhibited were nymphomania accompanied by a persistent passionate feeling, which was exaggerated by the smell of perfume, possibly because of the importance of the smell brain to emotion.

The accuracy of these anatomical correlations is, at this point, of less importance than their implications for MacLean. In his mind, all roads led to the smell brain as the seat of emotion. Noting that stimulation of the rhinencephalic areas, but not areas of the neocortex, typically produced autonomic responses (changes in respiration, blood pressure, heart rate, and other visceral functions), he renamed the rhinencephalon the visceral brain. MacLean suggested that, "although, in the ascension to higher forms, the rhinencephalon yields more and more control over the animal's movements to the neocortex, its persistent, strong connections with lower autonomic centers suggests that it continues to dominate in the realm of visceral activity." While the neocortex "holds sway over the body musculature and subserves the functions of the intellect," the visceral brain is the region involved in "ordering the affective behavior of the animal in such basic drives as obtaining and assimilating food, fleeing from or orally disposing of an enemy, reproducing, and so forth."[31]

MacLean's basic idea was that in primitive animals the visceral brain was the highest center available for coordinating behavior, since the neocortex had not yet evolved. In these primitive creatures, the visceral brain took care of all the instinctual behaviors and basic drives underlying the survival of the individual and the species. With the emergence of the neocortex in the mammals, the capacity for higher forms of psychological function, like thinking and reasoning, began to emerge and reached its zenith in man. But even in man, the visceral brain remains essentially unchanged and is involved in the primitive functions that it carried out in our distal evolutionary ancestors.

MacLean believed that emotional feelings involve the integration of sensations arising from the external environment with visceral sensations from within the body and proposed that these integrations take place in the visceral brain. His theory was in essence a feedback hypothesis about the nature of emotion, not unlike James'. That is, emotional stimuli in the external world produce responses in the visceral organs. Messages from these internal organs then return to the brain, where they are integrated with ongoing perceptions of the outside world. This integration of the internal and external worlds was viewed as the mechanism that generates emotional experience:

> The problem pertaining to emotional mechanisms is basically one of communication in the nervous system. It may be assumed that messages from both without and within the organism are relayed to the brain by nervous impulses traveling along nerve fibers and possibly by humoral agents carried in the blood stream. Ultimately, however, any correlation of these messages must be a function of a highly integrated body of neurones capable of sorting, selecting, and acting upon various patterns of bioelectrical activity. The indications are that both the experience and the expression of emotion are the resultant of the association and correlation of a wide variety of internal and external stimuli whose messages are transmitted as nervous impulses in cerebral analyzers.[32]

The cerebral analyzers underlying emotion were, in MacLean's view, located in the visceral brain, and particularly in the hippocampus, so-named for its sea horse shape—in Greek mythology, *hippokampos* was a horse-shaped (*hippo*) sea monster (*kampos*).

MacLean poetically described the large nerve cells in the hippocampus as an emotional keyboard. The keyboard idea comes from the fact that cells in this region are very orderly arranged, side by side. When the elements of the sensory world activate these cells, the tunes they play are the emotions we experience (see figure 4-9).

MacLean proposed that our emotions, in contrast to our

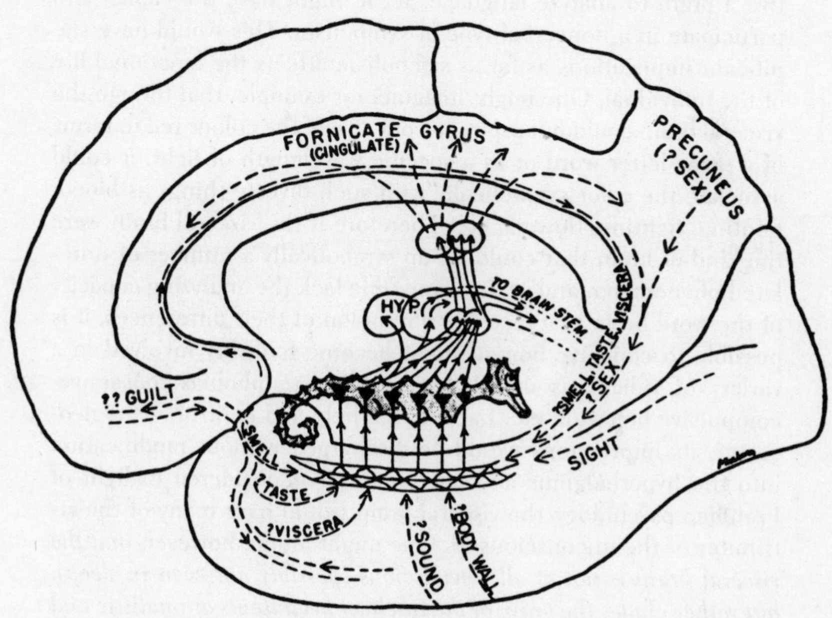

FIGURE 4-9

MacLean's Visceral Brain (Limbic System) Theory.

The centerpiece of the limbic system was the hippocampus (shown as a seahorse). It was believed to receive inputs from the external world (sight, smell, hearing, touch, taste) as well as from the internal or visceral environment. The integration of internal and external sensations was viewed as the basis of emotional experience. The pyramidal cells of the hippocampus (black triangle inside the seahorse) were viewed as the emotional keyboard (see text). (Reprinted from P. MacLean [1949], Psychosomatic disease and the "visceral brain." Recent developments bearing on the Papez theory of emotion. *Psychosomatic Medicine* 11, 338–53.)

thoughts, are difficult for us to understand precisely because of structural differences between the organization of the hippocampus, the centerpiece of the visceral brain, and the neocortex, the home of the thinking (word) brain: "the cortical cytoarchitecture of the hippocampal formation indicates that it would have little efficiency as an analyzer compared to the neocortex." Further elaborating this idea, he said:

> one might infer that the hippocampal system could hardly deal with information in more than a crude way, and was possibly too primitive a brain to analyze language. Yet it might have the capacity to participate in a nonverbal type of symbolism. This would have significant implications as far as symbolism affects the emotional life of the individual. One might imagine, for example, that though the visceral brain could not aspire to conceive of the colour red in terms of a three-letter word or as a specific wavelength of light, it could associate the color symbolically with such diverse things as blood, fainting, fighting, flowers, etc. Therefore if the visceral brain were the kind of brain that could tie up symbolically a number of unrelated phenomena, and at the same time lack the analyzing capacity of the word brain to a nice discrimination of their differences, it is possible to conceive how it might become foolishly involved in a variety of ridiculous correlations leading to phobias, obsessive-compulsive behavior, etc. Lacking the help and control of the neocortex, its impressions would be discharged without modification into the hypothalamus and lower centers. Considered in light of Freudian psychology, the visceral brain would have many of the attributes of the unconscious id. One might argue, however, *that the visceral brain is not at all unconscious (possibly not even in sleep), but rather eludes the grasp of the intellect because its animalistic and primitive structure makes it impossible to communicate in verbal terms.*[33] [MacLean's italics]

MacLean pursued the hypothesis, and a very radical one at the time, that psychiatric problems might be attributable to disorders of the visceral brain, and particularly that the visceral brain might be the source of pathology in patients with psychosomatic symptoms:

> so much of the information obtained from these patients has to do with material which in a Freudian sense is assigned to the oral and oral-anal level, or, as one might say all inclusively, the visceral level. In practically all the psychosomatic diseases such as hypertension,

> peptic ulcer, asthma, ulcerative colitis, that have been subject to fairly extensive psychiatric investigation, great emphasis has been placed on the "oral" needs, the "oral" dependencies, the "oral" drives, etc., of the patients. These oral factors have been related to rage, hostility, fear, insecurity, resentment, grief, and a variety of other emotional states. In certain circumstances, for example, eating food may be the symbolic representation of psychologic phenomena as diverse as 1) the hostile desire to eradicate an inimical person, 2) the need for love, 3) fear of some deprivation or punishment, 4) the grief of separation, etc. . . . Many of the seemingly paradoxical and ridiculous implications of the term "oral" result from a situation, most clearly manifest in children or primitive peoples, where there is a failure or inability to discriminate between the internal and external perceptions that make up the affective qualities of experience. . . . [The] emotional life [of psychosomatic patients] often becomes a matter of "inviscerating" or "exviscerating." It is as if the person never "learned to walk" emotionally. . . . In the psychosomatic patient it would almost seem that there was little direct exchange between the visceral brain and the word brain, and that emotional feelings built up in the hippocampal formation, instead of being relayed to the intellect for evaluation, found immediate expression through autonomic centers.[34]

He goes on to suggest that one would not expect to accomplish a great deal with words at the beginning of psychotherapy and that the most important first steps would be for the therapist to relate to the patient's visceral brain.[35]

In 1952, three years after publication of the visceral brain hypothesis, MacLean introduced the term "limbic system" as a new name for the visceral brain.[36] Limbic, you will recall, comes from Broca's description of the rim of medial cortex that later became the rhinencephalon. But in contrast to Broca, MacLean had function, not structure, on his mind when he packaged Broca's limbic cortex and related cortical and subcortical regions into the limbic system. In addition to the areas of the Papez circuit, MacLean included regions like the amygdala, septum, and prefrontal cortex in the limbic system. He then proposed that the structures of the limbic system comprise a phylogenetically early neural development that functions in an integrated way, in fact as a system, in maintaining the survival of the individual and the species. This system evolved to mediate visceral functions and affective behaviors, including feeding, defense, fighting, and reproduction. It underlies the visceral or emotional life of the individual.

MacLean has continued to develop and embellish the visceral brain/limbic system theory over the years. In 1970 he introduced his theory of the triune brain.[37] The forebrain, according to MacLean, has gone through three stages of evolution: reptilian, paleomammalian, and neomammalian. He notes, "there results a remarkable linkage of three cerebrotypes which are radically different in chemistry and structure and which in an evolutionary sense are eons apart. There exists, so to speak, a hierarchy of three-brains-in-one, or what I call, for short, a *triune brain*."[38] Each of the cerebrotypes, according to MacLean, has its own special kind of intelligence, its own special memory, its own sense of time and space, and its own motor and other functions. In humans, other primates, and advanced mammals, all three brains exist. Lower mammals lack the neomammalian brain, but have the paleomammalian and reptilian brain. All other vertebrate creatures (birds, reptiles, amphibians, and fishes) have only the reptilian brain. The paleomammalian brain, present in all mammals, is essentially the limbic system. The triune brain thus puts the limbic system into a broader evolutionary context to account for behaviors and mental functions of all levels of complexity.

Trouble in Paradise

What a synthesis! Reading MacLean's original writings, it is easy to see why the problem of the emotional brain seemed, by 1952, all wrapped up. The theory was far reaching and covered the latest in brain science, psychology, psychiatry, and even managed to talk about the new emerging ideas about computer modeling of neural activity. It was an amazing achievement. There have been few if any theories in neuroscience that have been as broad in their scope, as wide ranging in their implications, and as long-lived. The limbic system concept survives to this very day as the major view of the emotional brain. Textbooks of neuroanatomy routinely include a chapter on the structural organization and function of the limbic system. It is a household concept in the mind of every brain scientist. Lay dictionaries have an entry for the term, describing the limbic system as a circle of connections that mediates emotions.

Unfortunately, the idea that the limbic system constitutes the emotional brain is, for a variety of reasons, not acceptable. But before explaining why, I'd like to separate MacLean's fascinating and penetrating ideas about the nature of emotion and emotional disorders from the limbic system theory. I think he did an incredible job of conceptualizing the general way in which emotions might come out of brains. Like MacLean, and unlike many contemporary cognitive and social constructivist theorists, I believe that it is essential that the emotional brain be viewed from an evolutionary perspective.[39] I am very fond of his idea that the emotional brain and the "word brain" might be operating in parallel but using different codes and thus are not necessarily able to communicate with each other. And I also think that his idea that some psychiatric problems might represent the operation of the emotional brain independent of the "word brain" is on the mark. But these gems need to be separated from the rest of the limbic system theory.

The limbic system theory was a theory of localization. It proposed to tell us where emotion lives in the brain. But MacLean and later enthusiasts of the limbic system have not managed to give us a good way of identifying what parts of the brain actually make up the limbic system.

MacLean said that the limbic system is made up of phylogenetically old cortex and anatomically related subcortical areas. Phylogenetically old cortex is cortex that was present in very old (in an evolutionary sense) animals. Although these animals are long gone, their distal progeny are around and we can look in the brains of living fish, amphibians, birds, and reptiles and see what kinds of cortical areas they have and compare these to the kinds of areas that are present in newly evolved creatures—humans and other mammals. When anatomists did this early in this century, they concluded that the lowly animals only have the medial (old) cortex, but mammals have both the medial and lateral (new) cortex.[40]

This kind of evolutionary neurologic carried the day for a long time, and it was perfectly reasonable for Herrick, Papez, MacLean, and many others to latch on to it. But, by the early 1970s, this view had begun to crumble. Anatomists like Harvey Karten and Glenn Northcutt were showing that so-called primitive creatures do in fact have areas that meet the structural and functional criteria of neocortex.[41] What had been confusing was that these cortical areas were not exactly in the place that they are in mammals so it was not obvious that they were the same structures. As a result of these discoveries, it is no longer possible to say that some parts of the mammalian cortex were older than other parts. And once the distinction between old and new cortex breaks down, the whole concept of mammalian brain evolution is turned on its head.[42] As a result, the evolutionary basis of the limbic lobe, rhinencephalon, visceral brain, and limbic system concepts has become suspect.[43]

Another idea was that the limbic system might be defined on the basis of connectivity with the hypothalamus. After all, this is what led MacLean to the medial cortex in the first place. But with newer, more refined methods, it has been shown that the hypothalamus is connected with all levels of the nervous system, including the neocortex. Connectivity with the hypothalamus turns the limbic system into the entire brain, which doesn't help us very much.

MacLean also proposed that areas of the limbic system be identified on the basis of their involvement in visceral functions. While it is true that some areas traditionally included in the limbic system contribute to the control of the autonomic nervous system, other areas, like the hippocampus, are now believed to have relatively less involvement in autonomic and emotional functions than in cognition.[44] And other areas not included in the limbic system by anyone (especially areas in the lower brain stem) are primarily involved in autonomic regulation. Visceral regulation is a poor basis for identifying the limbic system.

Involvement in emotional functions is, obviously, another way the limbic system has been looked for. If the limbic system is the emotion system, then studies showing which brain areas are involved in emotion should tell us where the limbic system is. But this is backward reasoning. The goal of the limbic system theory was to tell us where emotion is in the brain on the basis of knowing something about the evolution of brain structure. To use research on emotion to find the limbic system turns this criterion around. Research on emotion can tell us where the emotion system is in the brain, but not where the limbic system is. Either the limbic system exists or it does not. Since there are no independent criteria for telling us where it is, I have to say that it does not exist.

But let's consider the issue of using research on emotion to define the limbic system a little further. MacLean had proposed that the limbic system was the kind of system that would be involved in primitive emotional functions and not in higher thought processes. Recent research, which we will discuss in detail later, is very problematic for this view. For example, damage to the hippocampus and some regions of the Papez circuit, like the mammillary bodies and anterior thalamus, have relatively little consistent effect on emotional functions but produce pronounced disorders of conscious or declarative memory—the ability to know what you did a few minutes ago and to store that information and retrieve it at some later time and to verbally describe what you remember. These were exactly the kinds of processes that MacLean proposed that the visceral brain and limbic system would not be involved with. The relative absence of involvement in emotion and the clear involvement in cognition are major difficulties for the view that the limbic system, however one chooses to define it, is the emotional brain.

How, then, has the limbic system theory of emotion survived so long if there is so little evidence for its existence or for its involvement in emotion? There are many explanations that one could come up with. Two seem particularly cogent. One is that, though imprecise, the limbic system term is a useful anatomical shorthand for areas located in the no-man's-land between the hypothalamus and the neocortex, the lowest and highest (in structural terms) regions of the forebrain, respectively. But scientists should be precise. The limbic system term, even when used in a shorthand structural sense, is imprecise and has unwarranted functional (emotional) implications. It should be discarded.[45]

Another explanation for the survival of the limbic system theory of emotion is that it is not completely wrong—some limbic areas have been implicated in emotional functions. Given that the limbic system is a tightly packaged concept (though not a tightly organized, well-defined system in the brain), evidence that one limbic area is involved in some emotional process has often been generalized to validate the idea that the limbic system as a whole is involved in emotion. And, by the same token, the demonstration that a limbic region is involved in one emotional process is often generalized to all emotional processes. Through these kinds of poorly reasoned associations, involvement of a particular limbic area in a very specific emotional process has tended to substantiate the view that the limbic system is the emotional brain.

Like other theories that preceded it, the limbic system theory of the emotional brain was meant to apply equally to all emotions. It was a general theory of how feelings come from the brain. This general explanation was based on a specific functional hypothesis—when information about the external world is integrated with sensations arising from within the body, we have feelings. While the general limbic system concept has been adopted by many subsequent researchers and theorists, the specific hypothesis about integration of internal and external senses that led MacLean to the general theory has been left behind. The general theory, that the brain has a limbic system and that our emotions come out of this system, took on a life of its own and has survived, and even thrived, independent of its conceptual origins. Even as research has shown that classical limbic areas are by no means dedicated to emotion, the theory has persisted. Implicit in such a view is that emotion is a single faculty of mind and that a single unified system of the brain evolved to mediate this faculty. While it is possible that this view is correct, there is little evidence that it is. A new approach to the emotional brain is needed.

One of MacLean's many important insights was his emphasis on the evolution of the brain as a key to understanding emotions. He saw emotions as brain functions involved in maintaining the survival of the individual and the species. From the vantage point of hindsight, it seems that his mistake was to package the entire emotional brain and its evolutionary history into one system. I think that his logic about emotional evolution was perfect, he just applied it too broadly. Emotions are indeed functions involved in survival. But since different emotions are involved with different survival functions—defending against danger, finding food and mates, caring for offspring, and so on—each may well involve different brain systems that evolved for different reasons. As a result, there may not be one emotional system in the brain but many.

5
THE WAY WE WERE

"Human subtlety . . . will never devise an invention more beautiful, more simple or more direct than does nature."

Leonardo da Vinci, *The Notebooks* (1508–1518)[1]

When engineers sit down to design machines, they start off with some function they want to implement and then figure out how to make a device that will accomplish the task. But biological machines aren't assembled from carefully engineered plans. The human brain, for example, happens to be the most sophisticated machine imaginable, or unimaginable, yet it wasn't predesigned. It is the product of evolutionary tinkering, where lots of little changes over extremely long periods of time have accumulated.[2]

Organisms, according to Stephen J. Gould, are Rube Goldberg devices, patchworks of quick fixes and partial solutions that shouldn't work but somehow do the trick.[3] Evolution works with what it has, rather than starting from scratch. As evolutionary biologist Richard Dawkins points out, this is horribly inefficient on small time scales—it would have been foolish to try and construct the first jet engine by modifying an existing gasoline engine.[4] But, as Dawkins notes, evolution's strategy of tinkering works pretty well over the huge spans of time on which it operates. Besides, there's no alternative.

The problem of figuring out how a brain works has been described by the linguist Steven Pinker as "reverse engineering."[5] We've got the product and we want to know how it functions. So we pick the brain apart in the hope that we will see what evolution was up to when it put the device together.

Although we often talk about the brain as if it has a function, the brain itself actually has no function. It is a collection of systems, sometimes called modules, each with different functions.[6] There is no equation by which the combination of the functions of all the different systems mixed together equals an additional function called brain function.

Evolution tends to act on the individual modules and their functions rather than the brain as a whole. For example, there is evidence that specific brain adaptations underlie certain capacities, like song learning in birds, memory for food location in foragers, sex differences, hand preferences, and language skills in humans.[7] It is true that at times evolution might act globally,[8] say increasing the size of the entire brain, but by and large most evolutionary changes in the brain take place at the level of individual modules. These modules accomplish such mental exotica as having thoughts or beliefs, but also activities as mundane as breathing. Evolutionary improvements in our ability to believe do not necessarily help us breathe any better. They may, but they don't necessarily.

Admittedly, breathing and believing are pretty distinct functions, clearly mediated by different brain regions. Breathing is controlled in the medulla oblongata, that utility station down in the subbasement of the brain, whereas believing, like all good higher cognitive functions, goes on up in the neocortical penthouse. Contrasting these is not so interesting.

So let's consider functions that are more similar, like different emotions. Do changes in our ability to detect danger and respond to it help us in our love life or make us less prone to anger or depression? They could, especially if there was a universal emotional system in the brain, a system dedicated to emotional functions and within which all emotional functions are mediated. A general improvement in the operation of this system would probably impact across the board on all emotions. We could certainly concoct an explanation for why feeling good and bad may have been beneficial to survival to our ancestors, and theirs, and therefore why an all-purpose emotion system might have evolved. In the last chapter, though, we saw that attempts to find a single unified brain system of emotion have not been very successful. It is possible that such a system exists and that scientists just haven't been clever enough to find it, but I don't think that's the case. Most likely, attempts to find an all-purpose emotion system have failed because such a system does not exist. Different emotions are mediated by different brain networks, different modules, and evolutionary changes in a particular network don't necessarily affect the others directly. There might of course be indirect effects—an increased ability to detect danger and defend against it might leave more time and resources for pursuing romantic interests—but this is a different matter.

If I'm correct, the only way to understand how emotions come out of brains is to study emotions one at a time. If there are different emotional systems and we ignore this diversity we will never make much sense of the brain's emotional secrets. If I'm wrong, though, we've lost nothing by taking this approach. We can mix the findings from fear, anger, disgust, and joy back together anytime.

For these reasons, my research on the emotional brain has been focused on the neural basis of one particular emotion—fear, and its various incarnations. Much of the remainder of this book is aimed at explaining what we know about the brain mechanisms of fear, especially what we've learned from research on fear behavior in nonhuman animals, and then seeing to what extent this knowledge can help us understand "emotion" in the broader sense of the term (especially human emotion). But before going further, I need to convince you that the study of fear behavior in animals is a good starting point. And before doing this, I need to go through some ideas about the evolution of emotions, some criticisms of them, and my take on where the balance lies.

To Change or Not to Change, That Is the (Evolutionary) Question

For some mental functions, like language, the job facing evolutionary theorists is to try to understand how the function came to be in humans. Our species seems to be the only one living now that is endowed with natural language.[9] So the big question, in terms of origins, is what did language evolve from—what were the intermediate phases that the brain passed through in the transition from non-speaking to speaking primates?

When it comes to emotions, though, we face a different problem. Contrary to the views of some humanists, I believe that emotions are anything but uniquely human traits and, in fact, that some emotional systems in the brain are essentially the same in many of the backboned creatures, including mammals, reptiles, and birds, and possibly amphibians and fishes as well. If this is true, and I'll try to convince you that it is, our first order of business is quite different from that of the evolutionary linguists. Rather than trying to figure out what is unique about human emotion, we need to examine how evolution stubbornly maintains emotional functions across species while changing many other brain functions and bodily traits.

If people had wings, William James would have posed his famous question about running away from the bear in terms of flying away. He would have asked whether fear is the result of flying away from danger or whether flying from danger causes us to be afraid. Stated this way, the question loses none of its meaning. Escaping from danger is something that all animals have to do to survive. Uniquely human traits, like the ability to compose poetry or solve differential equations, are irrelevant to what goes on when we are faced with a sudden and immediate threat to our existence. What is important is that the brain have a mechanism for detecting the danger and responding to it appropriately and quickly. The particular behavior that occurs is tailored to the species (running, flying, swimming), but the brain function underlying that response is the same—protection against the danger.[10] This is as true of a human animal as of a slimy reptile. And, as we will see, evolution has seen fit to pretty much leave well enough alone inside the brain when it comes to these functions.

Emotional Descent

The belief that at least some emotions might be shared by man and other creatures has been around for a long time, at least since Plato proclaimed that the passions are wild beasts trying to escape from the human body.[11] But an understanding of how and why aspects of mind and behavior might be commonly represented in humans and other species remained completely obscure until Charles Darwin conceived of the theory of evolution by natural selection in the last century.[12]

Darwin got his ideas by looking at life around him. He noted that children resemble their parents, but differ from them as well. And he was fascinated with the ability of breeders of domestic animals to build traits in offspring by carefully mixing and matching parents—cows could be made to produce more milk and horses to run faster by preselecting the parents. He reasoned that something similar might occur naturally. Armed with these observations, and others made on his famous voyage to the Galápagos Islands, Darwin proposed that, through heritability and variability, "descent with modification" occurs.

Stephen J. Gould tells us that Darwin did not use the term "evolution" to describe natural selection.[13] At the time, evolution had two other connotations, both of which were incompatible with Darwin's theory. One had to do with the notion that embryos grow from preformed homunculi enclosed in the egg and sperm (tiny preserved versions of Adam and Eve). The other was a vernacular usage that implied constant progress toward an ideal. Darwin felt that a so-called lower form of life, like an amoeba, could be as adapted to its environment as a human is to its—humans, in other words, are not necessarily closer to some evolutionary ideal than other animals. It was really Herbert Spencer, a contemporary of Darwin's, who transformed "descent with modification" into "evolution," the catchier term that we use today.[14]

In rough-and-ready terms, Darwin's theory of natural selection went something like this.[15] Those traits that were useful to the survival of a species in a particular environment became, over the long run, characteristic traits of the species. And, by the same token, the characteristic traits of current species exist because they contributed to the survival of distant ancestors. Because of limited food supplies, not all individuals that are born survive to the point of sexual maturity and procreate. The less fit get weeded out so that over time more and more of the better fit become parents and pass on their fitness to their offspring. But if the environment happens to change, and it does so constantly, then different traits become relevant to survival, and these eventually get selected for. Species that adapt in this way survive, whereas those that do not become extinct.

Darwin's theory is most often thought of as an explanation of how physical features of species evolved. However, he argued that mind and behavior are also shaped by natural selection. James Gould, a behavioral biologist, makes this point forcefully:

> Darwin's revolutionary insights into evolution . . . demonstrated for the first time the inextricable link between an animal's world and its behavior. His theory of natural selection made it possible to understand why animals are so well endowed with mysterious instincts—why a wasp, for example, gathers food she has never eaten to feed larvae she will never see. Natural selection, Darwin hypothesized, favors animals which leave the most offspring. Through countless generations the survivors of the unceasing struggle for a limited amount of food have to be ever more perfectly adapted to their worlds, both morphologically and behaviorally. . . . Carefully programmed behavior like that of the wasp must provide an enormous competitive advantage for animals.[16]

In *The Expression of the Emotions in Man and Animals,* Darwin proposed that "the chief expressive actions, exhibited by man and by the lower animals, are now innate or inherited,—that is, have not been learnt by the individual."[17] As evidence for emotional innateness, he noted the similarity of expressions both within and between species. In humans, Darwin was particularly impressed with the fact that the bodily expressions (especially of the face) occurring during emotions are similar in people around the world, regardless of racial origins or cultural heritage. He also pointed out that these same expressions are present in persons born blind, and thus lacking the opportunity to have learned the muscle movements from seeing them in others, and are also present in very young children, who also have had little opportunity to learn to express emotions by imitation.[18]

Darwin mustered instances of all sorts of bodily expressions that are similar in different species. Although the greatest similarities were found between closely related species, Darwin was able to identify some striking similarities, even within fairly dissimilar organisms. He pointed out how common it is for animals of all varieties, including humans, to urinate and defecate in the face of extreme danger. And many animals erect body hair in dangerous situations, presumably to make themselves look more vicious than they otherwise would. Piloerection, according to Darwin, is probably one of the most general of the emotional expressions, occurring in dogs, lions, hyenas, cows, pigs, antelopes, horses, cats, rodents, bats, to name a few.

Darwin suggested that goose bumps, a mild form of piloerection in humans, occur as a vestige of the more dramatic displays in our mammalian cousins. He points out that it is a remarkable fact that the thinly scattered hairs on the human body are erected in rage and terror, emotional states that cause body hair to stand on end in furry animals, where body piloerection has some purpose. But he noted that piloerection also occurs on the part of the human body that is well endowed with hair, the head, using Brutus' statement to the ghost of Caesar as evidence: "that mak'st my blood cold, and my hair stare."

Darwin gave many other examples of common emotional expression in different species. For example, he equated the snarl of an angry human with similar behaviors in other creatures. Again turning to literature for support, he quotes Dickens' description in *Oliver Twist* of a furious mob witnessing the capture of an atrocious murderer on the streets of London: "the people as jumping up one behind another, snarling with their teeth, and making at him like wild beasts." Darwin goes on to note that "Everyone who has had much to do with young children must have seen how naturally they take to biting, when in a passion. It seems as instinctive in them as in young crocodiles, who snap their little jaws as soon as they emerge from the egg." He also quotes from Dr. Maudsley, who specialized in human insanity and for whom the renowned Maudsley Hospital in London is named: "whence come 'the savage snarl, the destructive disposition, the obscene language, the wild howl, the offensive habits, displayed by some of the insane? Why should a human being, deprived of his reason, ever become so brutal in character, as some do, unless he has the brute nature within him?'" In response, Darwin says, "This question must, as it would appear, be answered in the affirmative."

For Darwin, an important function of emotional expressions is communication between individuals—they show others what particular emotional state one is in. Emission of vicious sounds and enlarging body parts (flashing of feathers, extension of fins or pointy spines, puffing up, and, as we have seen, erection of body hair), are used throughout the animal kingdom to dissuade an enemy from attacking. Sounds, smells, and various postures and displays of body parts or hidden colors serve as signals of sexual receptiveness as well. Sounds are also used to warn others that danger is near. While these signals are somewhat relevant to humans, in the passage below Darwin describes some emotional expressions that are particularly important to our species:

> The movements of expression in the face and body, whatever their origin may have been, are in themselves of much importance in our welfare. They serve as the first means of communication between the mother and her infant; she smiles approval, and thus encourages her child on the right path, or frowns disapproval. We readily perceive sympathy in others by their expression; our sufferings are thus mitigated and our pleasures increased; and mutual good feeling is thus strengthened. The movements of expression give vividness and energy to our spoken words. They reveal the thoughts and intentions of others more truly than do words, which may be falsified.

FIGURE 5-1
Commonality of Emotional Expression in the Faces of Animals and People.

Some emotional expressions are similar in humans and other animals. These two drawings illustrate angry facial expressions in a chimpanzee and human. In both species an expression of anger often involves a direct gaze and a partly opened mouth with lips retracted vertically so that the teeth show. (Drawings, by Eric Stoelting, are reprinted with permission from S. Chevalier-Skolnikoff [1973], Facial expression of emotion in nonhuman primates. In P. Ekman, *Darwin and Facial Expression.* New York: Academic Press.)

A picture may be worth a thousand words, but bodily expressions are priceless commodities in the emotional marketplace.

Darwin argued that although emotional expressions can sometimes be muted by willpower, they are usually involuntary actions. He pointed out how easy it is to tell the difference between a real, involuntary smile and one that is feigned. And he gives us an example from his own life to illustrate how difficult it is to suppress an emotional reaction that has been elicited naturally: "I put my face close to the thick glass-plate in front of a puff-ader in the Zoological Gardens, with the firm determination of not starting back if the snake struck at me; but, as soon as the blow was struck, my resolution went for nothing, and I jumped a yard or two backwards with astonishing rapidity. My will and reason were powerless against the imagination of a danger which had never been experienced."

Within the general class of innate emotions, Darwin suggested that some have older evolutionary histories than others. He noted that fear and rage were expressed in our remote ancestors almost as they are today in humans. Suffering, as in grief or anxiety, though, he placed closer to human origins. Nevertheless, Darwin was well aware of the pitfalls of such ideas about the time of origin of different emotions and noted: "It is a curious, though perhaps an idle speculation, how early in the long line of our progenitors the various expressive movements, now exhibited by man, were successively acquired."

Basic Instinct

A number of modern theorists carry on Darwin's tradition in their emphasis on a set of basic, innate emotions. For many, basic emotions are defined by universal facial expressions that are similar across many different cultures. In Darwin's day, the universality of emotional expression across cultures was presumed from casual observation, but modern researchers have gone into remote areas of the world to firmly establish with scientific methods that at least some emotions have fairly universal modes of expression, especially in the face. On the basis of this kind of evidence, the late Sylvan Tomkins proposed the existence of eight basic emotions: surprise, interest, joy, rage, fear, disgust, shame, and anguish.[19] These were said to represent innate, patterned responses that are controlled by "hardwired" brain systems. A similar theory involving eight basic emotions has been proposed by Carroll Izard.[20] Paul Ekman has a shorter list, consisting of six basic emotions with universal facial expression: surprise, happiness, anger, fear, disgust and sadness.[21] Other theorists, like Robert Plutchik[22] and Nico Frijda,[23] do not rely exclusively on facial expressions, but instead argue for the primacy of more global action tendencies involving many body parts. Plutchik points out that as one goes down the evolutionary scale there are fewer and fewer facial expressions, but still lots of emotional expressions involving other bodily systems. Plutchik's emotions list overlaps with the others, but also diverges to some extent—it is similar to Ekman's, with the addition of acceptance, anticipation, and surprise. Philip Johnson-Laird and Keith Oatley approach the problem of basic emotions by looking at the kinds of words we have for talking about emotions.[24] They come up with a list of five that overlaps with Ekman's six, dropping surprise. Jaak Panksepp has taken a different approach, using the behavioral consequences of electrical stimulation of areas of the rat brain to reveal four basic emotional response patterns: panic, rage, expectancy, and fear.[25] Other theorists have other ways of identifying basic emotions and their lists also overlap and diverge from the ones already described.[26]

Most basic emotions theorists assume that there are also nonbasic emotions that are the result of blends or mixes of the more basic ones. Izard, for example, describes anxiety as the combination of fear and two additional emotions, which can be either guilt, interest, shame, anger, or distress. Plutchik has one of the better developed theories of emotion mixes. He has a circle of emotions, analogous to a circle of colors in which mixing of elementary colors gives new ones. Each basic emotion occupies a position on the circle. Blends of two basic emotions are called dyads. Blends involving adjacent emotions in the circle are first-order dyads, blends involving emotions that are separated by one other emotion are second-order dyads, and so on. Love, in this scheme, is a first-order dyad resulting from the blending of adjacent basic emotions joy and acceptance, whereas guilt is a second-order dyad involving joy and fear, which are separated by acceptance. The further away two basic emotions are, the less likely they are to mix. And if two distant emotions mix, conflict is likely. Fear and surprise are adjacent and readily blend to give rise to alarm, but joy and fear are separated by acceptance and their fusion is imperfect—the conflict that results is the source of the emotion guilt.

The mixing of basic emotions into higher order emotions is typically thought of as a cognitive operation. According to basic emotions theorists, some if not all of the biologically basic emotions are shared with lower animals, but the derived or nonbasic emotions tend to be more uniquely human. Since the derived emotions are constructed by cognitive operations, they could only be the same to the extent that two animals share the same cognitive capacities. And since it is in the area of cognition that humans are believed to differ most significantly from other mammals, nonbasic, cognitively constructed emotions are more likely than basic emotions to differ between humans and other species. Richard Lazarus, for example, proposes that pride, shame, and gratitude might be uniquely human emotions.[27]

Plutchik's 8 Basic Emotions

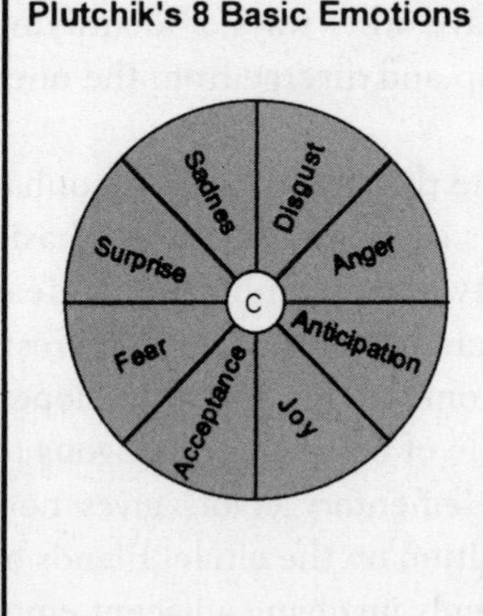

Some Psychosocially Derived Emotions

Primary Dyads (mix of adjacent emotions)
- joy + acceptance = friendliness
- fear + surprise = alarm

Secondary Dyads (mix of emotions, once removed)
- joy + fear = guilt
- sadness + anger = sullenness

Tertiary Dyads (mix of emotions, twice removed)
- joy + surprise = delight
- anticipation + fear = anxiety

FIGURE 5-2
Plutchik's Theory of Basic and Derived Emotions.

(Based on figure 11.4 and table 11.3 in R. Plutchik [1980], *Emotion: A Psychoevolutionary Synthesis.* New York: Harper and Row.)

Being a Wild Pig

The idea of biologically primitive emotions has many supporters but has also had its detractors. One challenge comes from various forms of cognitive emotion theory that propose that specific emotions, even those that are described as basic emotions, are psychological, not biological, constructions. Emotions, in this view, are due to the internal representation and interpretation (appraisal) of situations, not to the mindless workings of biological hardware.

We saw many examples of cognitive views of emotion in Chapter 3. Here, however, I want to focus on the social constructivist approach, which is even further removed from the biology of emotion than most other cognitive approaches. These theorists argue that emotions are products of society, not biology.[28] Cognitive processes play an important role in these theories by providing the mechanism through which the social environment is represented and, on the basis of past experience and future expectations, interpreted. Emotional diversity across cultures is used as evidence in support of this position.

James Averill, a major proponent of social constructivism, describes a behavior pattern, called "being a wild pig," that is quite unusual by Western standards, but is common and even "normal" among the Gururumba, a horticultural people living in the highlands of New Zealand.[29] The behavior gets its name by analogy. There are no undomesticated pigs in this culture, but occasionally, and for unknown reasons, a domesticated one will go through a temporary condition in which it runs wild. But the pig can, with appropriate measures, be redomesticated and returned to the normal pig life among the villagers. And, in a similar vein, Gururumba people can act this way, becoming violent and aggressive and looting and stealing, but seldom causing harm or taking anything of importance, and eventually returning to routine life. In some instances, after several days of living in the forest, during which time the stolen objects are destroyed, the person returns to the village spontaneously with no memory of the experience and is never reminded of the event by the villagers. Others, though, have to be captured and treated like a wild pig—held over a smoking fire until the old self returns. The Gururumba believe that being a wild pig occurs when one is bitten by the ghost of someone who recently died. As a result, social controls on behavior are lost and primitive impulses are set free. According to Averill, being a wild pig is a social, not a biological or even an individual, condition. Westerners are prone to think of this as psychotic, abnormal behavior, but for the Gururumba it is instead a way of relieving stress and maintaining community mental health in the village. Averill uses "being a wild pig" to support his claim that "most standard emotional reactions are socially constructed or institutionalized patterns of response" rather than biologically determined events (one wonders, though, where the wild impulses come from).

Another example of an emotional condition that is not common in Western cultures is the state of mind called "amae" in Japan.[30] Amae has no literal translation in Indo-European languages. The condition it represents is believed by some to be a key to understanding important aspects of Japanese personality structure.[31] It roughly means to presume upon another's love or to indulge in another's kindness. The Japanese psychiatrist Doi calls amae a sense of helplessness and the desire to be loved, to be a passive love object.[32] He believes that amae also occurs in Westerners, but in a much more limited way. The Japanese frequently amaeru (the verb form) but seldom talk about it because it is a nonverbal condition and it would be inappropriate to point it out in another. According to Doi, "those who are close to each other—that is to say, who are privileged to merge with each other—do not need words to express their feelings. One surely would not feel merged with another (that is amae), if one had to verbalize the need to do so!" Doi says that Americans feel encouraged and reassured by such verbal exchanges, but the Japanese neither need it nor find it desirable.

Display Rules

Social constructivists can produce endless lists of all sorts of ways that emotions differ in different cultures or social situations.[33] Certain emotion words from islands in the South Pacific or other remote areas have no translation in English. And even amongst Western cultures, there are differences in emotion words.[34]

But these kinds of observations are not enough to take down the basic emotions view. Basic emotions theorists do not deny that some differences exist in the way emotions are labeled and even expressed between cultures, or even between individuals within a culture. They simply say that some emotions and their expressions are fairly constant in all people. The social constructivists can then counter with the fact that a given individual may express a basic emotion, like anger, differently in different situations—overt anger is more likely to be displayed at those below than those above one in a social hierarchy.

In an attempt to reconcile theories that emphasize the similarity of facial expression across cultures and those that emphasize differences, basic emotions theorist Paul Ekman proposed a distinction between universal emotional expressions (especially facial expressions), which are common to all cultures, and other bodily movements (emblems and illustrators, for example) that vary from culture to culture.[35] Emblems are movements with a specific verbal meaning, such as head nodding to signify yes or no, or shrugging to indicate that you don't know the answer to a question. Emblems could be expressed in words but are not. Illustrators are closely tied to the content and flow of speech. They punctuate speech, help fill in when words cannot be found, or help explain what is being said. In some cultures people "talk with their hands" more than in others. Ekman suggests that social constructivists may be focusing on learned cultural differences in emotional expression, while the basic emotions theorists have been focused on the unlearned, universal expressions that occur in the movement of facial muscles during the occurrence of basic (innate) emotions in all cultures.

Ekman does not claim that basic emotional expressions always look exactly the same. He points out that even universal facial expressions can be regulated by learning and culture. They can be interrupted, diminished, or amplified by learned factors, or even masked by other emotions.[36] He uses the term "display rules" to refer to the conventions, norms, and habits that people develop to manage their emotional expressions. Display rules specify who can show what emotion to whom and when and how much. In Western cultures, there is a grief hierarchy at funerals. As Mark Twain said, "Where a blood relation sobs, an intimate friend should choke up, a distant acquaintance should sigh, a stranger should merely fumble sympathetically with his handkerchief.[37] According to Ekman, if the secretary looks sadder than the wife, suspicions may be aroused. Ekman also suggests that display rules can be personalized and override cultural norms. Some people end up being stoics and show little emotion, even in situations where society allows emotions to flow freely. In Ekman's view, the concept of basic emotions accounts for the similarity of basic emotional expression across individuals and cultures and display rules take care of many of the differences.

Ekman performed a powerful test of his hypothesis.[38] Starting with the assumption that Westerners are more emotionally expressive than Orientals, he studied the facial expressions of Japanese and American subjects while they watched an emotion-arousing film. The subjects were tested in their native countries, and they watched the film while either sitting alone in a room or sitting in a room with an authoritative-looking experimenter in a white coat. Their faces were secretly recorded on videotape throughout. Later, facial expressions were coded by observers who were ignorant of what the subjects were watching. In the private viewing condition, there was a tremendous similarity in the emotions expressed at various points in the film by the Japanese and American subjects. But when the white-coated experimenter was present, the facial movements were no longer the same. The Japanese looked more polite and showed more smiling and less emotional diversity than the Americans. Interestingly, slow-motion analysis of the film revealed that the smiles and other polite facial expressions of the Japanese subjects were superimposed over brief, prior-occurring facial movements that were, according to Ekman, the basic emotions leaking through.

Display rules are learned as part of one's socialization and become so ingrained that, like the basic emotional expressions themselves, they occur automatically, which is to say without conscious participation. At the same time, an individual may sometimes deliberately choose to conceal emotions for a particular advantage in a specific situation. This, however, is a skill that is hard to master—we aren't all good poker players.

Emotional Responses: Parts or Wholes?

The combination of universal emotional expressions and display rules goes a long way toward accounting for individual and cross-cultural variation in the expression of basic emotions, but has not completely inoculated the idea of basic emotions against further challenge. Cognitive scientists Andrew Ortony and Terrance Turner have raised important questions about whether basic emotions can be defined by universal facial expressions, or any other means.[39] They asked why, if basic emotions are so basic, there is so much disagreement about what the basic ones are, and why emotions that are considered basic by some theorists (like interest and desire) are not even considered to be emotions by others. Ortony and Turner then go on to argue that perhaps it is not the emotions and their expressions that are so basic. Instead, they propose that there might be basic, maybe even innate, response components that can be utilized in the expression of emotions, but that are used in other nonemotional situations as well. They note, "emotion expressions are built up by drawing on a repertoire of biologically determined components, and . . . many emotions are often, but by no means always, associated with the same limited subset of such components." They point out that bodily expressions similar to those in an emotion can arise independent of emotions and that the expression typical of one emotion can appear during a different emotional state. Shivering can occur because you are cold or because you are afraid. Crying can occur in extreme happiness as well as sadness. Frowning occurs in anger, but also in frustration, and eyebrows are raised in anger, but also in any condition that requires that we carefully attend to the environment.

For Ortony and Turner, emotion involves higher cognitive processes (appraisals) that organize the various responses that are appropriate to the situation faced by the organism. They accept that component responses can be biologically determined, but place emotion itself in the world of psychological rather than biological determinism. Fear, in their view, is not a biological package that is unwrapped by danger. It is a psychologically constructed set of responses and experiences that are tailored to the particular dangerous situation. There are no emotional responses, there are just responses, and these are put together on the spot when appraisals are made—the particular set of responses that occurs depends on the particular appraisal that occurs. As a result, the number of different emotions is limited only by the number of different appraisals that one can make. And because certain appraisals occur frequently and are often talked about by people, they are easily and reliably labeled with precise terms in most languages and this makes them seem basic (universal).

The reason that Ortony and Turner pushed for a difference between the innateness of emotional expressions and the innateness of response components is simple. If there are no universal expressions characteristic of certain emotions, then the evidence that some emotions, the so-called basic emotions, are biologically determined is brought to its knees. And if emotions are not biologically determined, then they must be psychologically determined. But Ortony and Turner seem to make two unacceptable assumptions. First, just because an appraisal is mental does not mean that it is not also biological. In fact, appraisals play a biological role in some basic emotions theories as the link between emotional stimuli and the characteristic responses they produce. Second, the innateness of individual response components does not preclude the possibility that higher levels of expression are also innate. Some innate behavioral patterns are known to involve hierarchically organized response components.[40] For example, reproductive behavior is often brought on by the presence of hormones that act on certain parts of the brain. When an organism is in the right hormonal state, either mating or fighting can occur, depending on whether a receptive female or a competitive male happens to be around. And these behaviors, though innate, involve many complex levels of control. Mating, for example, may start with a courtship dance, approach toward the partner, and ultimately end in copulation. Each phase, itself, has a complex hierarchy of events controlled by different levels of the nervous system, with the lower levels controlling the most specific components (individual patterns of muscle contraction and relaxation) and higher levels of the nervous system specifying more general aspects of behavior (the act of copulation).

Ortony and Turner caused quite a stir in the world of basic emotions. They made it painfully clear that basic emotions theorists could no longer continue to agree that basic emotions exist and at the same time disagree about what the basic ones are. But now that the dust has settled, it seems that Ortony and Turner were probably too hard on the basic emotions view. Some of the differences between the basic emotions lists of different investigators have to do with the words used rather than with the emotions implied by the words.[41] For example, joy and happiness, basic emotions in different lists, are probably just different names for the same emotion. If we allow these kinds of translations, there turns out to be a good deal of overlap of the different lists: many if not most of the lists include some version of fear, anger, disgust, and joy. Most of the remaining disagreement is over the fringe cases, like interest, desire, and surprise. The basic emotions theorists are not as divergent as they appeared, and, as we will see, at least for some emotions, the evidence for an innate, biological organization is quite strong.

If It Ain't Broke . . .

As has been apparent since Darwin's time (and even before) different animals can act in very similar ways under similar circumstances. This is what led Darwin to propose that certain human emotions have their roots in our animal ancestors. But behavioral commonalities between species can occur at several different levels, and not all of them involve responses that look the same.[42] In other words, the gold standard of whether two animals are doing the same thing is not necessarily whether the two things look exactly the same—emotional commonalities between species might be even broader than conceived by Darwin.

In order for behavior to occur, muscles have to move. So the reason that facial expressions of particular emotions look the same in different people is because everyone contracts and relaxes facial muscles in roughly the same way when exposed to a stimulus that characteristically evokes that emotion. And to the extent that different species show similar kinds of expressions it is because they are contracting and relaxing the same or similar muscle groups—the muscle movements required to furrow the brow and expose the teeth in anger are similar in a human and a chimp. At the same time, the behaviors may be similar at some broader level, but not at the level of individual muscles. People run from danger on two legs, but many other land mammals tend to do so on all four: although quadrupeds use more muscles and different patterns of muscle coordination than bipeds, the function performed is the same—escape. Most important, even when the behaviors are very different, the function achieved may be the same. Plutchik puts this nicely: "although a deer may run from danger, a bird may fly from it, and a fish may swim from it, there is a functional equivalence to all the different patterns of behavior; namely, they all have the common function of separating an organism from a threat to its survival."[43] Obviously, running, flying, and swimming are different behaviors involving different muscles, but each achieves escape.

The implication of Plutchik's notion is that certain basic functions that are necessary for survival have been conserved throughout evolution. They have been modified as needed, but the changes have occurred against a fairly consistent background. In an influential treatise on mother-child bonding in humans, the psychoanalytic theorist John Bowlby makes a similar point:

> The basic structure of man's behavioral equipment resembles that of infrahuman species but has in the course of evolution undergone special modifications that permit the same ends to be reached by a much greater diversity of means. . . . The early form is not superseded: it is modified, elaborated, and augmented but it still determines the overall pattern. . . . Instinctive behavior in humans . . . is assumed to derive from some prototypes that are common to other animal species.[44]

I don't mean to minimize the importance of species differences. The things that distinguish one species from another are often things that allowed its ancestors to survive their particular struggle for existence and pass on their traits to their offspring. The kind of body an animal has obviously limits the kinds of behaviors in which the animal can engage. Nevertheless, evolutionary solutions to problems that are common to many species may have some underlying functional equivalence that cuts across the behavioral differences imposed by the uniqueness of body forms.

The obvious question that this discussion raises is, how could a functional equivalence of behavior be maintained across species, especially across species in which the way the function is expressed behaviorally is radically different? The short answer to this very complicated issue is that the brain systems involved in mediating the function are the same in different species.

We know that there is a great deal of similarity in brain organization across the various vertebrate species. All vertebrates have a hindbrain, midbrain, and forebrain, and within each of the three divisions, one can find all of the basic structures and major neural pathways in all animals.[45] At the same time there are obvious differences between the brains of widely different groups of animals. Species differences can involve any brain region or pathway, due to particular brain specializations required for certain species-specific adaptations or to random changes. However, as one follows brain evolution from fish, through amphibians and reptiles, to mammals, and ultimately to humans, the greatest changes appear to have taken place in the forebrain.[46] But evolution should not be thought of as an ascending scale. It is more like a branching tree.[47] The long process of human brain evolution has not just been a matter of making the forebrain bigger and bigger; it has also become more diversified.[48] For example, as we saw in Chapter 4, it was long thought that the neocortex was a mammalian specialization, one that did not exist in other classes of animals (the designation "neo" reflects the supposed evolutionary newness of this part of the brain). However, it is now known that all vertebrates have areas of the cortex that correspond with what is called the neocortex in mammals—these are just located in a different place in nonmammalian species (birds and reptiles, for example) than in mammals, which caused anatomists to misjudge what these regions are.[49] Nevertheless, there are areas of the human neocortex that are apparently not present in the brains of other animals.[50] In spite of this diversification, though, brain evolution is basically conservative, and certain systems, especially those that have been generally useful for survival and have been around for a long time, have been preserved in their basic structure and function.

Circuits in the brain, like all other body parts, are assembled during embryonic development by processes encoded in our genes. If different animals indeed have circuits that achieve some common

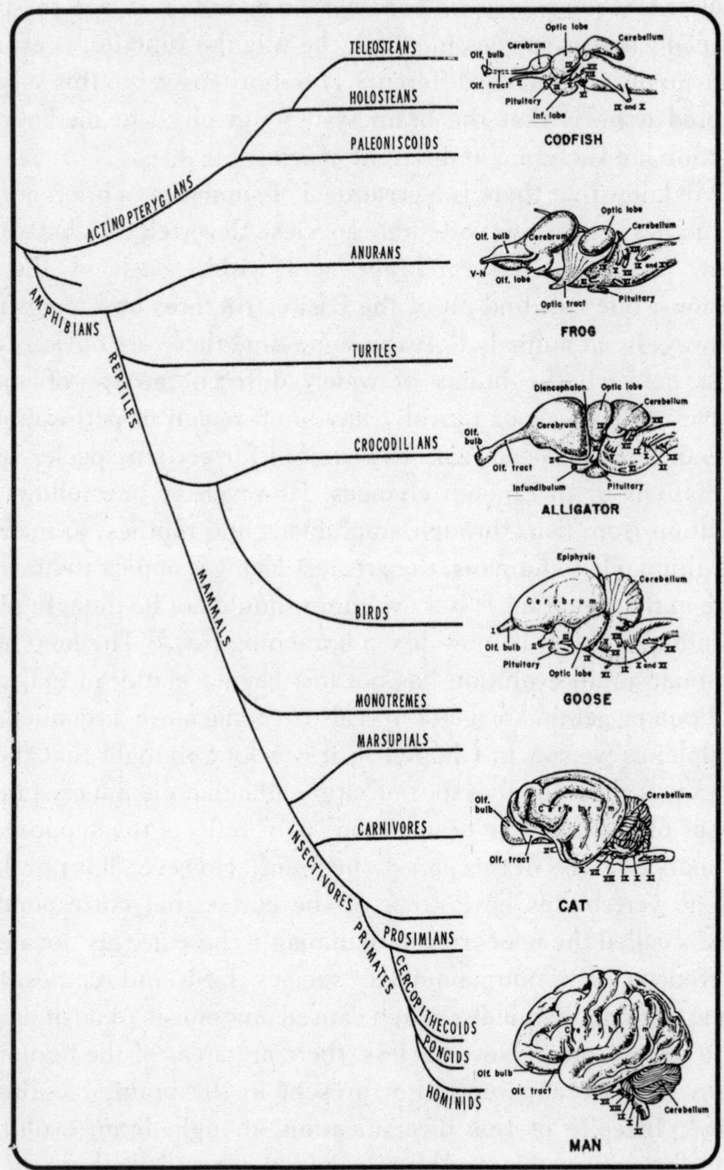

FIGURE 5-3
The Branching Tree of Brain Evolution.

(Modified from figure 5 of W. Hodos [1970], Evolutionary interpretation of neural and behavioral studies of living vertebrates. In F. O. Schmitt, ed., *The Neurosciences: Second Study Program.* New York: Rockefeller University Press. Used by permission of the Rockefeller University Press.)

function, but do it by controlling different behaviors, we would be led to the conclusion that the genetic code that controls the wiring of the functions in the brain during development is conserved across species in spite of the fact that the genetic code that constructs the body parts used to express those functions is different. Evolution, in other words, creates unique behavioral solutions to the problem of survival in different species, but it may do so by following a kind of "if it ain't broke don't fix it" rule for the underlying brain systems.

For now, I'm going ask you to take it on faith that the brain systems underlying certain emotional behaviors have been preserved throughout many levels of brain evolution. In the next chapter, though, I will present very strong evidence that this is true within the mammalian class, and also describe some hints that it may extend to existing reptiles and birds as well.

Throughout this discussion of the evolution of emotion, I've said nothing about what most people consider the most important, in fact, the defining feature of an emotion: the subjective feeling that comes with it. The reason for this is that I believe that the basic building blocks of emotions are neural systems that mediate behavioral interactions with the environment, particularly behaviors that take care of fundamental problems of survival.[51] And while all animals have some version of these survival systems in their brains, I believe that feelings can only occur when a survival system is present in a brain that also has the capacity for consciousness. To the extent that consciousness is a recent (in evolutionary time) development,[52] feelings came after responses in the emotional chicken-and-egg problem. I'm not going to say which animals are conscious (which ones have feelings) and which ones are not (which ones don't have feelings). But I will say that capacity to have feelings is directly tied to the capacity to be consciously aware of one's self and the relation of oneself to the rest of the world. We will return to these issues in Chapter 9. For now, I want to continue to pursue some ideas about the evolution of behaviors that are crucial to survival, or more precisely, some ideas about the evolution of the neural systems underlying these behaviors.

Specialized versus General-Purpose Neural Systems

Modern evolutionarily minded emotions theorists, like Ekman, argue that emotions deal with "fundamental life tasks."[53] A similar point is made by Johnson-Laird and Oatley, who say that each emotion "prompts us in a direction which in the course of evolution has done better than other solutions to recurring circumstances."[54] And Tooby and Cosmides argue that emotions involve situations that have occurred over and over throughout our evolutionary history (escaping from danger, finding food and mates) and cause us to appraise present events in terms of our ancestral past—that the structure of the past imposes an interpretive landscape on the present.[55]

In a sense, coming up with a list of the special adaptive behaviors that are crucial to survival would essentially be a list of basic emotions. I think starting with universal behavioral functions is a better way of producing a list of basic emotions than the more standard ways—facial expressions, emotion words in different languages, or conscious introspections. However, I'm not concerned with defining, from the start, what the different emotions are and I have no interest in producing yet another basic emotions list. Obviously, it is ultimately important to understand what all the biologically derived and socially constructed emotions are, and to determine where the line should be drawn between them. It is also important to draw the line between mental phenomena that are emotions and those that are not. However, for good reasons, efforts to identify what all of the emotions are frequently get bogged down in arguments over the fringe instances, as when Ortony and Turner took basic emotions theorists to task for their inability to agree about what the various basic emotions are, and especially for disagreeing about the fuzzy cases, like surprise, interest, and desire. I believe that once we've built up a core knowledge about the clear instances we will be in a better position to deal with the fuzzy ones, but we haven't reached that point yet.

To the extent that emotional responses evolved, they evolved for different reasons, and it seems obvious to me that there must be different brain systems to take care of these different kinds of functions. Lumping all of these together under the unitary concept of emotional behavior provides us with a convenient way of organizing things—for distinguishing behaviors that we call emotional (for example, those involved with fighting, feeding, sex, and social bonding) from those that reflect cognitive functions (like reasoning, abstract thinking, problem solving, and concept formation). However, the use of a label, like "emotional behavior," should not necessarily lead us to assume that all of the labeled functions are mediated by one system of the brain. Seeing and hearing are both sensory functions, but each has its own neural machinery.

I think that the most practical working hypothesis is that different classes of emotional behavior represent different kinds of functions that take care of different kinds of problems for the animal and have different brain systems devoted to them. If this is true, then different emotions should be studied as separate functional units.

At the neural level, each emotional unit can be thought of as consisting of a set of inputs, an appraisal mechanism, and a set of outputs. The appraisal mechanism is programmed by evolution to detect certain input or trigger stimuli that are relevant to the function of the network. We'll call these "natural triggers."[56] The sight of a predator is a good example. It is not uncommon for prey species to recognize predators the first time they see them. Evolution has programmed the prey brain so that certain features of the way the predator looks, sounds, or smells will be automatically appraised as being a source of danger. But the appraisal mechanism also has the capacity to learn about stimuli that tend to be associated with and predictive of the occurrence of natural triggers. These we'll call "learned triggers." The place where a predator was seen last, or the sound it made when it was charging toward the prey are good examples. When the appraisal mechanism receives trigger inputs of either type, it unleashes certain patterns of response that have tended to be useful in dealing with situations that have routinely activated the appraisal mechanism in ancestral animals. These networks evolved because they serve the function of connecting trigger stimuli with responses that are likely to succeed in keeping the organism alive. And because different kinds of problems of survival have different trigger stimuli and require different kinds of responses to deal with them, different neural systems are devoted to them.[57]

The particular functional unit that I have focused my research on is the fear system of the brain. In the next several chapters, we will look closely at the fear system, which is understood as well or better than other emotional systems. Once we see how this system is organized, we will be in a better position to consider the manner in which other emotions are organized in the brain, and how these relate to the fear system.

Why Fear?

I'm going to now lay out some of the reasons why I believe the fear system of the brain is a particularly good one to focus on as an anchoring point. However, I want to first be explicit about what I think the fear system is. The system is not, strictly speaking, a system that results in the experience of fear. It is a system that detects danger and produces responses that maximize the probability of surviving a dangerous situation in the most beneficial way. It is, in other words, a system of defensive behavior. As noted above, I believe that emotional behaviors, like defensive behaviors, evolved independent of, which is to say before, conscious feelings, and that we should not be too quick to assume that when an animal, other than a human one, is in danger it feels afraid. We should, in other words, take defensive behaviors at face value—they represent the operation of brain systems that have been programmed by evolution to deal with danger in routine ways. Although we can become conscious of the operation of the defense system, especially when it leads to behavioral expressions, the system operates independently of consciousness—it is part of what we called the emotional unconscious in the Chapter 3. Interactions between the defense system and consciousness underlie feelings of fear, but the defense system's function in life, or at least the function it evolved to achieve, is survival in the face of danger. Feelings of fear are a by-product of the evolution of two neural systems: one that mediates defensive behavior and one that creates consciousness. Either one alone is not enough to produce subjective fear. Feeling afraid can be very useful, but this is not the function programmed into the neural system of defense by evolution.

With the territory staked out in this way, let's now consider why the defensive system of the brain and its associated subjective emotion, fear, are attractive starting points for studying the emotional brain. I'll discuss three points below: fear is pervasive, fear is important in psychopathology, and fear is expressed similarly in man and many other animals. In the next chapter, another crucial point will be considered, namely that the neural basis of fear is similar in humans and other animals.

Fear Is Pervasive: William James once said that nothing marks the ascendancy of man from beast more clearly than the reduction of the conditions under which fear is evoked in humans.[58] By this, it would seem that James meant that man has managed to establish a less dangerous way of living. It is certainly true that in comparison to our primate ancestors, who lived in a world in which being someone's dinner was an ever-present possibility, humans have created a way of living in which the likelihood of encountering predators is greatly reduced. But not all dangers come in the form of bloodthirsty beasts. Snakes and tigers are rare in modern cities, except in zoos, where viewing dangerous animals in captivity reinforces our hope that life is safe. But in our quest to conquer nature we have created new forms of danger. Automobiles, airplanes, weapons, and nuclear energy give us a step up on the wild, but each is also a potential source of harm. We've traded in the dangers of a life amongst the wild things for other dangers that may, in the end, be far more harmful to our species than any natural predator. The dangers we face are not fewer or less significant than those of our animal ancestors, they're just different.

Even a casual analysis of the number of ways the concept of fear can be expressed in the English language reveals its importance in our lives: alarm, scare, worry, concern, misgiving, qualm, disquiet, uneasiness, wariness, nervousness, edginess, jitteriness, apprehension, anxiety, trepidation, fright, dread, anguish, panic, terror, horror, consternation, distress, unnerved, distraught, threatened, defensive.[59] The so-called ascent of man occurred in spite of the continued existence of fear rather than at its expense. As the renowned human ethologist Eibl-Eibesfeldt notes, "Perhaps man is one of the most fearful creatures, since added to the basic fear of predators and hostile conspecifics come intellectually based existential fears."[60] Indeed, for the existential philosophers (like Kierkegaard, Heidegger, and Sartre), dread, angst, and anguish are at the core of human existence.[61]

One can find evidence of fear lurking in the background of many kinds of emotions that on the surface might seem to be the antithesis of fear. Courage is the ability to overcome fear. Children learn to be moral to some extent by their fear of what will happen if they are not. Laws reflect our fear of social disorder and, by the same token, social order is maintained, however imperfectly, by fear of the consequences of breaking the rules. World peace is a desirable humanitarian goal, but in practice war is avoided, at least in part, because the weak fear the strong. These are bleak statements, hopefully overstatements, but even as partial truths they emphasize how deeply fear cuts into the mental fabric of persons and societies.

Fear Plays an Important Role in Psychopathology: While fear is a part of everyone's life, too much or inappropriate fear accounts for many common psychiatric problems. Anxiety, a brooding fear of what might happen, was at the core of Freud's psychoanalytic theory. Phobias are specific fears taken to extreme. Phobic objects (snakes, spiders, heights, water, open places, social situations) are often legitimately threatening, but not to the extent believed by the phobic person. Obsessive-compulsive disorder often involves extreme fear of something, like germs, and the patients will engage in compulsive rituals to avoid the feared object or event or to rid themselves of the fear object once it is encountered. Panic disorder involves the rapid onset of a host of physical symptoms and often the overwhelming fear that death is near. Post-traumatic stress disorder (PTSD), previously referred to as shell shock, often occurs in war veterans, who can be sent into intense distress by a stimulus that has some resemblance to events associated with battlefield trauma. Thunderclaps and the sound of a car backfiring are common examples. But PTSD extends to many other kinds of traumatic situations, including physical and sexual abuse. Fear is a core emotion in psychopathology.

Fear Is Expressed Similarly in Humans and Other Animals: It may not be the case that every form of emotional behavior has a long evolutionary history. Guilt and shame, for example, may be special human emotions.[62] Nevertheless, as we will see, human defensive behavior clearly seems to have a long evolutionary history. As a result, we can study fear responses in animals for the purpose of illuminating the mechanisms of human fear, including pathological fear. This is crucial, since for both ethical and practical reasons it is not possible to study brain mechanisms in much detail in humans.

All animals have to protect themselves from dangerous situations in order to survive, and there are only a limited number of strategies that animals can call upon to deal with danger. Isaac Marks, who has written extensively on fear, summarizes these as withdrawal (avoiding the danger or escape from it), immobility (freezing), defensive aggression (appearing to be dangerous and/or fighting back), or submission (appeasement).[63] The extent to which these strategies apply across the various vertebrates is striking.

Consider the following description of human defense by Caroline and Robert Blanchard, pioneers in fear research:

> If something unexpected occurs—a loud noise or sudden movement—people tend to respond immediately . . . stop what they are doing . . . orient toward the stimulus, and try to identify its potentiality for actual danger. This happens very quickly, in a reflex-like sequence in which action precedes any voluntary or consciously intentioned behavior. A poorly localizable or identifiable threat source, such as a sound in the night, may elicit an active immobility so profound that the frightened person can hardly speak or even breathe, i.e. freezing. However, if the danger source has been localized and an avenue for flight or concealment is plausible, the person will probably try to flee or hide. . . . Actual contact, particularly painful contact, with the threat source is also likely to elicit thrashing, biting, scratching, and other potentially damaging activities by the terrified person.[64]

Though anecdotal, the Blanchards' description goes a long way toward accounting for the way people behave when threatened. And different people tend to do roughly the same things in similar kinds of situations. This uniformity suggests that either we all learn to be fearful in the same way, or, more likely, that patterns of fear reactivity are genetically programmed into the human brain.

Research by the Blanchards and others has shown that the reaction pattern described above for a frightened human also occurs when rats are in danger.[65] For example, if a laboratory-reared rat (one that has never had the opportunity to see a cat or be threatened by one) is exposed to a cat, it stops what it is doing and turns toward the cat. Depending on whether the cat is close or far away and whether the two animals are in an enclosed or an open area, the rat will either freeze or try to escape. If trapped by the cat, the rat will vocalize and ultimately will attack the cat. This striking functional correspondence between human and rat fear responses holds for many mammals and other vertebrates: it is quite common to observe startle, orienting, then freezing or fleeing or attack, in the face of danger. We've already seen examples from Darwin about how hair erection is a common defense response in many animals, including people, and how it may be related to the flashing of feathers in birds and fin extensions in fish.

Not only are some general patterns of behavior similar in different animals, but so too are some of the underlying physiological changes that occur in dangerous or stressful situations. For example, it is well known that soldiers in battle fail to notice injuries that would, under less traumatic circumstances, be excruciatingly painful. Similarly, a rat, when exposed to a cat, will fail to notice painful heat applied to its tail.[66] The cat poses a greater overall threat than a wound to the tail, and pain suppression in the face of danger allows the organism to use its resources to deal with the most significant danger. In both humans and rats, stress-induced analgesia is a consequence of activation of the brain's natural opiate system.[67] When the brain detects danger, it also sends messages through the nerves of the autonomic nervous system to bodily organs and adjusts the activity of those organs to match the demands of the situation. Nerves reaching the gut, heart, blood vessels, and sweat and salivary glands give rise to the taut stomach, racing heart, high blood pressure, clammy hands and feet, and dry mouth that typify fear in humans. The cardiovascular responses associated with defensive behavior have been examined in birds, rats, rabbits, cats, dogs, monkeys, baboons, and people, to name a few of the better studied species, and the responses are controlled by similar kinds of brain networks and body chemistry in these different species.[68] Threatening stimuli also cause the pituitary gland to release adrenocorticotropic hormone (ACTH) that results in the release of a steroid hormone from the adrenal gland.[69] The adrenal hormone then travels back to the brain. Initially, these hormones help the body deal with the stress, but if the stress is prolonged the hormone can begin to have pathological consequences, interfering with cognitive functions and even causing brain damage.[70] This so-called stress response is ubiquitous amongst mammals, and also occurs in other vertebrates.[71] These various bodily responses are not random activities. They each play an important role in the emotional reaction and each functions similarly in diverse animal groups.

Nevertheless, it would be wrong to give the impression that all animals respond exactly the same in the face of danger. Obviously, they do not. Each animal is the product of its own evolutionary history. Within the general classes of defense reactions, much variation is possible. In fact, defense reactions should be thought of as constantly changing, dynamic solutions to the problem of survival. They are not static structures created in ancestral species and maintained unchanged. They change as the world in which they operate changes. For example, Richard Dawkins describes predators and prey as involved in evolutionary arms races, where a particular adaptation that makes a prey better at defending against a predator can lead to the selection of traits that then give the predator the edge up—the color of the prey may change so that it blends in better with the environment, and the predator may, in turn, evolve a more sensitive perceptual system for detecting the camouflaged prey.[72] But Dawkins also notes a certain imbalance in these arms races, what he calls the "life/dinner" principle. According to this notion, rabbits run faster than foxes because rabbits are running for their life but foxes are only running for their dinner. As a result, genetic mutations that make foxes run more slowly are more likely to survive in the gene pool than mutations that make rabbits run more slowly, since the penalty of slowness is more severe for rabbits than foxes—a fox may reproduce after being outrun by a rabbit, but no rabbit ever reproduced after being caught by a fox.

In spite of the fact that species can have their own special ways of responding to danger, commonality of functional patterns is the rule. In fact, what distinguishes fear reactions in humans and other animals is not so much the ways in which fear is expressed as the different kinds of trigger stimuli that activate the appraisal mechanism of the defensive system. Each animal has to be able to detect the particular things that are dangerous to it, but there is an evolutionary economy to using universal response strategies—withdrawal, immobility, aggression, submission—and universal physiological adjustments. Added cognitive power opens up the defensive hardware to new kinds of events, new learned triggers. Humans fear things that a rat could never conceptualize, but the human and rat body respond much the same to their special triggers.

The implications of this situation are enormous. For the purpose of understanding how fear is generated, it does not matter so much how we activate the system or whether we activate the system in a person or a rat. The system will respond in pretty much the same way using a limited set of given defense response strategies available to it. We can thus design experiments in rats (or other laboratory animals) for the purpose of understanding how the human fear system works.

Genetic Determinism and Emotional Freedom

All this talk about the evolution of emotional behavior is likely to make the imagination run wild with ideas about genetic determination of our emotions. After all, any characteristic that has evolved has done so because of the representation of that characteristic in the genes of the species. But I want to make clear two different implications of the genetics of emotional behavior.

On the one hand, there is the way genes maintain similar behavioral expressions of defense within a species and similar defensive functions across diverse species. This occurs, as I've argued, because the neural system of defense is conserved in evolution. As a result, all humans have the same general ways of expressing themselves when in danger, and these tend to be similar to the ways that other animals have for expressing themselves in the face of danger. This view of emotional genetics tries to find the common ground of emotional reactions across individuals and species—the stuff that particular emotional systems evolved to do.[73]

On the other hand, there is the question of how genes contribute to differences between individuals. Some people are good fighters, others are not. Some are adept in detecting dangers, others are oblivious to their surroundings. Differences between individuals in fearful behavior are due, at least in part, to genetic variation.

So far, I've emphasized the first implication—the way genes make emotional reactions similar amongst humans and between humans and other animals. But it is important to also consider in some detail the ways in which genes make us different from each other. We will then discuss whether, and if so to what extent, such differences predestine us to act in some particular way, again concentrating on the fear system.

Temperament runs through bloodlines. Some breeds of horses or dogs are jumpy, others complacent. These characteristics can sometimes be side effects of some other trait that was selected for, like running speed, but temperament can also be selected for itself. Indeed, selective breeding has been used to create strains of rats and mice that are particularly timid or courageous.[74]

For example, rats don't normally congregate in wide-open spaces. This makes a lot of sense evolutionarily—open places are unprotected from land and air predators and can be very dangerous for rodents. Those ancestral rodents that tended to hang out in an open area probably did not do so well in their struggle to survive, whereas those that hightailed it out to the nearest safe place did. Psychologists created an apparatus for testing this behavior—a large, well-lit circular arena called "the open field."[75] If you put your garden-variety rat in the center of an open field apparatus, it will make a beeline to the wall, which is the most protected place available. The rats also defecate—like people, rats can have the "$#!+" scared out of them. Defecation is controlled by the autonomic nervous system and the number of fecal pellets (poops) that are dropped is a reliable and measure of ANS activity. Defecation in the open field or other potentially dangerous situations has become a fairly standard measure of "fearfulness" in rodents.[76] But not all rats drop the same number of pellets in the open field, and the amount one rat drops tends to be fairly constant. If you divide a large group into those that drop more and fewer pellets in the open field, and then start breeding them on the basis of this selected trait, you can, in a few generations, create strains of timid and courageous rats—rats from the low pellet-dropping line act more courageous in the open field (they stay in the unprotected area longer) and in a variety of other tests. From this example, it is easy to imagine how personality traits might come to be part of a family, or even a culture. All you need is a few generations of inbreeding amongst a limited gene pool to begin to stabilize behaviorally significant characteristics.

In fact, considerable evidence shows that there is a genetic component to fear behavior in humans.[77] For example, identical twins (even those reared in separate homes) are far more similar in fearfulness than fraternal twins. This conclusion applies across many kinds of measurements, including tests of shyness, worry, fear of strangers, social introversion/extroversion, and others. Similarly, anxiety, phobic, and obsessive compulsive disorders tend to run in families and to be more likely to occur in both identical than in both fraternal twins.

The genetics of defensive behavior has been studied most extensively in bacteria.[78] Although not known as a particularly sophisticated organism from the psychological point of view, they do protect themselves from danger and there may be some biological lessons to be learned. Their defensive repertoire consists of moving away from substances assessed as harmful. The specific gene mutations controlling this behavior, which involves complex coding of chemical constituents in the immediate environment, have been identified. Similarly, much progress has been made in genetic analyses of fruit fly defensive behavior.[79] Through some ingenious experiments, Tim Tully has shown that these creatures can learn to avoid danger (electric shock) on the basis of stimulus cues (odors)—once shocked in the presence of a certain smell, they tend to avoid a chamber that has the smell. Using the modern tools of molecular biology and genetics, mutant flies have been created that are unable to use the smell cues to avoid the shock. They can smell just fine, they just can't link the smell with danger. It is admittedly a far leap from defensive responses in flies to humans, and at least a quantum leap from bacterial to human behavior. However, studies in these simple creatures may pave the way for future researchers to perform similar kinds of experiments in mammals, and these kinds of studies may shed light on the genetics of fear in humans. There is, after all, massive overlap in the genetic makeup of humans and chimpanzees, and a good deal of overlap in humans and other mammals as well.[80]

There's no denying that genes make each of us different from one another and explain at least part of the variability in the way different people act in dangerous and other situations. But we have to be very careful in interpreting differences in behavior between different people. As Richard Dawkins puts it, "If I am homozygous for a gene G, nothing save mutation can prevent my passing G on to all my children. So much is inexorable. But whether or not I, or my children, show the phenotypic effect normally associated with possession of G may depend very much on how we are brought up, what diet or education we experience, and what other genes we happen to possess."[81]

The bottom line is that our genes give us the raw materials out of which to build our emotions. They specify the kind of nervous system we will have, the kinds of mental processes in which it can engage, and the kinds of bodily functions it can control. But the exact way we act, think, and feel in a particular situation is determined by many other factors and is not predestined in our genes. Some, if not many, emotions do have a biological basis, but social, which is to say cognitive, factors are also crucially important. Nature and nurture are partners in our emotional life. The trick is to figure what their unique contributions are.

6

A FEW DEGREES OF SEPARATION

"The BRAIN—is wider than the sky—."

Emily Dickinson, *The Poems of Emily Dickinson*[1]

ONLY A FEW NEURAL links separate any particular set of neurons in the brain from most others. As a result, it is sometimes said that figuring out connections between brain areas is a waste of time, since information reaching one area can eventually influence many. But this criticism is misplaced. By way of a small number of acquaintances, we are each potentially connected to everyone else in the world.[2] Yet, we only get around to meeting a small subset of the earth's population in our lives. Communication between people, like the flow of information between neurons, is selective.

But how does one go about figuring out the selective channels of information flow in the brain? There are billions of neurons, and each gives rise to one or more axons (nerve fibers that allow neurons to communicate with one another). The axons themselves branch, so that the number of synapses (the connection made by an axon from one neuron with another) is far greater than the number of neurons. And each neuron has multiple dendrites that receive thousands of synaptic contacts from many others. Can we ever hope to relate this intricate mesh of interconnected neural elements to emotion, a term that itself refers to an enormously complex set of phenomena?

The field of neuroscience has a vast arsenal of techniques for figuring out how the brain is organized—how it is wired together.

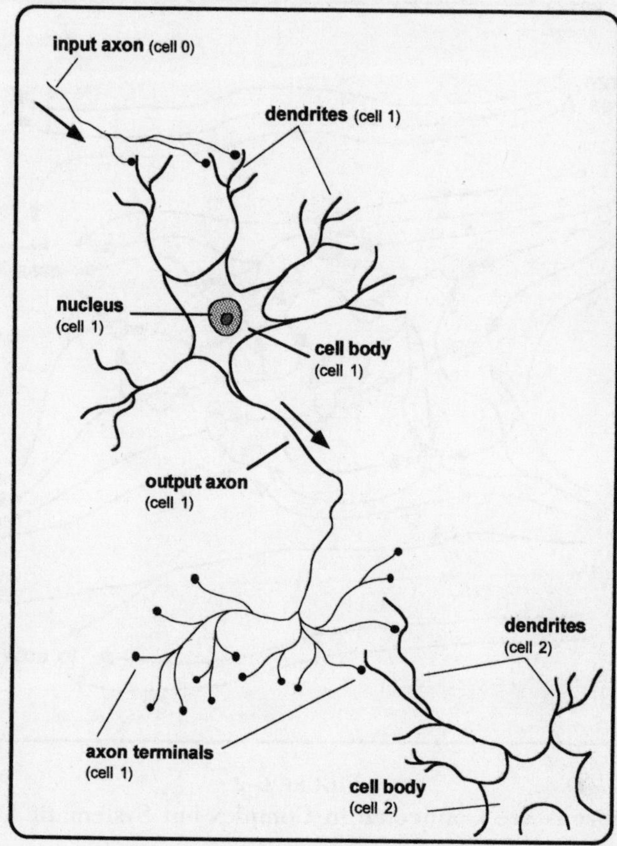

FIGURE 6-1
A Neuron.

Neurons (brain cells) have three parts: a cell body, an axon, and some dendrites. Typically, information from other neurons comes into a brain cell by way of the dendrites (but the cell body or axon can also receive inputs). Each cell receives inputs from many others. When a neuron receives enough inputs at the same time, it will fire an action potential (a wave of electrical charge) down the axon. Although a neuron usually has only one axon, it branches extensively, allowing many other neurons to be influenced. When the action potential reaches the axon terminals, a chemical, called a neurotransmitter, is released. The neurotransmitter diffuses from the terminal to the dendrites of adjacent neurons and contributes to the firing of action potentials in these. The space between the axon terminal of one cell and its neighbor is called the synapse. For this reason, communication between neurons is referred to as synaptic transmission. (Based on figure 1 in B. Katz [1966], *Nerve, Muscle, and Synapse.* New York: McGraw-Hill.)

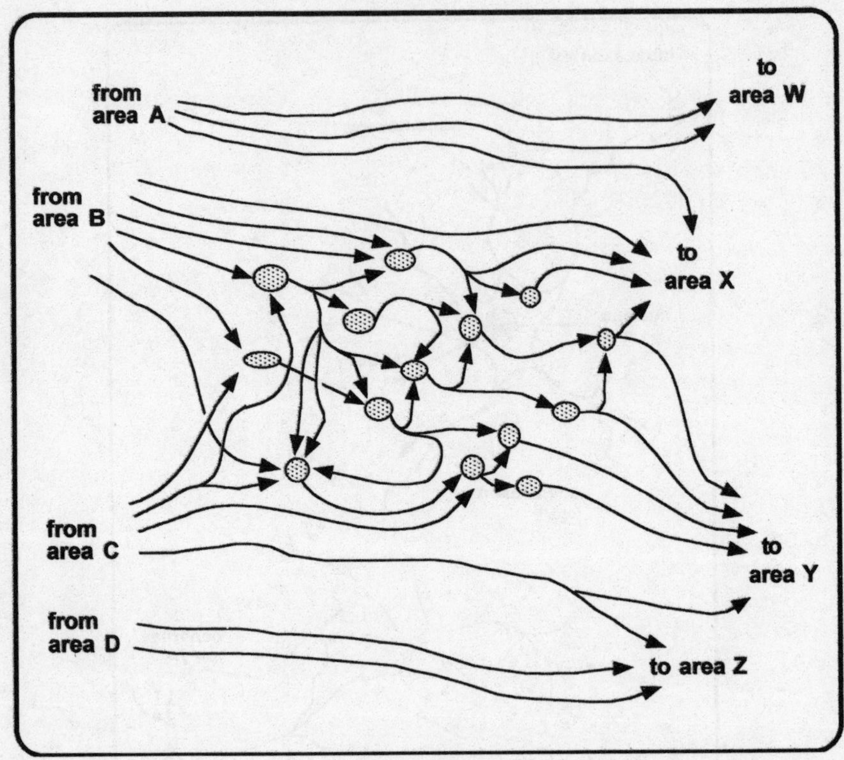

FIGURE 6-2

Neurons Are Connected in Complex but Systematic Ways.

Although the trillions of connections made by the billions of neurons in the brain may seem to constitute a hopelessly complex web of relations, very systematic patterns of interactions exist between neurons in various brain areas. The interconnected brain network shown in the center receives inputs from areas B and C, but not from A or D, and gives rise to outputs that reach areas X and Y, but not W or Z. Further, area C communicates with Y both directly and by way of links in the central network. In real brains, these relations are figured out by tracing axonal connections between areas, as illustrated in figure 6-8.

Though these techniques make important, even crucial, contributions to the effort to understand the emotional brain, they are not enough. In order to figure out how emotional functions are mediated by specific patterns of neural wiring, we also need good ways of telling when the brain is in an emotional state. For this we depend on behavioral tools, ways of determining, from what an animal or person is doing, that the brain is engaged in emotional activity. And if the picture of the emotional brain that I have been painting is accurate, the particular behavioral tools we need will be dictated by the kind of emotional function we are interested in understanding: the tools that allow us to reliably measure responses that depend upon the system that underlies fear behavior will probably not be very useful for studying aggressive or sexual behavior or mother-infant relations. Armed with a good behavioral task, together with the bag of tricks that modern brain science offers, we can go searching for the brain networks that mediate specific emotional functions and actually expect to find them. But without good behavioral tools, the effort to understand emotional networks is doomed.

Fortunately, there is an incredibly good task for studying fear mechanisms. It is called fear conditioning. Below I will explain what fear conditioning is and why it is so useful. I will then describe how, using fear conditioning, it has been possible to isolate, from the billions of neurons and trillions of connections, those that are important in fear behavior.

For Whom the Bell Tolls

If your neighbor's dog bites you, you will probably be wary every time you walk by his property. His house and yard, as well as sight and sound of the beast, have become emotional stimuli for you because of their association with the unpleasant event. This is fear conditioning in action. It turns meaningless stimuli into warning signs, cues that signal potentially dangerous situations on the basis of past experiences with similar situations.

In a typical fear conditioning experiment, the subject, say a rat, is placed in a small cage. A sound then comes on, followed by a brief, mild shock to the feet. After very few such pairings of the sound and the shock, the rat begins to act afraid when it hears the sound: it stops dead in its tracks and adopts the characteristic freezing posture—crouching down and remaining motionless, except for the rhythmic chest movements required for breathing. In addition, the rat's fur stands on end, its blood pressure and heart rate rise, and stress hormones are released into its bloodstream. These and other conditioned responses are expressed in essentially the same way in every rat, and also occur when a rat encounters its perennial arch-enemy, a cat, strongly suggesting that, as a result of fear conditioning, the sound activates the neural system that controls responses involved in dealing with predators and other natural dangers.

Fear conditioning is a variation on the procedure discovered by Ivan Pavlov around the turn of the century.[3] As everyone knows, the great Russian physiologist observed that his dogs salivated when a bell was rung if the sound of the bell had previously occurred while the dog had a juicy morsel of meat in its mouth. Pavlov proposed that the overlap in time of the meat in the mouth with the sound of the bell resulted in the creation of an association (a connection in the brain) between the two stimuli, such that the sound was able to substitute for the meat in the elicitation of salivation.

Pavlov abhorred psychological explanations of behavior and

FIGURE 6-3
Pavlov and His Dog.

Photograph of I. P. Pavlov demonstrating classical conditioning to students and visitors at the Russian Army Medical Academy sometime around 1904. (Caption from figure on p. 177 of C. Blakemore and S. Greenfield [1987], *Mindwaves.* Oxford: Basil Blackwell.)

sought to account for the anticipatory salivation physiologically, without having to "resort to fantastic speculations as to the existence of any subjective state in the animal which may be conjectured on analogy with ourselves." He thus explicitly rejected the idea that salivation occurred because the hungry dogs began to think about the food when they heard the bell. In this way Pavlov, like William James (see Chapter 3), removed subjective emotional states from the chain of events leading to emotional behavior.

Pavlov called the meat an unconditioned stimulus (US), the bell a conditioned stimulus (CS), and the salivation elicited by the CS a conditioned response (CR). This terminology derives from the fact that the capacity of the bell to elicit salivation was conditional upon its relation to the meat, which elicited salivation naturally, which is to say, unconditionally. Applying these terms to the fear conditioning experiment described above, the tone was the CS, the shock was the US, and the behavioral and autonomic expressions were the CRs. And in the language used in the previous chapter to describe the stimuli that initiate emotional behaviors, a US is a *natural trigger* while a CS is a *learned trigger.*

Fear conditioning does not involve response learning. Although rats freeze when they are exposed to a tone after but not before conditioning, conditioning does not teach the rats how to freeze. Freezing is something that rats do naturally when they are exposed to danger. Laboratory-bred rats who have never seen a cat will freeze if they encounter one.[4] Freezing is a built-in response, an innate defense response, that can be activated by either *natural or learned triggers.*

Fear conditioning opens up channels of evolutionarily shaped responsivity to new environmental events, allowing novel stimuli that predict danger (like sounds made by an approaching predator or the place where a predator was seen) to gain control over tried-and-true ways of responding to danger. The danger predicted by these *learned trigger stimuli* can be real or imagined, concrete or abstract, allowing a great range of external (environmental) and internal (mental) conditions to serve as CSs.

Conditioned fear learning occurs quickly, and can occur after a single CS-US pairing. An animal in the wild does not have the opportunity for trial-and-error learning. Evolution has arranged things

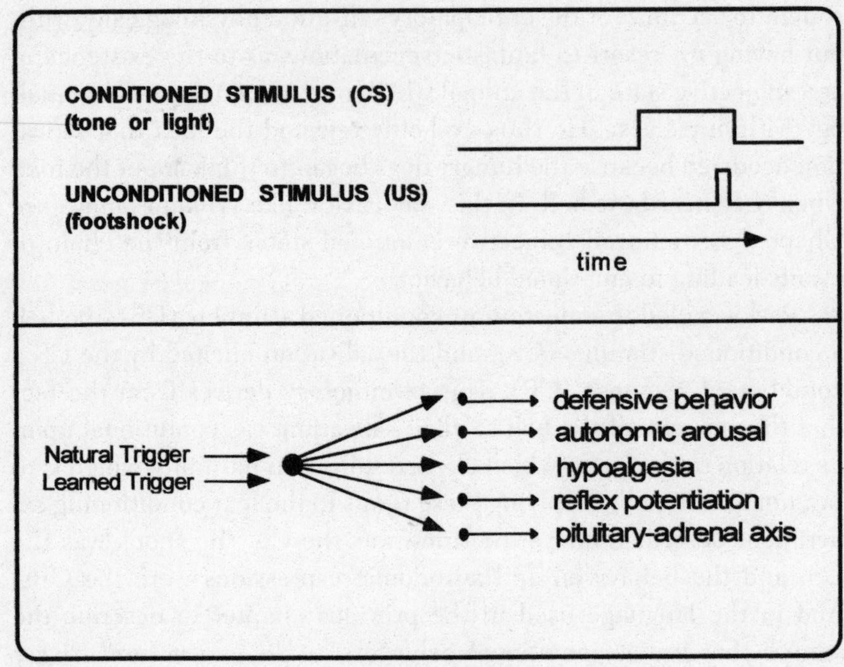

FIGURE 6-4
Fear Conditioning.

In fear conditioning an unconditioned stimulus (typically a brief, mild footshock) is delivered at the end of the conditioned stimulus (usually a tone or light). After a few pairings, the conditioned stimulus acquires the capacity to elicit a wide variety of bodily responses. Similar responses occur in the presence of natural dangers that are innately programmed into the brain. For example, in the presence of either a conditioned fear stimulus or a cat, rats will freeze and exhibit blood pressure and heart rate changes, alterations in pain responsivity, more sensitive reflexes, and elevation of stress hormones from the pituitary gland. Because rats do not require prior exposure to cats to exhibit these responses, the cat is a natural trigger *of defense responses for rats. And because the tone only elicits these responses after fear conditioning, it is a* learned trigger. *Similar patterns of defense responses occur in humans and other animals when exposed to fear triggers (natural and learned). Studies of nonhuman animals can thus illuminate important aspects of fear reactivity in humans.*

so that if you survive one encounter with a predator you can use your experience to help you survive in future situations. For example, if the last time a rabbit went to a certain watering hole it encountered a fox and barely escaped, it will probably either avoid that watering hole in the future or the next time it goes there it will approach the scene with trepidation, taking small cautious steps, searching the environment for any clue that might signal that a fox is near.[5] The watering hole and fox have been linked up in the rabbit's brain, and being near the watering hole puts the rabbit on the defensive.

Not only is fear conditioning quick, it is also very long lasting. In fact, there is little forgetting when it comes to conditioned fear. The passing of time is not enough to get rid of it.[6] Nevertheless, repeated exposure to the CS in the absence of the US can lead to "extinction." That is, the capacity of the CS to elicit the fear reaction is diminished by presentation of the CS over and over without the US. If our thirsty but fearful rabbit has only one watering hole to which it can go, and visits it day after day without again encountering a fox, it will eventually act as though it never met a fox there.

But extinction does not involve an elimination of the relation between the CS and US. Pavlov observed that a conditioned response could be completely extinguished on one day, and on the next day the CS was again effective in eliciting the response. He called this "spontaneous recovery."[7] Recovery of extinguished conditioned responses can also be induced. This has been nicely demonstrated in studies by Mark Bouton.[8] After rats received tone-shock pairings in one chamber, he put them in a new chamber and gave them the tone CS over and over until the conditioned fear responses were no longer elicited—the conditioned fear reaction was completely extinguished. He then showed that simply placing the animals back in the chamber where the CS and US were previously paired was enough to *renew* the conditioned fear response to the CS. Extinguished conditioned fear responses can also be *reinstated* by exposing the animals to the US or some other stressful event.[9] Spontaneous recovery, renewal, and reinstatement suggest that extinction does not eliminate the memory that the CS was once associated with danger but instead reduces the likelihood that the CS will elicit the fear response.

These findings in rats fit well with observations on humans with pathological fears (phobias).[10] As a result of psychotherapy, the fear of the phobic stimulus can be kept under control for many years. Then, after some stress or trauma, the fear reaction can return in full force. Like extinction, therapy does not erase the memory that ties fear reactions to trigger stimuli. Both processes simply prevent the stimuli from unleashing the fear reaction. I'll have much more to say about this in Chapter 8.

The indelibility of learned fear has an upside and a downside. It is obviously very useful for our brain to be able to retain records of those stimuli and situations that have been associated with danger in the past. But these potent memories, which are typically formed in traumatic circumstances, can also find their way into everyday life, intruding into situations in which they are not especially useful, and such intrusions can be quite disruptive to normal mental functioning. We'll consider traumatic memory again in Chapters 7 and 8.

Although most of the research on the neural basis of conditioned fear has been conducted in animals, fear conditioning procedures can be used in identical ways in humans.[11] Numerous studies of humans have conditioned autonomic nervous system responses, such as changes in heart rate or in sweat gland activity (so-called galvanic skin responses), by pairing tones or other neutral stimuli with mild shocks. Because conditioned fear responses are not dependent on verbal behavior and conscious awareness, they have often been used to study unconscious (subliminal) emotional processing in humans, as described in Chapter 3.

When a human is presented with a consciously perceptible CS that predicts the imminent delivery of a painful stimulus, he or she typically feels fearful or anxious during the CS.[12] We might therefore be inclined to say that the CS elicits a state of fear that then causes the responses. In fact, a number of psychologists and neuroscientists who study fear conditioning assume that "fear" connects the CS to the CR.[13] However, like Pavlov and James, I find it neither necessary nor desirable to insert a conscious state of fear into the chain of events connecting trigger stimuli to fear responses. Here are my reasons why. First, fear conditioning procedures can be used to couple defensive responses to neutral stimuli in worms, flies, and snails, as well as in fish, frogs, lizards, pigeons, rats, cats, dogs, monkeys, and people.[14] I doubt that all of these animals consciously experience fear in the presence of a CS that predicts danger. This is admittedly a slippery slope to slide down, and one that I'm going to delay detailed discussion on until Chapter 9. But if for the time being we assume I'm correct that we don't need conscious fear to explain fear responses in some species, then we don't need it to explain fear responses in humans.[15] Second, even in humans, the one species in which we can study conscious processes with some confidence, fear conditioning can be achieved without conscious awareness of the CS or the relation between the CS and US.[16] The conscious fear that can come with fear conditioning in a human is not a cause of the fear responses; it is one consequence (and not an obligatory one) of activating the defense system in a brain that also has consciousness.

One of the key aspects of fear conditioning that makes it so valuable as a tool for studying the brain mechanisms of fear is that the fear responses come to be coupled to a specific stimulus. This offers

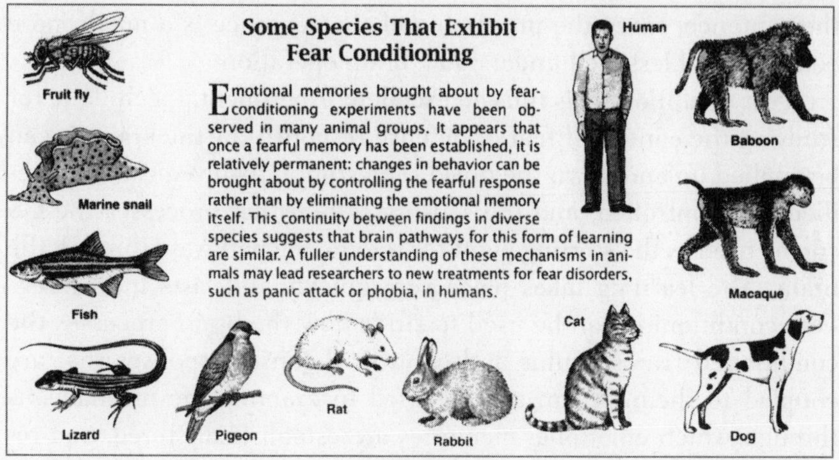
Some Species That Exhibit Fear Conditioning
Emotional memories brought about by fear-conditioning experiments have been observed in many animal groups. It appears that once a fearful memory has been established, it is relatively permanent: changes in behavior can be brought about by controlling the fearful response rather than by eliminating the emotional memory itself. This continuity between findings in diverse species suggests that brain pathways for this form of learning are similar. A fuller understanding of these mechanisms in animals may lead researchers to new treatments for fear disorders, such as panic attack or phobia, in humans.
Fruit fly
Marine snail
Fish
Lizard
Pigeon
Rat
Rabbit
Cat
Human
Baboon
Macaque
Dog

FIGURE 6-5
Animals Throughout the Phyla Can Be Fear Conditioned.

Fear conditioning is an evolutionarily old solution to the problem of acquiring and storing information about harmful or potentially harmful stimuli and situations. It has been studied in several invertebrate species, and in a variety of vertebrates. Within the vertebrates, the behavioral expression of fear conditioning and its neural basis appear very similar in all species that have been examined in detail. (From J.E. LeDoux, Emotion, memory and the brain. *Scientific American* [June, 1994], vol 270, p.39. © 1994 by Scientific American Inc., all rights reserved.)

several important advantages. First, once the stimulus is established as a learned trigger of fear, it will lead to the expression of fear responses each time it occurs. The expression of the fear response is thus under the control of the experimenter, which is very convenient. Second, we can begin to build the emotional processing circuit on the shoulders of the known organization of the sensory system engaged by the CS. Since the sensory systems are understood better than other aspects of the brain, we can use these as a launching pad, tracing the fear processing circuit forward from them. Third, the CS can be a very simple sensory stimulus that is processed with minimal brain power, allowing us to bypass much of the cognitive machinery in the study of fear. We can, in other words, study how the brain appraises the danger implied by a stimulus without getting too bogged down in how the stimulus itself is processed. While it is possible to use either a simple tone or a spoken sentence as a CS, it will be much more difficult to trace the pathways involved in fear conditioning to the sentence, since the processing of the sentence is a much more complex, and less well understood, brain operation.

Fear conditioning is thus an excellent experimental technique for studying the control of fear or defense responses by the brain. It can be applied up and down the phyla. The stimuli involved can be specified and controlled, and the sensory system that processes the CS can be used as the starting point for tracing the pathways through the brain. The learning takes place very quickly and lasts indefinitely. Fear conditioning can be used to study how the brain processes the conditioned fear stimulus and controls defense responses that are coupled to them. It can also be used to examine the mechanisms through which emotional memories are established, stored, and retrieved, and, in humans, the mechanisms underlying conscious fear.

Fear conditioning is not the only way to study fear behavior[17] and it may not be a valid model of all of the many phenomena that are referred to by the term "fear."[18] Nevertheless, it is a quite powerful and versatile model of fear behavior and has been very effectively used to trace brain pathways. Fear conditioning may not tell us everything we need to know about fear, but it has been an excellent way to get started.

Measure for Measure

Once the meaning of a stimulus has been modified by fear conditioning, the next occurrence of the stimulus unleashes a whole host of bodily responses that prepare the organism to deal with the impending danger about which the stimulus warns. Any of these can be used to measure the effects of conditioning.

For example, when a conditioned fear stimulus occurs, the subject will typically stop all movement—it will freeze.[19] Many predators respond to movement[20] and withholding movement is often the best thing to do when danger is near.[21] Freezing can also be thought of as preparatory to rapid escape when the coast clears, or to defensive fighting if escape is not possible. Since the muscle contractions that underlie freezing require metabolic energy, blood has to be sent to those muscles. Indeed, the autonomic nervous system is strongly activated by a conditioned fear stimulus, producing a variety of cardiovascular and other visceral responses that help support the freezing response. These also help the body prepare for the escape or fighting responses that are likely to follow.[22] Additionally, stress hormones are released into the bloodstream to further help the body cope with the threatening situation.[23] Reactivity to pain is also suppressed, which is useful since the conditioned stimulus often announces a situation in which the probability of bodily harm is high.[24] And reflexes (like eyeblink or startle responses) are potentiated, allowing quicker, more efficient reactions to stimuli that normally elicit protective movements.[25]

These various responses are part of the body's overall adaptive reaction to danger and each has been used to examine the brain systems involved in conditioned fear responses. For example, David Cohen[26] has studied the brain pathways of fear conditioning in pigeons using heart rate responses, and Bruce Kapp,[27] Neil Schneidermann and Phil McCabe[28] and Don Powell[29] have used heart rate responses in rabbits. Michael Fanselow[30] has used freezing and pain suppression in rats as measures, while Michael Davis[31] has exploited the potentiation of reflexes by a fear eliciting conditioned stimulus, also in rats. Orville Smith[32] has studied fear conditioning in baboons, measuring a variety of cardiovascular responses in conjunction with measures of movement inhibition. And in my research on the brain mechanisms of fear conditioning, I've made simultaneous measurements of freezing and blood pressure responses in rats.[33]

The amazing fact is that it has not really mattered very much how conditioned fear has been measured, or what species has been studied, as all of the approaches have converged on a common set of brain structures and pathways that are important. Although there are some minor differences and controversies over some of the details, in broad outline there is remarkable consensus. This contrasts with studies of the neural basis of many other behaviors, where slight changes in the experimental procedure or the species can result in profound differences in the neural systems involved. Fear conditioning is so important that the brain does the job the same way no matter how we ask it to do it.

Highways and Byways

Imagine being in a unfamiliar land. You are handed a piece of paper on which the locations of a starting point and a destination are indi-

FIGURE 6-6
A Rat Undergoing Fear Conditioning.

The rat is first exposed to the sound alone. It orients toward the sound, but after several occurrences, the sound is ignored. Next, the sound and the brief, relatively mild shock occur together several times. Later, the sound, when presented alone, will elicit conditioned fear responses. The sound, by association with the shock, has become a learned trigger of fear responses. This is similar to what goes on in humans when they are exposed to dangers or trauma. The stimuli associated with the danger or trauma become learned triggers *that unleash emotional reactions in us. Studies of fear conditioning in rats can thus reveal important aspects of the way human emotional (fear) learning occurs.* (From J.E. LeDoux, Emotion, memory and the brain. *Scientific American* [June 1994], vol 270, p. 34. © 1994 by Scientific American Inc., all rights reserved.)

cated. There are lots of other points marked on the paper. There are also some lines between some of the points, indicating possible ways to get from one to another. But you are told that the lines between the points may or may not indicate real roads, and also that not all of the roads that exist between points are marked. Your job is to get in your car at the starting point and find the best way to the destination, and to make an accurate map along the way.

This is essentially the problem that we faced when we began to try to figure out how networks in the brain make it possible for a novel acoustic stimulus to come to elicit defensive responses as a result of fear conditioning. We knew the starting point (the ear and its connections into the brain) and the end point (the behavioral defense responses and their autonomic concomitants), but the points that linked the inputs and outputs in the brain were unclear. Many of the relevant connections in the brain had been demonstrated with older techniques that were prone to lead to false results—identifying nonexistent connections between two points or failing to find real ones.[34] Relatively little work on the neural basis of fear had used fear conditioning.[35] And while research on fear using techniques other than fear conditioning had suggested some ideas about which brain areas might be involved, it wasn't clear whether these were essential way stations, interesting detours, or just plain wrong turns.

Go with the Flow: Much of the earlier work on the emotional brain had started in the middle of the brain, not surprisingly, in the limbic system.[36] This work showed that lesions of limbic areas can interfere with some emotional behaviors, and that stimulation of limbic areas can elicit emotional responses. But these studies left unclear how the lesioned or stimulated area relates to the rest of the brain. Also, most of the earlier work used techniques that lacked a discrete eliciting stimulus and thus could not benefit from the advantages, described above, that a conditioned stimulus offers.

My approach was to let the natural flow of information through the brain be my guide.[37] In other words, I started at the beginning, at the point that the auditory-conditioned stimulus enters the brain, and tried to trace the pathways forward from this system toward the final destinations that control the conditioned fear responses. I thought that this strategy would be the best and most direct way of figuring out the road map of fear. In retrospect, this strategy worked pretty well.

I began by asking a simple question: which parts of the auditory system are required for auditory fear conditioning (fear conditioning tasks in which an auditory stimulus serves the CS)?[38] The auditory system, like other sensory systems, is organized such that the cortical component is the highest level; it is the culmination of a sequence of information processing steps that start with the peripheral sensory receptors, in this case, receptors located in the ear. I reasoned that damaging the ear would be uninteresting, since a deaf animal is obviously not going to be able to learn anything about a sound. So, instead, I started by damaging the highest part of the auditory pathway. If auditory cortex lesions interfered with fear conditioning, I would be able to conclude that the auditory stimulus had to go all the way through the system in order for conditioning to occur, and that the next step in the pathway should be an output connection of the auditory cortex. If, however, auditory cortex lesions did not disrupt conditioning, I would have to make lesions in lower stations to find the highest level that the auditory stimulus has to reach in order for conditioning to take place.

Damage to the auditory cortex, in fact, turned out to have no effect at all on the conditioning of either the freezing or the blood pressure responses. I then lesioned the next lower station, the auditory thalamus, and these lesions completely prevented fear conditioning. So did lesions of the next lower auditory station in the midbrain. On the basis of these studies I concluded that the auditory stimulus has to rise through the auditory pathway from the ear to the thalamus, but does not have to go the full distance to the auditory cortex. This presented me with a paradox.

Traditionally, the sensory processing structures below the cortex are viewed as slaves to the cortical master. Their job is to get the information to the cortex, where all of the interesting things are done to the stimulus, like assembling neural bits and pieces of the input into the perceptions of the external world that we experience. According to neuroanatomy textbooks, the auditory cortex was the main if not the only target of the auditory thalamus. Where, then, was the auditory stimulus going after it left the thalamus in its journey toward emotional reactivity, if not to the cortex?

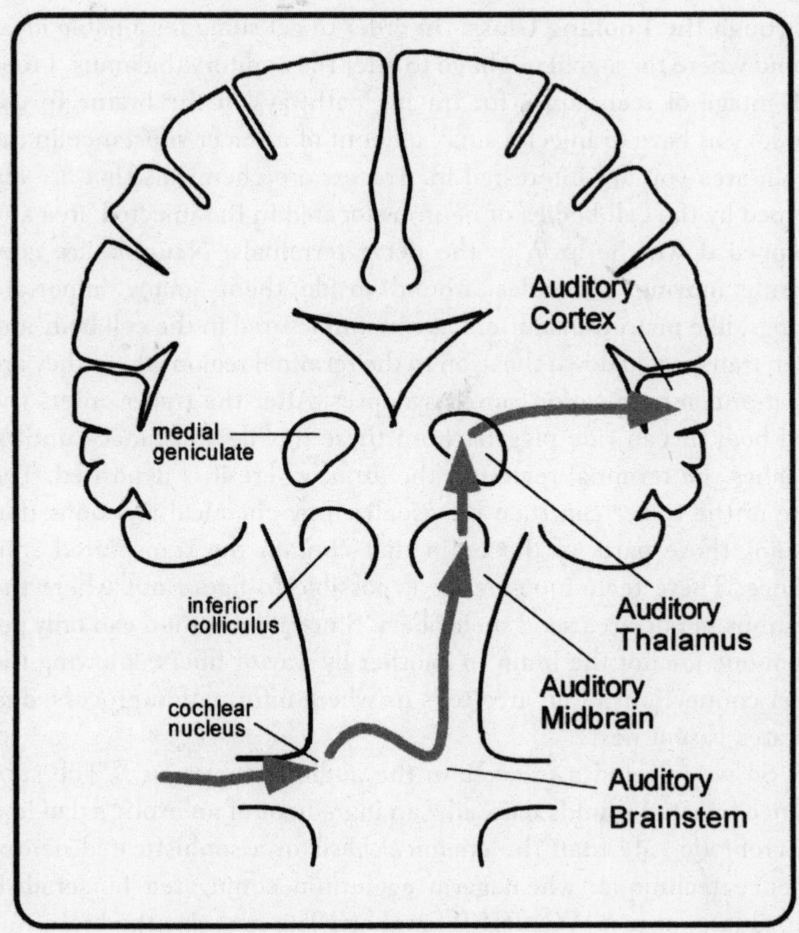

FIGURE 6-7
Auditory Processing Pathways.

This is a highly simplified depiction of auditory pathways in the human brain. A similar organization plan holds for other vertebrate species. Acoustic signals in the environment are picked up by special receptors in the ear (not shown) and transmitted into the brain by way of the auditory nerve (arrow at bottom left), which terminates in the auditory brainstem nuclei (cochlear nucleus and related regions). Axons from these regions then mostly cross over to the other side of the brain and ascend to the inferior colliculus of the midbrain. Inferior collicular axons then travel to the auditory thalamic relay nucleus, the medial geniculate body, which provides the major inputs to the auditory cortex. The auditory cortex is composed of a number of regions and subregions (not shown).

Through the Looking Glass: In order to get some reasonable ideas about where the signal might go to after the auditory thalamus, I took advantage of techniques for tracing pathways in the brain. To use these, you have to inject a small amount of a tracer substance in the brain area you are interested in. Tracers are chemicals that are absorbed by the cell bodies of neurons located in the injected area and shipped down the axon to the nerve terminals. Neurons are constantly moving molecules around inside them—many important things, like neurotransmitters, are manufactured in the cell body and then transported down the axon to the terminal region where they are used in communication across synapses. After the tracer enters the cell body, it can ride piggyback on these mobile substances until it reaches the terminal region of the axon, where it is deposited. The fate of the tracer can then be visualized by chemical reactions that "stain" those parts of the brain that contain the transported substance. These techniques make it possible to figure out where the neurons in one area send their fibers. Since information can only get from one area of the brain to another by way of fibers, knowing the fiber connections of an area tells us where information processed in an area is sent next.

So we injected a tracer into the auditory thalamus.[39] The substance injected sounds more like an ingredient of an exotic salad in a macrobiotic café than the chemical basis of a sophisticated neuroscience technique: wheat germ agglutinin conjugated horseradish peroxidase, or just WGA-HRP for short. The next day the brain was removed and sectioned, and the sections were stained by reacting them with a special chemical potion. We put the stained sections on slides and then looked at them with a microscope set up for dark-field optics, which involves shining indirect light onto the slide—this makes it easier to see the tracer reaction in the sections.

I'll never forget the first time I looked at WGA-HRP with dark-field optics. Bright orange particles formed streams and speckles against a dark blue-gray background. It was like looking into a strange world of inner space. It was incredibly beautiful and I stayed glued to the microscope for hours.

Once I got past the sheer beauty of the staining, I turned to the task at hand, which was to find out where, if anywhere, the auditory thalamus projected to besides the auditory cortex. I found four sub-

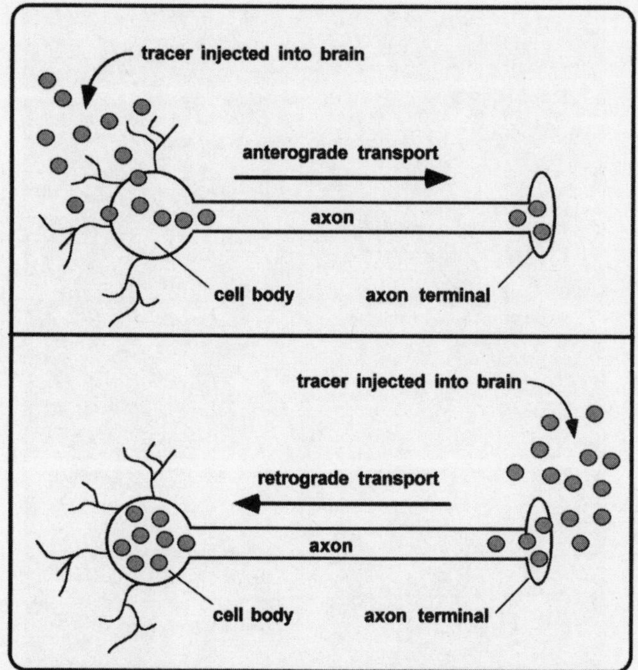

FIGURE 6-8
Tracing Pathways in the Brain with Axonal Transport.

In order to figure out whether neurons in two different brain regions are interconnected, tracers are injected into one of the regions. The tracer is then picked up by neurons that are bathed by the injection. Once the tracer is inside the neuron, it is transported through the axon. Some tracers are picked up by cell bodies and transported to axon terminals (anterograde transport), whereas other tracers are picked up by terminals and transported to cell bodies (retrograde transport).

cortical regions that contained heavy sprinkling with the tiny orange dots, suggesting that these regions receive projections from the auditory thalamus. This was surprising, given the well-received view that sensory areas of the thalamus project mainly, if not exclusively, to the cortex.

It seemed likely that one of the four labeled regions might be the crucial next step in the fear conditioning pathway—the place where the stimulus goes after the thalamus. So, I designed a lesion study that would interrupt the flow of information from the auditory thalamus to each of these regions.[40] Three of the lesions had absolutely no

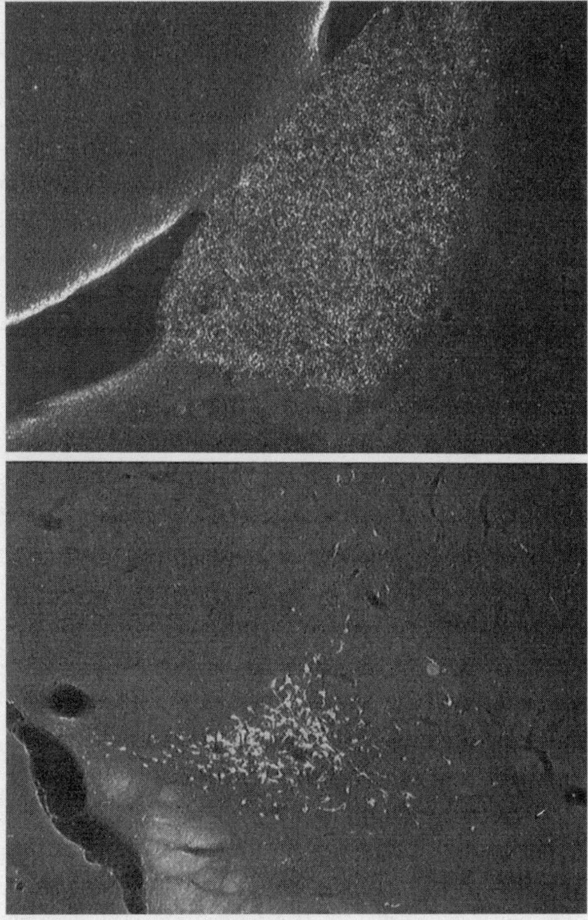

FIGURE 6-9
Examples of Anterograde and Retrograde Transport in the Thalamo-Amygdala Pathway.

The top photograph shows anterograde labeling of terminals in the lateral amygdala after an injection of a tracer in the auditory thalamus. These terminals in the lateral amygdala thus originate from cell bodies in the auditory thalamus. Note the fine, punctate nature of anterograde terminal labeling. The bottom photograph shows cell bodies in the auditory thalamus that were retrogradely labeled by an injection of a tracer in the lateral nucleus of the amygdala. The labeled cells are the bright white structures that cluster together in a triangular region. The cells in the auditory thalamus thus send their axons to the lateral amygdala. Note the large size of the labeled cell bodies, as compared to the terminals above. The two images are black-and-white photographs of dark-field illuminated brain sections taken through a microscope.

effect. But disconnection of auditory thalamus from the fourth area—the amygdala—prevented conditioning from taking place.

Almond Joy: The amygdala is a small region in the forebrain, named by the early anatomists for its almond shape (amygdala is the Latin word for almond). It was one of the areas of the limbic system and had long been thought of as being important for various forms of emotional behavior—earlier studies of the Klüver-Bucy syndrome had pointed to it (see Chapter 4), as had electrical stimulation studies (see below).

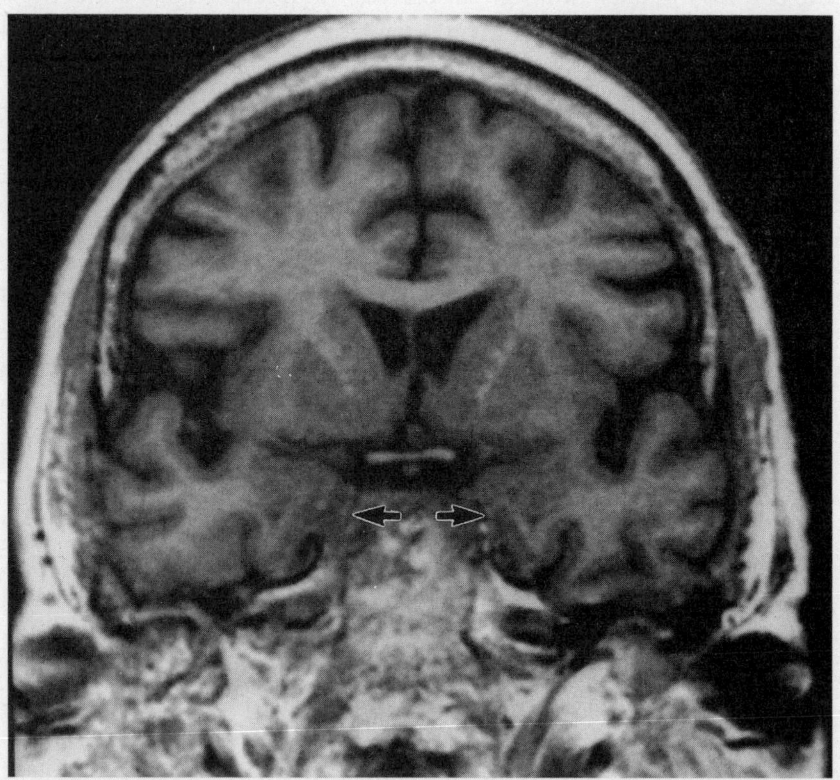

FIGURE 6-10
Magnetic Resonance Imaging Scan Showing the Location of the Amygdala in the Human Brain.

The amygdala on each side of the brain is indicated by the arrows. (Image provided by E. A. Phelps of Yale University.)

The discovery of a pathway that could transmit information directly to the amygdala from the thalamus suggested how a conditioned fear stimulus could elicit fear responses without the aid of the cortex. The direct thalamic input to the amygdala simply allowed the cortex to be bypassed. The brain is indeed a complex mesh of connections, but anatomical findings were taking us on a delightful journey of discovery through this neuronal maze.

I wasn't really looking for the amygdala in my work. The dissection of the brain's pathways just took me there. But my studies, when they first started coming out, fit nicely with a set of findings that Bruce Kapp had obtained concerning a subregion of the amygdala—the central nucleus. Noting that the central nucleus has connections with the brain stem areas involved in the control of heart rate and other autonomic nervous system responses, he proposed that this region might be a link in the neural system through which the autonomic responses elicited by a conditioned fear stimulus are expressed. And when he lesioned the central nucleus in the rabbit, his hypothesis was confirmed—the lesions dramatically interfered with the conditioning of heart rate responses to a tone paired with shock.[41]

Kapp went on to show that stimulation of the central amygdala produced heart rate and other autonomic responses, strengthening his idea that the central nucleus was an important forebrain link in the control of autonomic responses by the brain stem. However, he also found that stimulation of the central nucleus elicited freezing responses, suggesting that the central amygdala might not just be involved in the control of autonomic responses, but might be part of a general-purpose defense response control network.

Indeed, subsequent research by several laboratories has shown that lesions of the central nucleus interfere with essentially every measure of conditioned fear, including freezing behavior, autonomic responses, suppression of pain, stress hormone release, and reflex potentiation.[42] It was also found that each of these responses are mediated by different outputs of the central nucleus.[43] For example, I demonstrated that lesions of different projections of the central nucleus separately interfered with freezing and blood pressure conditioned responses—lesions of one of the projections (the periaqueductal gray) interfered with freezing but not blood pressure responses, whereas lesions of another (the lateral hypothalamus) interfered with the blood pressure but not the freezing response.[44] And while lesions of a third projection (the bed nucleus of the stria terminalis) had no effect on either of these responses, other scientists later showed that lesions of this region interfere with the elicitation of stress hormones by the CS.[45]

Journey to the Center of the Amygdala: The studies of the central amygdala and its outputs seemed to clear up how the responses get expressed, but some mysteries still remained about how the stimulus reaches the central nucleus in its quest to gain control over the responses. Again using the WGA-HRP tracing techniques, I examined whether the auditory stimulus might be sent to the central amygdala directly from the auditory thalamus.[46]

I injected the tracer WGA-HRP into the central nucleus. This time, though, I was tracing connections in the reverse direction, from the area of termination of a pathway back to the cell bodies that give rise to it—the tracer does its piggyback ride in this direction as well. When I examined the sections under the microscope, I found bright orange cells containing the tracer in thalamic areas adjacent to the auditory thalamus but not the auditory thalamus itself. As a result, it seemed unlikely that an auditory stimulus is sent directly to the central nucleus in the process of controlling fear responses.

But when I made injections in another amygdala subregion, the lateral nucleus, there were orange cell bodies in the auditory thalamus.[47] And when I aimed injections for the region of the auditory thalamus that contained these labeled cells, I found the fine orange speckles characteristic of terminals in the lateral nucleus (see Figure 6-9). It seemed that the auditory stimulus might travel from the thalamus to the lateral nucleus of the amygdala. To test this hypothesis, I made lesions of the lateral nucleus. Like central amygdala lesions, these interfered with fear conditioning.[48]

On the basis of these lesion studies, together with the results of anatomical tracing experiments, the lateral nucleus came to be thought of as the region of the amygdala that receives the CS inputs in fear conditioning and the central nucleus as the interface with response control systems. The inputs and outputs had been mapped.

Still, an important set of linkages remained uncharted. If the CS inputs enter the amygdala by way of the lateral nucleus and the CR

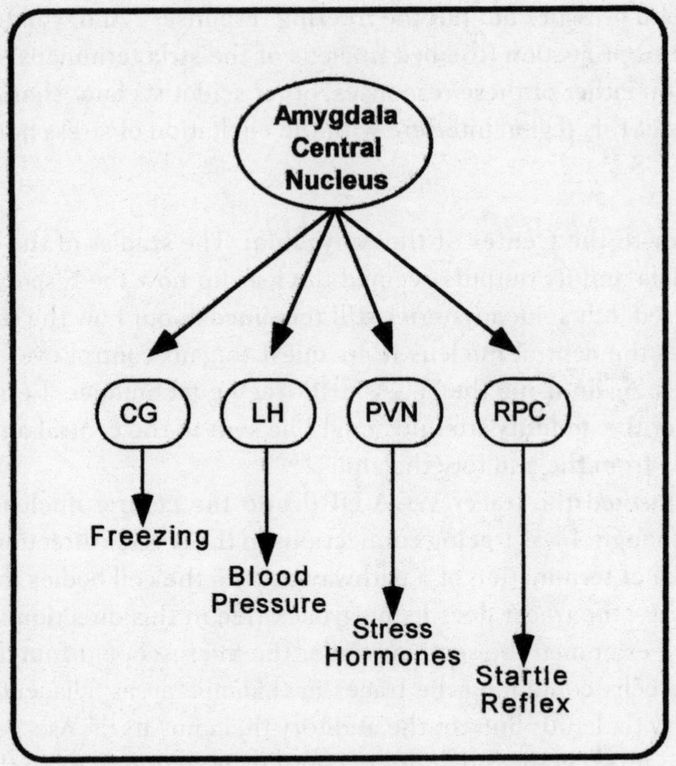

FIGURE 6-11
Different Outputs of the Amygdala Control Different Conditioned Fear Responses.

In the presence of danger or stimuli that warn of danger, behavioral, autonomic, and endocrine responses are expressed, and reflexes are modulated. Each of these responses is controlled by a different set of outputs from the central nucleus of the amygdala. Lesions of the central nucleus block the expression of all these responses, whereas lesions of the output pathways block only individual responses. Selected examples of central amygdala outputs are shown. Abbreviations: CG, central gray; LH, lateral hypothalamus; PVN, paraventricular hypothalamus (which receives inputs from the central amygdala directly and by way of the bed nucleus of the stria terminalis); RPC, reticulopontis caudalis.

outputs leave through the central nucleus, how does information received by the lateral nucleus reach the central nucleus? Although this question has not yet been answered completely, anatomical findings have provided us with some clues.[49] The lateral nucleus has some direct projections to the central nucleus, and also can influence the central nucleus by way of projections to two other amygdala nuclei (the basal and accessory basal), each of which gives rise to strong projections to the central nucleus. There are thus several ways for information entering the lateral nucleus to reach the central nucleus, but exactly which is most crucial is not yet known.

The amygdala is composed of about a dozen or so subregions, and not all or even most are involved in fear conditioning. Only lesions that damage amygdala regions that are part of the fear conditioning circuitry should be expected to disrupt fear conditioning. The lateral and central nuclei are, without doubt, crucially involved, but the role of other amygdala regions is still under study.

The Low and the High Road: The fact that emotional learning can be mediated by pathways that bypass the neocortex is intriguing, for it suggests that emotional responses can occur without the involvement of the higher processing systems of the brain, systems believed to be involved in thinking, reasoning, and consciousness. But before we pursue this notion, we need to further consider the role of the auditory cortex in fear conditioning.

In the experiments described so far, a simple sound was paired with a shock. The auditory cortex is clearly not needed for this. But suppose the situation is somewhat more complex. Instead of just one tone paired with a shock, suppose the animal gets two similar tones, one paired with the shock and the other not, and has to learn to distinguish between them. Would the auditory cortex then be required? Neil Schneidermann, Phil McCabe, and their colleagues looked at this question in a study of heart rate conditioning in rabbits.[50] With enough training, the rabbits eventually only expressed heart rate responses to the sound that had been associated with the shock. And when the auditory cortex was lesioned, this capacity was lost. Interestingly, the auditory cortex lesions did not interfere with conditioning by blocking responses to the stimulus paired with the shock. Instead, the cortically lesioned animals responded to both stimuli as if they had each been paired with the shock.

These findings make sense given what we know about the neurons in the thalamus that project to the amygdala as opposed to those that provide the major inputs to the auditory cortex.[51] If you put an

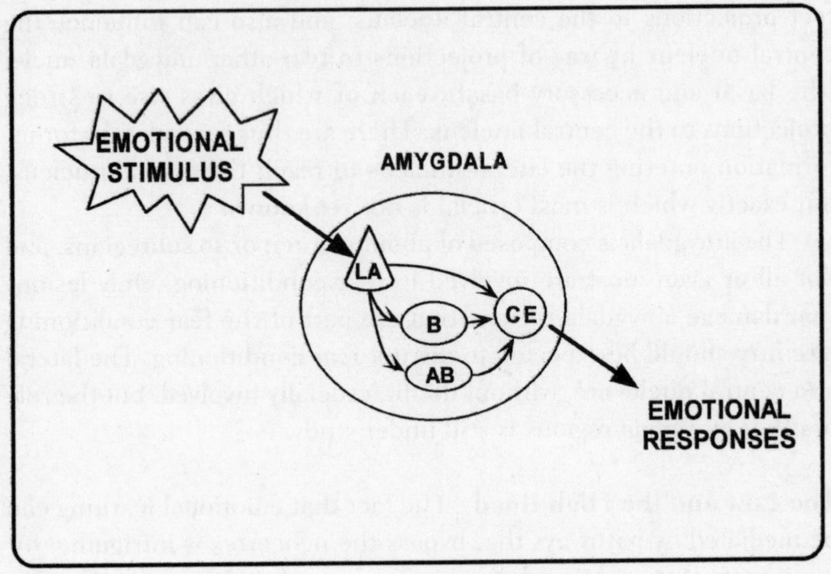

FIGURE 6-12
Organization of Information-Processing Pathways in the Amygdala.

The lateral nucleus (LA) is the gateway into the amygdala. Stimuli from the outside world are transmitted to LA, which then processes the stimuli and distributes the results to other regions of the amygdala, including the basal (B), accessory basal (AB), and central nuclei (CE). The central nucleus is then the main connection with areas that control emotional responses. As shown in figure 6-11, different outputs of the central nucleus regulate the expression of different responses.

electrode in the brain, you can record the electrical activity of individual neurons in response to auditory stimulation. Neurons in the area of the thalamus that projects to the primary auditory cortex are narrowly tuned—they are very particular about what they will respond to. But cells in the thalamic areas that project to the amygdala are less picky—they respond to a much wider range of stimuli and are said to be broadly tuned. The Beatles and Rolling Stones (or, if you like, Oasis and the Cranberries) will sound the same to the amygdala by way of the thalamic projections but quite different by way of the cortical projections. So when two similar stimuli are used in a conditioning study, the thalamus will send the amygdala essentially the same information, regardless of which stimulus it is processing, but when the cortex processes the different stimuli it will send the amygdala different signals. If the cortex is damaged, the animal has only the direct thalamic pathway and thus the amygdala treats the two stimuli the same—both elicit conditioned fear.

The Quick and the Dead: Why should the brain be organized this way? Why should it have the lowly thalamic road when it also has the high cortical road?

Our only source of information about the brains of animals from long ago is the brains of their living descendants. Studies of living fish, amphibians, and reptiles suggest that sensory projections to rudimentary cortical areas were probably relatively weak compared to projections to subcortical regions in primordial animals.[52] In contemporary mammals, the thalamic projections to cortical pathways are far more elaborate and important channels of information processing. As a result, it is possible that in mammals the direct thalamic pathway to the amygdala is simply an evolutionary relic, the brain's version of an appendix. But I don't think this is the case. There's been ample time for the direct thalamo-amygdala pathways to have atrophied if they were not useful. But they have not. The fact that they have existed for millions and millions of years side by side with thalamo-cortical pathways suggests that they still serve some useful function. But what could that function be?

Although the thalamic system cannot make fine distinctions, it has an important advantage over the cortical input pathway to the amygdala. That advantage is time. In a rat it takes about twelve milliseconds (twelve one-thousandths of a second) for an acoustic stimulus to reach the amygdala through the thalamic pathway, and almost twice as long through the cortical pathway. The thalamic pathway is thus faster. It cannot tell the amygdala exactly what is there, but can provide a fast signal that warns that something dangerous may be there. It is a quick and dirty processing system.

Imagine walking in the woods. A crackling sound occurs. It goes straight to the amygdala through the thalamic pathway. The sound also goes from the thalamus to the cortex, which recognizes the sound to be a dry twig that snapped under the weight of your boot, or that of a rattlesnake shaking its tail. But by the time the cortex has

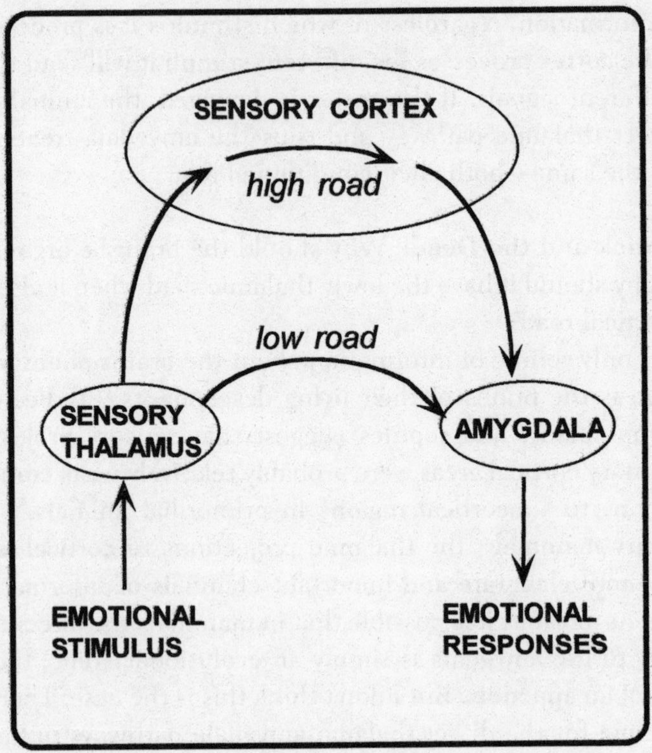

FIGURE 6-13
The Low and the High Roads to the Amygdala.

Information about external stimuli reaches the amygdala by way of direct pathways from the thalamus (the low road) as well as by way of pathways from the thalamus to the cortex to the amygdala. The direct thalamo-amygdala path is a shorter and thus a faster transmission route than the pathway from the thalamus through the cortex to the amygdala. However, because the direct pathway bypasses the cortex, it is unable to benefit from cortical processing. As a result, it can only provide the amygdala with a crude representation of the stimulus. It is thus a quick and dirty *processing pathway. The direct pathway allows us to begin to respond to potentially dangerous stimuli before we fully know what the stimulus is. This can be very useful in dangerous situations. However, its utility requires that the cortical pathway be able to override the direct pathway. It is possible that the direct pathway is responsible for the control of emotional responses that we don't understand. This may occur in all of us some of the time, but may be a predominant mode of functioning in individuals with certain emotional disorders (discussed in more detail in Chapter 8).*

figured this out, the amygdala is already starting to defend against the snake. The information received from the thalamus is unfiltered and biased toward evoking responses. The cortex's job is to prevent the inappropriate response rather than to produce the appropriate one. Alternatively, suppose there is a slender curved shape on the path. The curvature and slenderness reach the amygdala from the thalamus, whereas only the cortex distinguishes a coiled up snake from a curved stick. If it is a snake, the amygdala is ahead of the game. From the point of view of survival, it is better to respond to potentially dangerous events as if they were in fact the real thing than to fail to respond. The cost of treating a stick as a snake is less, in the long run, than the cost of treating a snake as a stick.

So we can begin to see the outline of a fear reaction system. It involves parallel transmission to the amygdala from the sensory thalamus and sensory cortex. The subcortical pathways provide a crude image of the external world, whereas more detailed and accurate representations come from the cortex. While the pathway from the thalamus only involves one link, several links are required to activate the amygdala by way of the cortex. Since each link adds time, the thalamic pathway is faster. Interestingly, the thalamo-amygdala and cortico-amygdala pathways converge in the lateral nucleus of the amygdala. In all likelihood, normally both pathways transmit signals to the lateral nucleus, which appears to play a pivotal role in coordinating the sensory processes that constitute the conditioned fear stimulus. And once the information has reached the lateral nucleus it can be distributed through the internal amygdala pathways to the central nucleus, which then unleashes the full repertoire of defensive reactions. Although I have mainly discussed my own work, research by others (especially Michael Davis, Michael Fanselow, Norman Weinberger, and Bruce Kapp) has also contributed significantly to our understanding of the neural basis of fear conditioning.[53]

A Sea Horse for All Occasions: Consider another example. You are walking down the street and notice someone running toward you. The person, upon reaching you, hits you on the head and steals your wallet or purse. The next time someone is running toward you, chances are a set of standard fear responses will be set into play. You will probably freeze and prepare to defend yourself, your blood pressure and

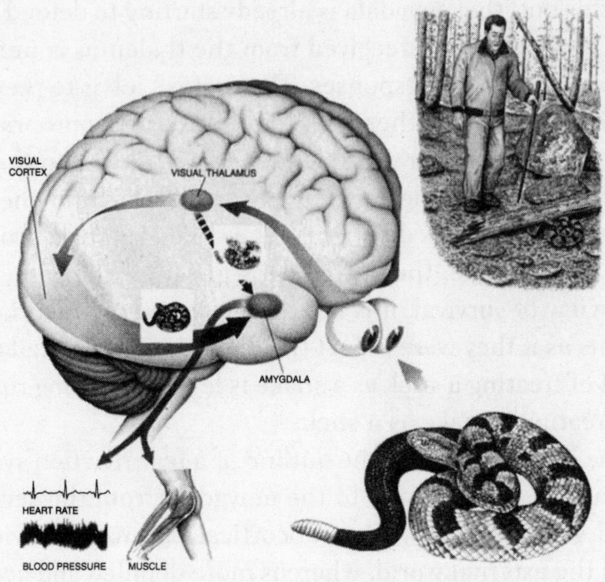

FIGURE 6-14
Brain Pathways of Defense.

As the hiker walks through the woods, he abruptly encounters a snake coiled up behind a log on the path (upper right inset). The visual stimulus is first processed in the brain by the thalamus. Part of the thalamus passes crude, almost archetypal, information directly to the amygdala. This quick and dirty transmission allows the brain to start to respond to the possible danger signified by a thin, curved object, which could be a snake, or could be a stick or some other benign object. Meanwhile, the thalamus also sends visual information to the visual cortex (this part of the thalamus has a greater ability to encode the details of the stimulus than does the part that sends inputs to the amygdala). The visual cortex then goes about the business of creating a detailed and accurate representation of the stimulus. The outcome of cortical processing is then fed to the amygdala as well. Although the cortical pathway provides the amygdala with a more accurate representation than the direct pathway to the amygdala from the thalamus, it takes longer for the information to reach the amygdala by way of the cortex. In situations of danger, it is very useful to be able to respond quickly. The time saved by the amygdala in acting on the thalamic information, rather than waiting for the cortical input, may be the difference between life and death. It is better to have treated a stick as a snake than not to have responded to a possible snake. Most of what we know about these pathways has actually been learned by studies of the auditory as opposed to the visual system, but the same organizational principles seem to apply. (From J.E. LeDoux, Emotion, memory and the brain. *Scientific American* [June 1994], vol 270, p. 38.)

heart rate will rise, your palms and feet will sweat, stress hormones will begin to flow through your bloodstream, and so on. The sight of someone running toward you has become a conditioned fear stimulus. But suppose you later find yourself on the street where you were mugged. Although there is no one running toward you, your body may still be going through its defense motions. The reason for this is that not only did you get conditioned to the immediate stimulus directly associated with the trauma (the sight of the mugger running toward you), but also to the other stimuli that just happen to have been there. These made up the occasion or context in which the mugging took place, and like the sight of the mugger they too were conditioned by the traumatic experience.

Psychologists have studied contextual conditioning extensively. If you place a rat in a box and give it a few exposures to a mild shock in the presence of a tone, the rat will become conditioned to the tone, as we've already seen, but will also get conditioned to the box. So the next time the rat is placed in the box, the conditioned fear responses—freezing, autonomic and endocrine arousal, pain suppression, reflex potentiation—will occur, even in the absence of the tone. The context has become a CS.

In a contextual fear conditioning experiment, the context is made up of all of the stimuli present, other than the explicit CS. In other words, the CS is in the foreground—it is the most salient and predictive stimulus with respect to the shock. All other stimuli are in the background of the CS and constitute the context. The context is always there, but the CS only comes on sometimes. For this reason, it is often necessary to test the effects of a CS in a novel context, one that has not been associated with the shock, since fear responses elicited by the ever-present context can prevent the detection of responses that occur to the occasionally occurring CS.

In a sense contextual conditioning is incidental learning. During conditioning, the subject is paying attention to the most obvious stimulus (the tone CS) but the other stimuli get bought for the same purchase price. This is very useful from an evolutionary point of view. Our rabbit that escaped from the fox got conditioned not only to the stimuli that were immediately and directly associated with the arrival of the fox—its sight and smell and the sounds it made when attacking—but also to the place where the fox encounter took place—the watering hole and its surroundings. These extra stimuli are very useful in expanding the impact of conditioning beyond the most obvious and direct stimuli, allowing the organism to use even remotely related cues to avoid or escape from danger.

The interesting thing about a context is that it is not a particular stimulus but a collection of many. For some time it has been thought that the integration of individual stimuli into a context that no longer contains the individual elements is a function of the hippocampus.[54] Unlike the amygdala, the hippocampus does not get information from brain regions that process individual sensory stimuli, like lights and tones.[55] Instead, the sights and sounds of a place are pooled together before reaching the hippocampus, and one job of this brain region is to create a representation of the context that contains not individual stimuli but relations between stimuli.[56]

With this view of the hippocampus in mind, Russ Phillips and I, as well as Mike Fanselow and colleagues, examined whether the hippocampus might play a crucial role in the conditioning of fear responses to background contextual events.[57] In other words, we examined whether damage to the hippocampus might interfere with the conditioning of fear responses to the chamber in which tone-shock pairings occurred. Normal rats froze as soon as they were placed in the conditioning box. Rats with hippocampal lesions showed little freezing to the conditioning box. But as soon as the tone came on, the lesioned rats started freezing. The hippocampal lesion, in other words, selectively eliminated fear responses elicited by contextual stimuli without affecting fear responses elicited by a tone. The tone still worked because the tone could get to the amygdala directly. We reasoned that the hippocampal lesioned animals showed no fear responses to the box because they couldn't form the contextual representation and send it to the amygdala. Indeed, amygdala damage interfered with contextual conditioning just as it did with tone conditioning.[58]

A Hub in the Wheel of Fear: The amygdala is like the hub of a wheel. It receives low-level inputs from sensory-specific regions of the thalamus, higher level information from sensory-specific cortex, and still higher level (sensory independent) information about the general situation from the hippocampal formation. Through such connections, the amygdala is able to process the emotional significance of individual stimuli as well as complex situations. The amygdala is, in essence, involved in the appraisal of emotional meaning. It is where trigger stimuli do their triggering.

It is not unreasonable to suggest that by knowing what the different inputs to the amygdala are, and having some idea of what function those areas play in cognition, we can get some reasonable hypotheses about what kinds of cognitive representations can arouse fear responses. And by the same token, if we know how the brain achieves some cognitive function, and we can determine how the brain regions involved in that function are connected with the amygdala, we can come up with some plausible ideas about how fear might be aroused by that kind of cognition.

It is easy to imagine how malfunctions of the amygdala and its neural partners might lead to emotional disorders. If in some individuals (for genetic or acquired reasons) thalamic pathways are dominant or otherwise uncoupled from the cortical pathways, these persons might form emotional memories on the basis of stimulus events that do not coincide with their ongoing conscious perceptions of the world mediated by the cortex. That is, because thalamic pathways to the amygdala exit the sensory system before conscious perceptions are created at the cortical level, the processing that occurs through these subcortical pathways, which can only represent features and fragments of stimuli, does not necessarily coincide with the perceptions occurring in the cortex. Such people would have very poor insight into their emotions. At the same time, if the hippocampal system were uncoupled from the thalamic and cortical projections to the amygdala, we might have persons who express emotions that are inappropriate to the immediate context, including possibly the social context. These are purely speculative suggestions at this point, but they are consistent with the facts now available.

Same as It Ever Was

Through studies of fear conditioning in rats, we have been able to map out in great detail the brain mechanisms that underlie fear reactions. The reason we study fear in rats is obvious—we want to learn

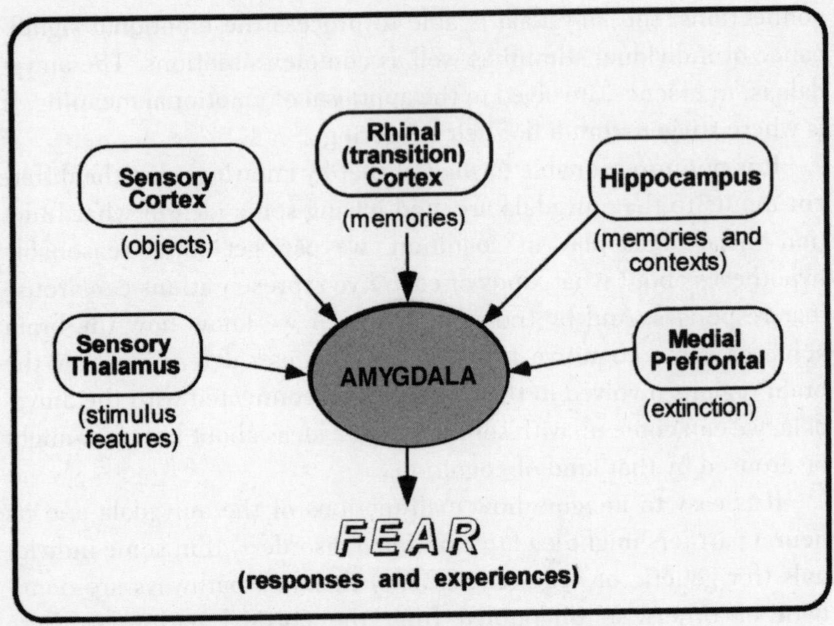

FIGURE 6-15

The Amygdala: Hub in the Wheel of Fear.

The amygdala receives inputs from a wide range of levels of cognitive processing. By way of inputs from sensory areas of the thalamus, the emotional functions of the amygdala can be triggered by low-level stimulus features, whereas inputs from cortical sensory processing systems (especially the late stages of processing in these systems) allow more complex aspects of stimulus processing (objects and events) to activate the amygdala. Inputs from the hippocampus play an important role in setting the emotional context. In addition, as we'll see in Chapter 7, the hippocampus and related areas of the cortex (including the rhinal or transitional cortical areas) are involved in the formation and retrieval of explicit memories, and inputs to the amygdala from these areas may allow emotions to be triggered by such memories. The medial prefrontal cortex has been implicated in the process known as extinction, whereby the ability of conditioned fear stimuli to elicit conditioned fear responses is weakened by repeated exposure to the conditioned stimulus without the unconditioned stimulus. Inputs to the amygdala from the medial prefrontal cortex appear to contribute to this process (see Chapter 8). By knowing which cortical areas project to the amygdala, and knowing the functions in which those areas participate, we can make predictions about how those functions might contribute to fear reactions. Anatomy can, in other words, illuminate psychology.

how human fear works. Less obvious, perhaps, is whether this is a reasonable approach. Can we really learn something about human fear by studying the brain of a rat? I believe we can.

Although no other creature has been studied as thoroughly with fear conditioning as the rat, and though no other technique has been used to study fear more extensively than fear conditioning, if we compile the evidence across species and experimental approaches we reach the inescapable conclusion that the basic brain mechanisms of fear are essentially the same through many levels of evolutionary development.

Let's start with our basic model of fear, fear conditioning. The effects of amygdala lesions on fear conditioning have been studied in birds, rats, rabbits, monkeys, and people using autonomic nervous system activity as the conditioned response. In each of these species, damage to the amygdala interferes with conditioned fear reactions—the CS fails to elicit the CR when the amygdala is damaged.

Pigeons are the only nonmammalian species in which the effects of amygdala lesions on fear conditioning have been examined. The similarity of the effects in pigeons and mammals means either that the amygdala was selected as a key component of the defense system of the vertebrate brain before birds and mammals separated from reptiles, or that the amygdala evolved to perform this function separately in the two post-reptilian lines. The best way to resolve this issue would be to know whether amygdala lesions interrupt fear conditioning in reptiles. Unfortunately, this experiment has not been performed. As a result, we need to turn to some other kind of evidence in search of an answer.

Another technique that has been used to map the brain pathways of fear or defensive behavior is brain stimulation. These techniques have been applied to reptiles as well as mammals and birds, and might thus be able to help us piece together an answer as to whether the amygdala has been involved in defense since at least the time when birds and mammals diverged from reptiles.

The first step we need to take, though, is to be certain that brain stimulation identifies the same pathways of fear reactivity that studies of fear conditioning have in the mammalian brain, where fear conditioning has been most clearly related to brain pathways. There is a long and interesting history to studies of brain stimulation in mammals, which we will only be able to touch on here.[59] Our main concern is whether stimulation of the amygdala, the heart and soul of the fear system revealed by fear conditioning studies, gives rise to defense responses in mammals. Clearly this occurs. It is well established that stimulation of the amygdala in anesthetized mammals elicits autonomic nervous system responses, and in awake mammals such stimulations elicit freezing, escape, and defensive attack responses, in addition to autonomic changes.[60] These kinds of studies have been performed in rats, cats, dogs, rabbits, and monkeys, all with similar results. Further, defense responses can be elicited from the central nucleus of the amygdala, the region by which the amygdala communicates with brain stem areas that control conditioned fear responses. And interruption of the pathways connecting the amygdala with these brain stem pathways interferes with the expression of the defense responses. Studies of fear conditioning and brain stimulation reveal similar output pathways in the expression of fear responses.

Let's now descend the phyletic tree and see what happens when we stimulate the amygdala of reptiles. It's tricky business to use living reptiles as examples of what reptiles might have been like when mammals diverged, as current reptiles themselves come from lines that have diverged from the ancestral lines. Nevertheless, since brain and behavior are not preserved in fossil records, this is the only way comparative studies of brain function can be conducted. Stimulation of the amygdala in lizards elicits the defensive behaviors these animals characteristically show when they are threatened by a predator, and lesions of the same regions reduce the expression of these behaviors in response to natural trigger stimuli.[61]

Now going up the branching evolutionary tree, we can consider the effects of stimulation of the human amygdala.[62] Such studies are performed in conjunction with brain surgery for otherwise untreatable epilepsy. Since the stimuli are delivered to the amygdala while the subjects are awake, it is possible to not only record expressive responses that are elicited, but also to ask the subjects to report on their experiences. Interestingly, the most common experience reported is a sense of foreboding danger, of fear. Fear is also the most commonly reported experience occurring in association with epileptic seizures, which are in essence spontaneous electrical stimulations that originate in the amygdala.

Recent studies of humans with amygdala damage also suggest that it plays a special role in fear. It is extremely rare to encounter patients with damage to only the amygdala, but it is not that rare to come across patients with damage that includes the amygdala. This is particularly common in patients who undergo surgery to remove epileptic regions of their temporal lobe. Kevin LaBar, Liz Phelps, and I conducted a study of fear conditioning in patients of this type.[63] Because we were studying humans rather than rats, we chose to use a very loud obnoxious noise as the US instead of electric shock. This worked just fine for conditioning autonomic nervous system responses to a softer, non-noxious sound in the control subjects. Importantly, we found that autonomic conditioned responses were reduced in the temporal lobe lesioned group. Interestingly, the patients consciously "knew" the relationship between the CS and US: when asked what went on in the experiment, they typically said, "Oh, there was a sound followed by this other really loud sound." This knowledge was not enough to transform the meaningless sound into a trigger stimulus. Although the lesions included areas other than the amygdala, we know from the animal studies that of all the areas included in the lesion, damage to the amygdala is the likely cause of the deficit in fear conditioning. This is a good example of why animal studies are so important. Without the animal studies the human experiment would be uninterpretable.

Although damage restricted to the human amygdala is very rare, Antonio Damasio and his colleagues at the University of Iowa have come across such a patient.[64] They have performed some extremely important and fascinating studies on her. For example, in one study they examined her ability to detect the emotional expression on faces. She was able to correctly identify most classes of expressions, except when the faces showed fear. And most importantly they have recently examined whether the capacity for fear conditioning is interfered with. Indeed, it was. Unlike the temporal lobe lesioned patients, this case unequivocally implicates the amygdala. Again, though, this study was inspired by the body of animal research that had already implicated the amygdala. If this study had been performed twenty years ago, before any of the animal conditioning studies had been done, we would have little understanding of the pathways through which the amygdala contributes to fear conditioning. In point of fact, though, the human studies might not have even been performed had the animal studies not set the stage for them—without the known effects of amygdala damage on conditioned fear in experimental animals, why would anyone consider doing such a study in humans with amygdala pathology?

The point of this discussion is to illustrate that the amygdala seems to do the same thing—take care of fear responses—in all species that have an amygdala. This is not the only function of the amygdala,[65] but it is certainly an important one. The function seems to have been established eons ago, probably at least since dinosaurs ruled the earth, and to have been maintained through diverse branches of evolutionary development. Defense against danger is perhaps an organism's number one priority and it appears that in the major groups of vertebrate animals that have been studied (reptiles, birds, and mammals) the brain performs this function using a common architectural plan.

The remarkable fact is that at the level of behavior, defense against danger is achieved in many different ways in different species, yet the amygdala's role is constant. It is this neural correspondence across species that no doubt allows diverse behaviors to achieve the same evolutionary function in different animals. This functional equivalence and neural correspondence applies to many vertebrate brains, including human brains. When it comes to detecting and responding to danger, the brain just hasn't changed much. In some ways we are emotional lizards.[66] I am quite confident in telling you that studies of fear reactions in rats tell us a great deal about how fear mechanisms work in our brains as well.

Beyond Evolution

By way of the amygdala and its input and output connections, the brain is programmed to detect dangers, both those that were routinely experienced by our ancestors and those learned about by each of us as individuals, and to produce protective responses that are most effective for our particular body type, and for the ancient environmental conditions under which the responses were selected.

Prepackaged responses have been shaped by evolution and occur automatically, or as Darwin pointed out, involuntarily.[67] They take place before the brain has had the chance to start thinking about what to do. Thinking takes time, but responding to danger often needs to occur quickly and without much mulling over the decision. Recall Darwin's encounter with the puff adder at the Zoological Gardens—the snake struck and Darwin recoiled back quick as a flash. If the snake had not been behind glass, Darwin's life would have been at the mercy of his involuntary responses—if they were quick enough, he would have survived; if they were too slow, he would have perished. He certainly had no time to decide whether or not to jump once the snake started to strike. And even though he had resolved not to jump, he could not stop himself.

While many animals get through life mostly on emotional automatic pilot, those animals that can readily switch from automatic pilot to willful control have a tremendous extra advantage. This advantage depends on the wedding of emotional and cognitive functions. So far we've emphasized the role of cognitive processes as a source of signals that can trigger prepackaged emotional reactions. But cognition also contributes to emotion by giving us the ability to make decisions about what kind of action should occur next, given the situation in which we find ourselves now. One of the reasons that cognition is so useful a part of the mental arsenal is that it allows this shift from *reaction* to *action.* The survival advantages that come from being able to make this shift may have been an important ingredient that shaped the evolutionary elaboration of cognition in mammals and the explosion of cognition in primates, especially in humans.

In responding first with its most-likely-to-succeed behavior, the brain buys time. This is not to say that the brain responds automatically first for the purpose of buying time. The automatic responses came first, in the evolutionary sense, and cannot exist for the purpose of serving responses that evolved later. Buying time is a fortunate byproduct of the way information processing is constrained by brain organization.

Imagine that you are a small mammal, say a prairie dog. You come out of your burrow to look for dinner. You begin exploring around, and all of a sudden you spot a bobcat, which you know to be a serious enemy. You immediately stop all movement. Freezing is evolution's gift to you. You do it without having to weigh decisions. It just happens. The sight or sound of the bobcat goes straight to your amygdala and out comes the freezing response. If you had to make a deliberate decision about what to do, you would have to consider the likelihood of each possible choice succeeding or failing and could get so bogged down in decision making that you might be eaten before you made the choice. And if you started fidgeting around or pacing while trying to decide, you would surely attract the predator's attention and certainly decrease your likelihood of surviving. Freezing, of course, is not the only automatic response. But it is a fairly universal initial response to detection of danger throughout the animal kingdom (see Chapter 5). Automatic responses like freezing have the advantage of having been test-piloted through the ages; reasoned responses do not come with this kind of fine-tuning.

Presumably, evolution could work toward making cognition faster, so that thought could always precede action, eliminating involuntary action altogether from the behavioral repertoire. But this would be quite costly. There are many things that we are better off not having to think about, like putting one foot in front of the other when we walk, blinking when objects come near the eye, getting the glove to just the right spot to catch a fly ball, inserting the subject and verb in the correct place when we speak, responding quickly and appropriately to danger, and so forth. Behavioral and mental functions would slow down to a crawl if every response had to be preceded by a thought.

But no matter how useful automatic reactions are, they are only a quick fix, especially in humans. Eventually you take control. You make a plan and carry it out. This requires that your cognitive resources be directed to the emotional problem. You have to stop thinking about whatever you were thinking about before the danger occurred and start thinking about the danger you are facing (and already responding to automatically). Robert and Caroline Blanchard call this behavior "risk assessment."[68] This is something we do all the time. We're always sizing up situations and planning how to maximize our gains and minimize our losses. Surviving is not just something we do in the presence of a wild beast. Social situations are often survival encounters.

We don't really fully understand how the human brain sizes up a situation, comes up with a set of potential courses of action, predicts possible outcomes of different actions, assigns priorities to possible actions, and chooses a particular action, but these activities are unquestionably amongst the most sophisticated cognitive functions. They allow the crucial shift from reaction to action. From what we currently know, it seems likely that regions like the prefrontal cortex may be involved.[69] The prefrontal cortex is the part of the cerebral cortex that has expanded the most in primates, and it may not even exist in other mammals.[70] When this region is damaged in people, they have great difficulty in planning what to do.[71] So-called frontal lobe patients tend to do the same thing over and over again. They are glued to the present and unable to project themselves into the future. Some regions of the prefrontal cortex are linked with the amygdala, and together these regions, and possibly others, may play key roles in planning and executing emotional actions. We'll again consider the role of the prefrontal cortex in emotion when we turn to the topic of emotional consciousness in Chapter 9. Another brain region that may be involved is the basal ganglia, a collection of areas in the subcortical forebrain. These regions have long been implicated in controlling movement, and recent work has shown that interactions between the amygdala and the basal ganglia may be important in instrumental emotional behavior, which is essentially what I am calling emotional actions.[72]

Emotional plans are a wonderful addition to emotional automaticity. They allow us to be emotional *actors,* rather than just *reactors.* But the capacity to make this switch has a price. Once you start thinking, not only do you try to figure the best thing to do in the face of several possible next moves that a predator (including a social predator) is likely to make, you also think about what will happen if the plan fails. Bigger brains allow better plans, but for these you pay in the currency of anxiety, a topic that we'll return to in Chapter 8.

The appraisal theorist Lazarus has talked about emotional coping.[73] In the scheme presented here, emotional coping represents the cognitive planning of voluntary *actions* once we find ourselves in the midst of an involuntarily elicited emotional *reaction*. Evolutionary programming sets the emotional ball rolling, but from then on we are very much in the driver's seat. How effectively we deal with this responsibility is a matter of our genetic constitution, past experience, and cognitive creativity, to name but a few of the many factors that are important. And while we will need to understand all of these before we understand "emotion," it seems to me that the way to start understanding emotion is by elucidating the first step in the sequence—the elicitation of prepackaged emotional *reactions* by innate or learned trigger stimuli. We clearly need to go beyond evolution in order to understand emotion, but we should get past it by understanding its contribution rather than ignoring it. I think we have now done that, at least for the emotion fear, or at least for those aspects of the emotion fear that are captured by studies of fear conditioning.

7
REMEMBRANCE OF EMOTIONS PAST

"Every man has reminiscences which he would not tell to everyone but only to his friends. He has other matters in his mind which he would not reveal even to his friends, but only to himself, and that in secret. But there are other things which a man is afraid to tell even to himself, and every decent man has a number of such things stored away in his mind."

Fyodor Dostoevsky, *Notes from the Underground*[1]

BICYCLING. SPEAKING ENGLISH. The Pledge of Allegiance. Multiplication by 7s. The rules of dominoes. Bowel control. A taste for spinach. Immense fear of snakes. Balancing when standing. The meaning of "halcyon days." The words to "Subterranean Homesick Blues." Anxiety associated with the sound of a dentist drill. The smell of banana pudding.

What do all of these have in common? They are each things I've learned and stored in my brain. Some I've learned to do, or learned to expect; others are remembered personal experiences; and still others are just rote facts.

For a long time, it was thought that there was one kind of learning system that would take care of all the learning the brain does. During the behaviorist reign, for example, it was assumed that psychologists could study any kind of learning in any kind of animal and find out how humans learn the things we learn. This logic was not only applied to those things that humans and animals both do, like finding food and avoiding danger, but also to things that humans do easily and animals do poorly if at all, like speaking.

It is now known that there are multiple memory systems in the brain, each devoted to different memory functions. The brain system that allowed me to learn to hit a baseball is different from the one that allows me to remember trying to hit the ball and failing, and this is different still from the system that made me tense and anxious when I stepped up to the plate after having been beaned the last time up. Though these are each forms of long-term memory (memory that lasts more than a few seconds), they are mediated by different neural networks. Different kinds of memory, like different kinds of emotions and different kinds of sensations, come out of different brain systems.

In this chapter we are going to be concerned with two learning systems that the brain uses to form memories about emotional experiences. The separate existence of these two kinds of memories in the brain is nicely illustrated by considering a famous case study in which one of these systems was damaged, but the other continued to function normally.

Is That a Pin in Your Hand or Are You Just Glad to See Me?

In the early part of this century, a French physician named Edouard Claparede examined a female patient who, as a result of brain damage, had seemingly lost all ability to create new memories.[2] Each time Claparede walked into the room he had to reintroduce himself to her, as she had no recollection of having seen him before. The memory problem was so severe that if Claparede left the room and returned a few minutes later, she wouldn't remember having seen him.

One day, he tried something new. He entered the room, and, as on every other day, he held out his hand to greet her. In typical fashion she shook his hand. But when their hands met, she quickly pulled hers back, for Claparede had concealed a tack in his palm and had pricked her with it. The next time he returned to the room to greet her, she still had no recognition of him, but she refused to shake his hand. She could not tell him why she would not shake hands with him, but she wouldn't do it.

Claparede had come to signify danger. He was no longer just a man, no longer just a doctor, but had become a stimulus with a specific emotional meaning. Although the patient did not have a conscious memory of the situation, subconsciously she learned that shaking Claparede's hand could cause her harm, and her brain used this stored information, this memory, to prevent the unpleasantness from occurring again.

These instances of memory sparing and loss were not easily interpreted in Claparede's time and until recently were thought of as reflecting the survival and breakdown of different aspects of one learning and memory system. But modern studies of the brain mechanisms of memory have given us a different view. It now seems that Claparede was seeing the operation of two different memory systems in his patient—one involved in forming memories of experiences and making those memories available for conscious recollection at some later time, and another operating outside of consciousness and controlling behavior without explicit awareness of the past learning.

Conscious recollection is the kind of memory that we have in mind when we use the term "memory" in everyday conversation: to remember is to be conscious of some past experience, and to have a memory problem (again, in everyday parlance) is to have difficulty with this ability. Scientists refer to conscious recollections as declarative or explicit memories.[3] Memories created this way can be brought to mind and described verbally. Sometimes we may have trouble dredging up the memory, but it is potentially available as a conscious memory. As a result of brain damage, Claparede's patient had a problem with this type of memory.

But the patient's ability to protect herself from a situation of potential danger by refusing to shake hands reflects a different kind of memory system. This system forms implicit or nondeclarative memories about dangerous or otherwise threatening situations. Memories of this type, as we saw in the last chapter, are created through the mechanisms of fear conditioning—because of its association with the painful pinprick, the sight of Claparede became a *learned trigger* of defensive behavior (a conditioned fear stimulus). We also saw that conditioned fear responses involve implicit or unconscious processes in two important senses: the learning that occurs does not depend on conscious awareness and, once the learning has taken place, the stimulus does not have to be consciously perceived in order to elicit the conditioned emotional responses. We may become aware that fear conditioning has taken place, but we do not have control over its occurrence or conscious access to its workings. Claparede's patient shows us something similar: as a result of brain damage, she had no conscious memory of the learning experience through which the conditioned fear stimulus implicitly acquired the capacity to protect her from being pricked again.

Through brain damage we can thus see the operation of an implicit emotional memory system in the absence of explicit conscious memory of the emotional learning experience. Normally, though, in the undamaged brain, explicit memory and implicit emotional memory systems are working at the same time, each forming their own special brand of memories. So if you met Claparede today and he was, after all these years, still up to his old tricks, you would form an explicit conscious memory of being pricked by the old codger, as well as an implicit or unconscious memory. We are going to call the implicit, fear-conditioned memory an "emotional memory" and the explicit declarative memory a "memory of an emotion." Having already explored how fear conditioning works, we will now examine the neural organization of the explicit or declarative memory system, and also take a look at interactions between this conscious memory network and the unconsciously functioning fear conditioning system.

Henry Mnemonic: The Life and Times of Case H.M.

Karl Lashley, the father of modern physiological psychology and one of the most influential brain researchers in the first half of the twentieth century, conducted an extensive series of investigations attempting to find the locus of memory in the rat brain.[4] His conclusion, that memory is not mediated by any particular neural system but is instead diffusely distributed in the brain, was widely accepted. By mid-century researchers had quit looking for the location of memory in the brain—it seemed that this was a fruitless and even a misguided quest. However, the tides began to shift when a young man suffering from an extreme case of epilepsy was operated on in Hartford, Connecticut, in 1953.[5]

Known to legions of brain scientists and psychologists as H.M.,[6] this patient has single-handedly, though unwittingly, shaped the course of research on the brain mechanisms of explicit (conscious) memory over the past forty years. At the time of the operation he was twenty-seven and had been experiencing convulsive epileptic attacks since sixteen. All attempts to control the seizures with medications available at the time had failed. Because of the severity and intractability of his epilepsy, H.M. was deemed an appropriate candidate for a radical, last-resort, experimental procedure in which the brain tissue containing the major sites or "foci" of the disease are removed. In his case, it was necessary to remove large regions of the temporal lobes on both sides of his brain.

Measured by the extent to which its medical goal was achieved, the surgery was a great success—the epileptic seizures came to be controllable by anticonvulsant medications. On the other hand, there was one unfortunate and unanticipated consequence. H.M. lost his memory. More specifically, he lost his capacity to form explicit, declarative, or conscious long-term memories. However, the distinction between explicit and implicit memories did not arise until much later, and in fact was based in part on the studies of H.M. So we will forsake the distinction for a while until we've considered H.M. and his problems in more detail.

H.M.'s memory disorder, his amnesia, has been studied and written about extensively over the years. Neal Cohen and Howard Eichenbaum, two leading memory researchers, recently summarized H.M.'s condition: "Now, nearly 40 years after his surgery, H.M. does not know his age or the current date; does not know where he is living; does not know the current status of his parents (who are long deceased); and does not know his own history. . . ."[7] And Larry Squire, another leader in the field, describes it this way: "Although his epileptic condition was markedly improved, he could accomplish little, if any, new learning. . . . His impairment in new learning is so pervasive and severe that he requires constant supervisory care. He does not learn the names or faces of those who see him regularly. Having aged since his surgery, he does not now recognize a photograph of himself."[8] But probably the most straightforward and telling characterization of H.M.'s condition appeared in the first publication that described his unfortunate state. William Scoville, the surgeon, and Brenda Milner, the psychologist who studied H.M. initially, noted that H.M. forgot the events of daily life as quickly as they occurred.[9]

One of the facts that was clear from Milner's studies was that H.M.'s memory problem had nothing to do with a loss of intellectual ability. H.M.'s IQ after the surgery was in the normal range, in fact on the high side of normal, and remained there over the years. The black holes of knowledge in his mind did not reflect some general breakdown in his ability to think and reason. He was not stupid. He simply couldn't remember.

In many ways, H.M.'s memory deficit was quite similar to the disorder in Claparede's patient. However, for two reasons, H.M. is a more important case for understanding memory. The first is that H.M. was extensively examined from the mid-1950s until only a few years ago. Probably no patient in the history of neurology has been studied in such detail and over such an extended period. Throughout, he was a willing and able subject, but in recent years, as age took its toll, he became less capable of participating in these studies. The result of all this work is that we know exactly which aspects of his memory were compromised. The second reason that H.M. has been so important for understanding memory is that we know the location of the damage in his brain. His lesion was the result of a precise surgical removal (rather than an accident of nature). The surgical record thus indicates where the damage is. It has also been possible to look inside his skull with modern brain imaging techniques and confirm the location of the damage. By combining this exacting neurological information about the locus of brain damage with the detailed information about which aspects of memory are disturbed and intact, researchers studying H.M. have obtained important insights into the way memory is organized in the brain.

The Long and the Short of It[10]

Today, it is widely accepted that memory can be divided into a short-term store, which lasts seconds, and a long-term one lasting from minutes to a lifetime.[11] What you are conscious of now is what is mometarily in your short-term memory (especially what is called working memory, a special kind of short-term memory that will be discussed in Chapter 9), and what goes into your short-term memory is what can go into your long-term memory.[12] This distinction had been around since the late nineteenth century, having been proposed (with different terms) by William James (who else?),[13] but the most conclusive evidence that short- and long-term memory are really different processes mediated by distinct brain systems probably came from Milner's early studies of H.M.

Although H.M. seemed to forget almost everything that happened to him (he was unable to form long-term memories), he could nevertheless hold on to information for a few seconds (he had short-term memory). For example, if he was shown a card with a picture on it and the card was put away, he could say what was on the card if he was asked immediately, but if a minute or so passed he was completely unable to say what he had seen, or even whether he had seen anything. From the results of many kinds of tests, it became clear that removal of regions of the temporal lobe in H.M. interfered with long- but not short-term memory, suggesting that the formation of long-term memories is mediated by the temporal lobe, but that short-term memory involves some other brain system.[14]

H.M. has also taught us that the brain system involved in forming new long-term memories is different from the one that stores old long-term memories. H.M. could remember events from his childhood and early adult life quite well. In fact, his memory of things before the operation was good, up to a couple of years prior to the surgery. Consequently, Milner pointed out that H.M. had a very severe anterograde amnesia (an inability to put new information into long-term memory) but only a mild retrograde amnesia (an inability to remember things that happened before the surgery). H.M.'s major deficit was thus one of depositing new learning into the long-term memory bank, rather than withdrawing information placed there earlier in life.

The findings from H.M. thus clearly distinguished short-term and long-term memory, and also suggested that long-term memory involves at least two stages, an initial one requiring the temporal lobe regions that were removed, and a later stage involving some other brain regions, most likely areas of the neocortex.[15] The temporal lobe is needed for forming long-term memories, but gradually, over years, memories become independent of this brain system. These are powerful concepts that remain central to our understanding of the brain mechanisms of memory.

In Search of a Model

The areas of the temporal lobe that were damaged in H.M. included major portions of the hippocampus and amygdala, and surrounding transitional areas. Some of these were areas that MacLean had identified as components of the limbic system, which, as we saw earlier, was supposed to constitute the emotional system of the brain. H.M. provided some of the first difficulties for the limbic system theory of emotion, suggesting that some regions of the limbic system are involved at least as much in cognitive functions (like memory) as in emotion.

Although several temporal lobe regions were damaged in H.M., the view emerged that damage to the hippocampus was primarily responsible for the memory disorder. Other patients were operated on, in addition to H.M., and when these were all considered together it seemed that the extent of the memory disorder was directly related to the amount of the hippocampus that had been removed. On the basis of these observations, the hippocampus emerged as the leading candidate brain region for the laying down of new memories. Surgeons now make every effort to leave the hippocampus and related brain regions intact, at least on one side of the brain, when operating on the temporal lobes so that devastating effects on memory can be prevented.

By the late 1950s, the task at hand for memory researchers seemed clear and straightforward: turn to studies of experimental animals to figure out how the hippocampus accomplishes its mnemonic job. In animal studies, memory is tested not by asking the subject whether he or she remembers, but by determining whether behavioral performance is affected by prior learning experiences. Countless studies of the effects of hippocampectomy (hippocampal removal) on memory were performed in a variety of animals. The results were inconsistent and disappointing. Lesions sometimes interfered with the ability of animals to remember what they learned, and sometimes did not. It seemed that either the human and nonhuman brain have different mechanisms of memory, or that the researchers had just not found the right way to test memory in animals.

But in the early 1970s David Gaffan, an Oxford psychologist, came up with a way of testing memory in monkeys that proved to be a reliable measure of hippocampal-dependent functions.[16] It was a task called delayed nonmatching to sample. The monkey was shown a stimulus, say a toy soldier. The stimulus was then removed. After a delay, two stimuli appeared, the toy soldier and a toy car of about the same size. The monkey could get a treat (like a raisin or Froot Loop) by picking the stimulus that had not appeared before (the stimulus that did not match the sample), which in this case was the car. If the sample (the soldier) was picked, no treat followed.

Having a sweet tooth, monkeys are very willing to play these kinds of games. Normal monkeys do fine, even at relatively long delays between the sample and the two test stimuli. Monkeys with hippocampal damage also perform reasonably well at short delays. But as the delay increases, they perform miserably—they respond randomly to the two stimuli, choosing the stimulus that matches the sample as often as the nonmatch.[17] This breakdown at the long delays cannot be due to a simple failure to learn the rule (pick the stimulus that does not match the sample). They learn the rule before the hippocampus is removed, so they already know it and just have to apply it to the stimuli that appear in a particular test. Most important, they use the rule well at short delays. The problem is really one of holding on to the memory of the sample long enough to choose the nonmatching item.

Delayed nonmatching to sample does not exactly resemble the kinds of tasks used to test memory in amnesic or normal humans.[18] Humans are given verbal instructions about how to perform the task, whereas animals are given weeks or months of behavioral training so that they can learn the rule. Humans are typically tested on verbal material or are required to give verbal responses even with nonverbal test stimuli. Animals always express their memory through behavioral performance. Humans are not given sweets every time they get an answer right. The important thing about delayed nonmatching to sam-

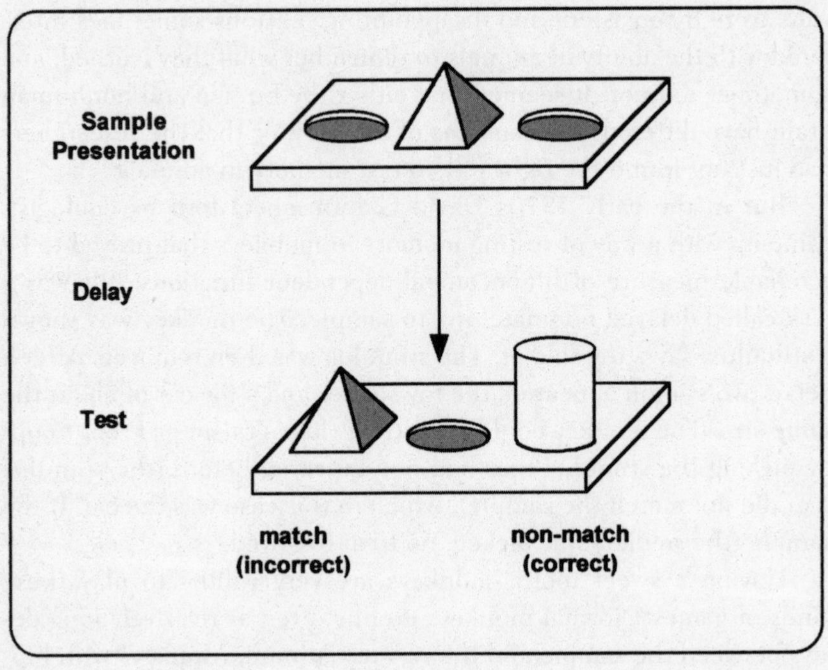

FIGURE 7-1
Delayed Nonmatching to Sample Procedure.

In this procedure, a monkey is shown a sample stimulus (in the center of the tray). After a delay, the sample is presented along with a novel stimulus. If the monkey picks the novel stimulus (the nonmatch) he will find a reward under it (peanut, Fruit Loop, or raisin). This procedure has been used extensively to study the role of the hippocampus and related cortical areas in memory processes in animals. (Based on an illustration provided by E. A. Murray of the National Institute of Mental Health.)

ple was thus not that it perfectly corresponded to the kinds of tests used to reveal memory problems in H.M. but that it proved to be a reliable means of revealing a hippocampal-dependent memory in animals. For this reason, delayed nonmatching to sample became the gold standard for modeling human temporal lobe amnesia in monkeys.

Delayed nonmatching to sample was also tried in studies of other species, particularly rats, and was found to be a good way of testing hippocampal-dependent memory in these animals as well.[19] But through studies of rats, other kinds of tasks were also discovered that reliably implicated the hippocampus. These mostly involved various forms of learning and memory that depend on the use of spatial cues. In one task, rats are tested in a maze in which different alleys radiate out from a central platform.[20] The rat is put in the center and has to choose one of the alleys. His job is to remember which alleys he's not gone down before. If he picks one of the previously unvisited alleys, he gets a treat. If he goes down a previously visited one, he gets nothing. The only way to solve this task is to use spatial cues, such as the location of an alley with respect to the location of other items in the room in which the maze is. In another task, rats are put into a tank containing milky water.[21] They are decent swimmers, but don't really care for it, and will swim to safety as soon as possible. Initially, there is a platform above the water. Once they've learned where it is, the platform is submerged just below the surface. The rats have to remember where the platform was and use spatial cues around the room to guide their swim to safety. Lesions of the hippocampus interfere with spatial memory in both the radial maze and the water maze.

By the late 1970s, the ducks seemed to be lining up. Studies of animals and humans were finally both pointing to the hippocampus as a key player in the game of memory. But then Mortimer Mishkin of the National Institute of Mental Health noted a problem with this neat and tidy story about the role of the hippocampus in memory and amnesia.[22] He pointed out that all of the patients, including H.M., who became amnesic as a result of temporal lobe lesions had damage to the amygdala as well as the hippocampus. Might the amygdala also be important? Mishkin tested this idea by examining the effects of combined lesions of the hippocampus and amygdala versus separate lesions of these two areas in monkeys. The findings seemed crystal clear. Damage to the amygdala and hippocampus together was more detrimental than damage to either alone on delayed nonmatching to sample. The idea that limbic areas, like the amygdala and hippocampus, are more involved in emotion than cognition was already challenged by the discovery that the hippocampus contributes to cognition (memory). The possibility that the amygdala was part of the memory system blurred this distinction between cognitive and emotional functions of limbic areas even more.

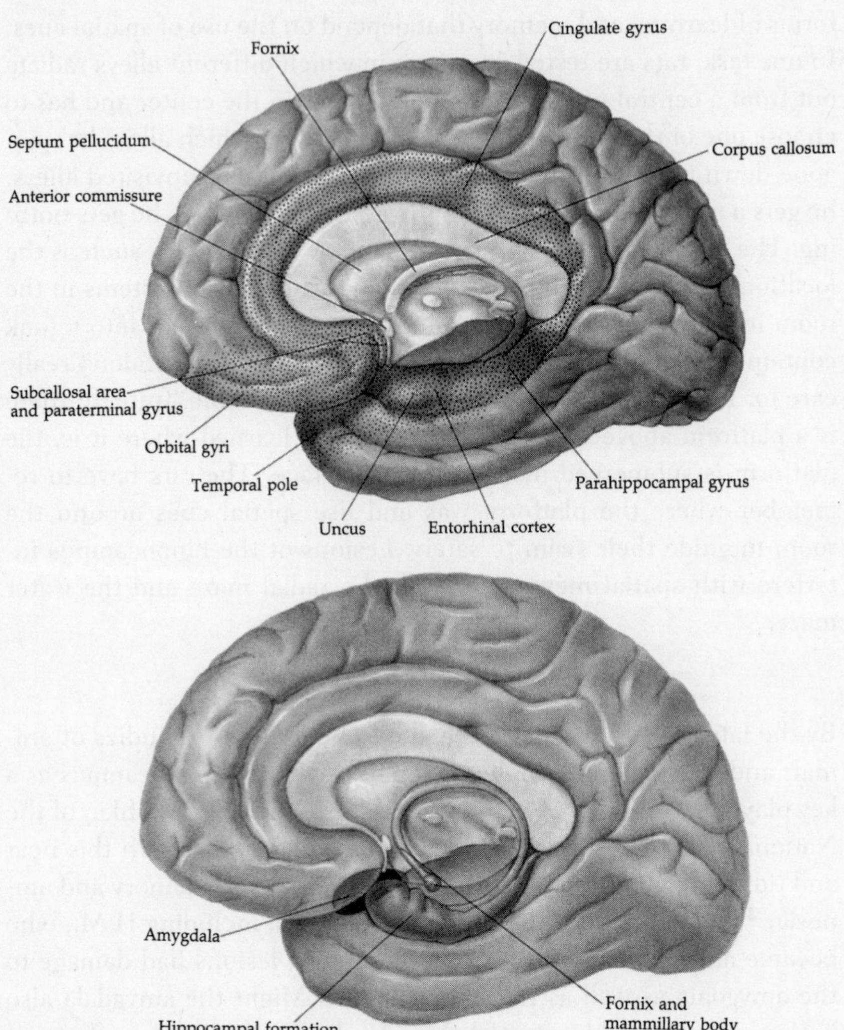

FIGURE 7-2

Location of the Hippocampus and Amygdala and Surrounding Cortical Areas.

The two diagrams show the medial or inside wall of the human cerebrum. The stippled area is the classic limbic lobe (see Chapter 4). The amygdala and hippocampus are found deep inside the medial part of the temporal lobe, underneath the uncus, entorhinal cortex, and parahippocampal gyrus (top). These cortical areas have been stripped away to show the location of the hippocampus and amygdala in the bottom illustration. (Reprinted from figures 15-1 and 15-2 in J. H. Martin [1989], *Neuroanatomy: Text and Atlas*. New York: Elsevier. Copyright © [1989] by Appleton and Lange.)

Other researchers, however, did not fully accept the view that the amygdala was part of the memory system and by the late 1980s the tide was shifting back toward the hippocampus as the core of the long-term memory system. Larry Squire, Stuart Zola-Morgan, and David Amaral of San Diego examined a patient with a severe memory disorder, not unlike H.M.'s.[23] Not too long afterward, the patient died, and the brain was made available for analysis. This patient turned out to have a pure hippocampal lesion. There was no detectable damage anywhere else. This selective lesion resulted from anoxia, a reduction in oxygen supply to the brain, which especially affects cells in the hippocampus. Amnesia, it seemed, could result from damage to only the hippocampus.

Why, then, did the combined hippocampal and amygdala lesion produce more of a deficit on delayed nonmatching to sample than the hippocampal lesions alone in monkeys? The San Diego team next took on this issue. They noted that in the process of removing the amygdala, surgeons often damage cortical areas that provide an important linkage between the neocortex and the hippocampus. Perhaps, Mishkin's effect was not due to amygdala damage but to an interruption of the flow of information back and forth between the neocortex and the hippocampus. The San Diego researchers figured out how to remove the amygdala without disturbing the cortical areas related to the hippocampus. This pure amygdala lesion had no effect on delayed nonmatching to sample.[24] Importantly, though, the pure amygdala lesion did produce the emotional concomitants of the Klüver-Bucy syndrome, especially reduced fear.[25] The role of the hippocampus in memory seemed rescued, and the burden of cognition was again lifted from the amygdala.

What, then, is the relative contribution of the hippocampus as opposed to those pesky cortical areas surrounding the amygdala and hippocampus? Mishkin and Betsy Murray showed that damage to the surrounding cortex also produces deficits in delayed nonmatching to sample; in fact, these lesions produced more of a deficit than the hippocampal damage.[26] On the basis of this finding, Murray and Mishkin questioned the premier role of the hippocampus in memory and argued instead that the surrounding cortex is particularly crucial. Other researchers, however, point out that this is too strong a conclusion to base solely on delayed nonmatching to sample, which may

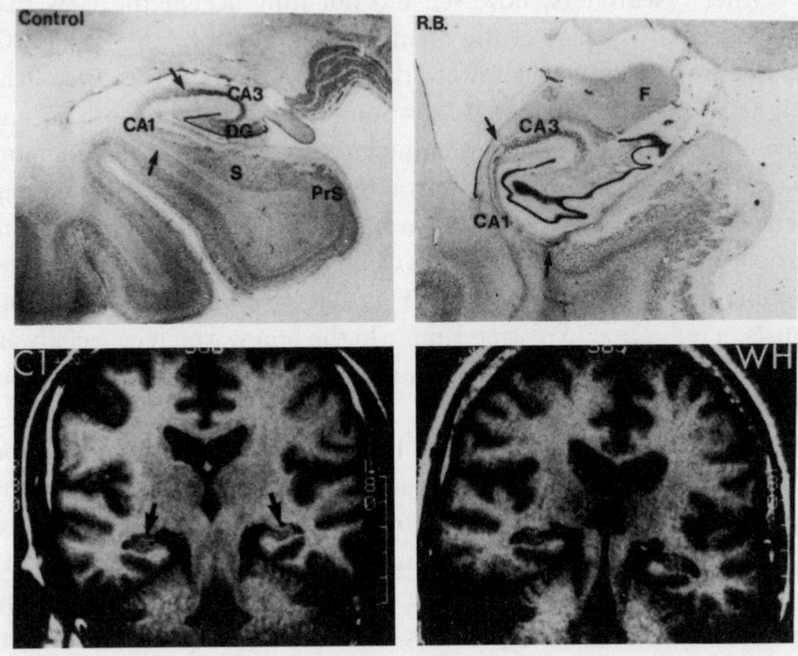

FIGURE 7-3
Magnetic Resonance Imaging Scan of the Normal and Damaged Hippocampus.

Section through the hippocampus of a normal human (top left) and an amnesic patient (top right). The CA1 region of the hippocampus is indicated by the arrow in the normal brain. In the amnesic patient the CA1 region is damaged. Magnetic resonance scans of a normal human (bottom left) and an amnesic patient (bottom right). The hippocampal formation is greatly reduced in the amnesic patient. (Top panels reprinted with permission from L. R. Squire [1986], Mechanisms of Memory, *Science* 232, 1612–1619, © 1986 American Association for the Advancement of Science. Bottom panels reprinted with permission from G. Press, D. G. Amaral, and L. R. Squire [1989], Hippocampal abnormalities in amnesic patients revealed by high-resolution magnetic resonance imaging. *Nature* 341, p. 54, © Macmillan Magazines Ltd.)

not be the magic bullet that it has been made out to be.[27] After all, there is pretty solid evidence that pure hippocampal damage in humans can lead to amnesia (recall the anoxia case above). Delayed nonmatching to sample may be a better test of the function of the surrounding cortex than of the hippocampus, which would suggest that these two areas each contribute uniquely to memory.[28]

This debate over the details will no doubt continue. However, most researchers in the field are in agreement about the broad outline of how the temporal lobe memory system works.[29] Sensory processing areas of the cortex receive inputs about external events and create perceptual representations of the stimuli. These representations are then shuttled to the surrounding cortical regions, which, in turn, send further processed representations to the hippocampus. The hippocampus then communicates back with the surrounding regions, which communicate with the neocortex. The maintenance of the memory over the short run (a few years) requires that the temporal lobe memory system be intact, either because components of this system store the memory trace or because the trace is maintained by interactions between the temporal lobe system and the neocortex. Gradually, over years, the hippocampus relinquishes its control over the memory to the neocortex, where the memory appears to remain as long as it is a memory, which may be a lifetime.

This model of memory, which has emerged from studies of amnesic animals and humans, gives us a way of understanding the mental changes that take place over time in Alzheimer's disease.[30] The disease begins its attack on the brain in the temporal lobe, particularly in the hippocampus, thus explaining why forgetfulness is the first warning sign. But the disease eventually creeps into the neocortex, suggesting why, as the disease progresses, all aspects of memory (old and new) are compromised, along with a variety of other cortically dependent cognitive functions. Without years of research on amnesia in both humans and animals the cognitive dissolution that occurs as Alzheimer's disease spreads through the forebrain would not be so readily interpreted. And having these insights about how the disease is compromising the mind and brain in tandem may be one of the best aids in figuring out approaches to preventing, arresting, or reversing the cognitive meltdown that occurs.

Pockets of Memory

In the early days, H.M. failed to form new long-term memories in the vast majority of memory tasks that he was given.[31] It did not matter so much whether he was tested with words, pictures, or sounds, he

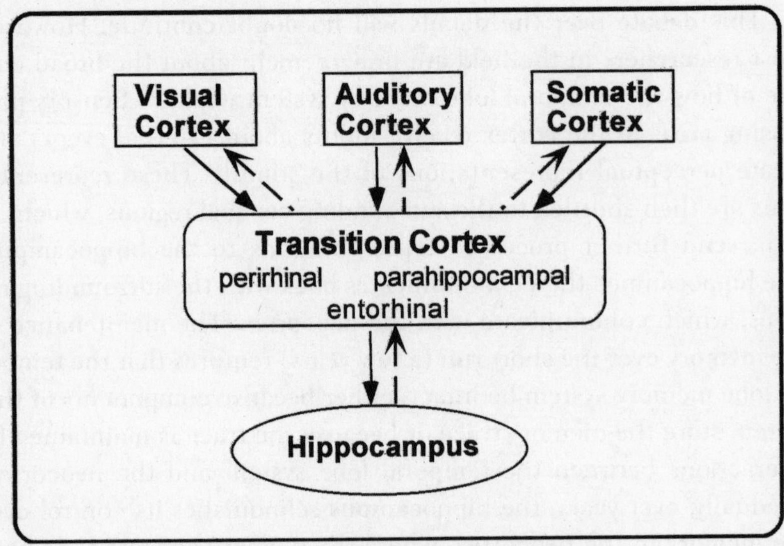

FIGURE 7-4
Cortical Inputs to the Hippocampus.

Each of the major sensory processing systems of the neocortex gives rise to projections to areas that are transitional between the neocortex and hippocampus (that is, the perirhinal and parahippocampal areas). These send their outputs to the entorhinal cortex. The entorhinal cortex then provides the main source of inputs to the hippocampal formation. The hippocampus projects back to the neocortex by way of the same pathways through the transitional cortex.

could not remember. His memory problem was, accordingly, described as a "global amnesia." The few odd things that he could retain seemed, at first, like isolated, unrelated snippets of memory. However, as more and more findings came in from different kinds of memory tests, it became clear that there were well-defined pockets of memory sparing in H.M. As a result of these discoveries, temporal lobe amnesia is no longer considered to be a global memory disorder affecting all forms of new learning. By figuring out what kinds of learning functions are spared and disrupted in amnesia, researchers have been able to characterize the contribution of the temporal lobe system to memory.

One of the first observations of spared learning resulted when Milner asked H.M. to try to copy a picture of a star while only watching a mirror reflection of his hand.[32] This required that he learn to control his hand on the basis of the abnormal (visually reversed) feedback coming to his brain about where his hand was in space. He did poorly the first time, but with practice he improved, and he retained the ability to express the improved performance over time. Suzanne Corkin of MIT then found that H.M. also improved with practice in another manual skill learning task—one in which he was required to keep a handheld stylus on a small dot that was spinning around on a turntable.[33] As with the mirror drawing task, the more times he did it the better he got. Interestingly, and importantly, on both manual skill learning tasks, he improved in spite of the fact that he had no conscious memory of the earlier experiences that led to the improved performance. These findings suggested that the learning and remembrance of manual skills might be mediated by some system other than the temporal lobe system.

Neal Cohen later examined whether spared skill learning ability in amnesia might extend to what is called cognitive skills, the ability to get better at doing mental tasks.[34] He showed that the ability of amnesics to read mirror images of words improved with practice (e.g., "egral" is the mirror image of "large"). He also showed that the patients could learn some complicated rule-based strategies required to solve certain mathematical problems or puzzles. One much discussed instance is a puzzle game called the Tower of Hanoi. To solve this problem, discs of different sizes have to be moved in a certain way between three pegs without ever letting a smaller peg be under a larger one. Even normal subjects have difficulty finding the "optimal" solution to the puzzle. However, with lots of practice, the amnesic patients, including H.M., were able to achieve this solution. As with the other learning tasks, they had no recollection of playing the game.

Research by Elizabeth Warrington and Larry Weiskrantz in England showed that "priming" is preserved in amnesic patients.[35] Like the other instances of spared learning and memory in amnesia, priming involves the demonstration of learning by the effects of prior experience on later behavioral performance rather than by the subject's knowledge of the prior learning. For example, in one version of priming, the subject is given a list of words. Later, if asked to recall the items that were on the list, amnesic patients perform miserably. However, if instead of being asked to recall the items they are given fragments of words and asked to complete them, like normal subjects they do better on the fragments that go with words in the study list than they do for fragments that have no match in the study list.

Weiskrantz and Warrington also showed that the classical conditioning of eyeblink responses is preserved in amnesia.[36] In this task, a tone is paired with an aversive stimulus (usually an air puff to the eye). After hundreds of trials, the tone elicits eyelid closure immediately before the onset of the air puff. This precisely timed response protects the delicate tissues of the eye from the air puff. Amnesic patients show normal eyeblink conditioning. This is not surprising given the fact that we now know from animal studies that eyeblink conditioning involves circuits in the brain stem and is unaffected by removal of all brain tissue above the midbrain.[37] The patients nevertheless later have no memory of having seen the conditioning apparatus.

The Multiplicity of Memory

What do all these spared learning and memory functions have in common, and how do they differ from the functions that are disrupted in temporal lobe amnesia? Cohen and Squire put all of the findings together and came up with an answer.[38] They proposed that damage to the temporal lobe memory system interferes with the ability to consciously recollect, but leaves intact the ability to learn certain skills. They called these two processes declarative and procedural memory. A similar dichotomy was offered by Daniel Schacter of Harvard, who distinguished between explicit and implicit memory.[39] Conscious awareness of the basis of performance occurs in explicit memory, but in implicit memory performance is guided by unconscious factors. Skill learning, priming, and classical conditioning are all examples of implicit or procedural learning. These are each intact in temporal lobe amnesia and involve brain areas other than the temporal lobe memory system. Other memory dichotomies have been proposed over the years,[40] but the distinction between conscious, explicit, declarative memory, on the one hand, and unconscious, implicit, procedural memory, on the other, has had the greatest impact on current thinking and will be emphasized here.

The distinction between explicit and implicit memory is dramatically illustrated by a study performed by Squire and his colleagues.[41] They showed that amnesics could be made to either succeed or fail a memory test, simply by changing the instructions—some instructions took the patients down the explicit memory path, which gave rise to failure, whereas other instructions led them on a successful stroll through implicit memory land. In all conditions the stimuli were the same, and only the memory instruction changed. The subjects were first given a list of words to study. Then a few minutes later, they received one of three sets of instructions: recall as many words as you can from the list; use the following cues to help you remember as many words on the list as you can; or, say the first word that comes to mind when you see the following cues. The cues in the latter two conditions were three-letter stems of words that had been on the list: MOT for MOTEL, ABS for ABSENT, INC for INCOME, and so on. Each stem could come from many other words: MOT could be from MOTHER or MOTLEY as well as MOTEL. Not surprisingly, the amnesics did poorly when they had to recall without any cues. They also did poorly when told to use the cues to help them remember the words. But they were as good as normal subjects when the instruction was to say the first word that comes to mind after seeing a cue. In the latter case, the cues were primes rather than recall aids. When performing the priming task, and thus using an implicit memory system, they functioned fine, but damage to the temporal lobe memory system prevented them from consciously recalling the items, even with the aid of cues. Howard Eichenbaum, in studies of rats, found something similar: depending on the instructions given to rats (through training experiences), he showed that it was possible to make a learning situation either dependent upon or independent of the hippocampus.[42]

Cohen and Squire were quick to point out that explicit, declarative memory is mediated by a single memory system, the temporal lobe memory system, but that there are multiple implicit or procedural memory systems. Thus, the brain system that mediates priming is different from the systems involved in skill learning or classical conditioning. Further, different forms of classical conditioning are also mediated by different neural systems—eyeblink conditioning by brain stem circuits and fear conditioning by the amygdala and its connections. The brain clearly has multiple memory systems, each devoted to different kinds of learning and memory functions.

In retrospect, the multiplicity of memory systems should have been apparent from the fact that it was so hard to find memory tasks that depend on the hippocampus in animals. Although a few such tasks were found, the vast majority of the memory tasks used to study animal memory are performed just fine in the absence of the hippocampus. If performance on some tasks depends on the hippocampus and performance on others does not, it must be the case that memory is not a unitary phenomenon and that different memory systems exist in the brain. But in the 1960s and 1970s, there wasn't a clear framework for understanding these variable effects. They led to confusion rather than clarity. The idea of multiple memory systems helped it all make sense.

So What Does the Hippocampus Represent?

We can get a pretty good idea about what makes the hippocampus so important for its brand of memory by examining the kinds of inputs that the hippocampus receives from the neocortex.[43] As we mentioned above, the major link between the hippocampus and the neocortex is the transition cortex (see Figure 7-4). This region receives inputs from the highest stages of neocortical processing in each of the major sensory modalities. So once a cortical sensory system has done all that it can do with a stimulus, say a sight or a sound, it ships the information to the transition region, where the different sensory modalities can be mixed together. This means that in the transition circuits we can begin to form representations of the world that are no longer just visual or auditory or olfactory, but that include all of these at once. We begin to leave the purely perceptual and enter the conceptual domain of the brain. The transition region then sends these conceptual representations to the hippocampus, where even more complex representations are created.

One of the first clues as to the way the hippocampus accomplishes its job came from a study performed in the early 1970s by

John O'Keefe at University College, London.[44] He found that cells in the hippocampus of a rat became very active when the rat moved into a certain part of a test chamber. The cells then became inactive when the rat moved elsewhere. He found lots of these cells, and each one became active in a different place. O'Keefe called these "place cells." The chamber was topless and rats could see out into the room. O'Keefe showed that the firing of the cells was controlled by the rat's sense of where it was in the room, for if the various cues around the room were removed, the firing patterns changed dramatically. Importantly, though, the place cells were not strictly responding to visual stimuli, since they maintained their "place fields" (the location where they became active) in complete darkness. O'Keefe and colleague Lynn Nadel published an influential book in 1978 called *The Hippocampus as a Cognitive Map* in which they proposed that the hippocampus forms sensory-independent spatial representations of the world.[45] One important function of these spatial representations, according to O'Keefe and Nadel, is to create a context in which to place memories. Context makes memories autobiographical, locating them in space and time, and this, they say, accounts for the role of the hippocampus in memory. They proposed an early multiple memory system account that distinguished a locale (spatial) memory system mediated by the hippocampus from several other systems mediated by other brain regions. O'Keefe and Nadel were mainly concerned with the locale system and made no attempt to identify the brain systems underlying the other forms of learning.

O'Keefe's observations, and the book with Nadel, created a whole industry devoted to understanding the role of the hippocampus in processing spatial cues. The demonstration of hippocampal-dependent memory in the radial maze[46] and the water maze[47] were direct outgrowths of the place cell findings and many experiments were conducted to figure out exactly how the hippocampus encodes space. In addition to O'Keefe's continued research,[48] particularly notable has been the work of Bruce McNaughton and Carol Barnes in Tucson,[49] the late David Olton in Baltimore,[50] Richard Morris in Edinburgh,[51] and Jim Ranck, John Kubie, and Bob Muller in Brooklyn.[52]

But not all investigators have accepted the idea that the hippocampus is a spatial machine. Howard Eichenbaum, for example, questions the role of the hippocampus in spatial processing per se, arguing that what the hippocampus is especially good at and important for is creating representations that involve the multiple cues at once, with space being a particular example of this rather than the primary instance of it.[53] Others, like Jerry Rudy and Rob Sutherland, have argued that the hippocampus creates representations that involve configurations (blends) of cues that transcend the individual stimuli making up the configuration.[54] This differs from Eichenbaum's hypothesis, which argues that the hippocampal representation involves the relation between individual cues rather than a representation in which the cues are fused into a newly synthesized configuration.

More work is needed to choose between spatial, configural, and relational hypotheses. The ultimate verdict will rest with the one that eventually explains how the sights, smells, and sounds of an experience, as well as the arrangement of all of the various stimuli and events in space and time, are represented in the hippocampus.

When Paul MacLean was putting forth his limbic system theory, he proposed that the hippocampus was an ideal place for emotion to reside. He suggested that because of its primitive, simple architecture, the hippocampus wouldn't be able to make fine distinctions between stimuli and would easily mix things up.[55] This, MacLean suggested, might account for the irrationality and confusion of our emotional life. But today, the pendulum has swung in the other direction. The hippocampus is thought to have an exquisite design that leads to sophisticated computational power[56] rather than a primitive organization that leads to confusion. The hippocampus has indeed come to be thought of as a key link in one of the most important cognitive systems of the brain, the temporal lobe memory system.

Tweedledee and Tweedledum: Emotional Memories and Memories of Emotions

Let's now explore the implications of the distinction between explicit and implicit memory for our understanding of how memories are formed in an emotional situation. Suppose you are driving down the road and have a terrible accident. The horn gets stuck on. You are in pain and are generally traumatized by the experience. Later, when you hear the sound of a horn, both the implicit and explicit memory systems are activated. The sound of the horn (or a neural representation of it), having become a conditioned fear stimulus, goes straight from the auditory system to the amygdala and implicitly elicits bodily responses that typically occur in situations of danger: muscle tension (a vestige of freezing), changes in blood pressure and heart rate, increased perspiration, and so on. The sound also travels through the cortex to the temporal lobe memory system, where explicit declarative memories are activated. You are reminded of the accident. You consciously remember where you were going and who you were with. You also remember how awful it was. But in the declarative memory system there is nothing different about the fact that you were with Bob and the fact that the accident was awful. Both are just facts, propositions that can be declared, about the experience. The particular fact that the accident was awful is not an emotional memory. It is a declarative memory about an emotional experience. It is mediated by the temporal lobe memory system and it has no emotional consequences itself. In order to have an aversive emotional memory, complete with the bodily experiences that come with an emotion, you have to activate an emotional memory system, for example, the implicit fear memory system involving the amygdala (see Figure 7-5).

There is a place, though, where explicit memories of emotional experiences and implicit emotional memories meet—in working memory and its creation of immediate conscious experience (working memory and consciousness will be discussed in Chapter 9). The sound of the horn, through the implicit emotional memory system, opens the floodgates of emotional arousal, turning on all the bodily responses associated with fear and defense. The fact that you are aroused becomes part of your current experience. This fact comes to rest side by side in consciousness with your explicit memory of the accident. Without the emotional arousal elicited through the implicit system, the conscious memory would be emotionally flat. But the co-representation in awareness of the conscious memory and the current emotional arousal give an emotional flavoring to the conscious memory. Actually, these two events (the past memory and the present arousal) are seamlessly fused as a unified conscious experience of the

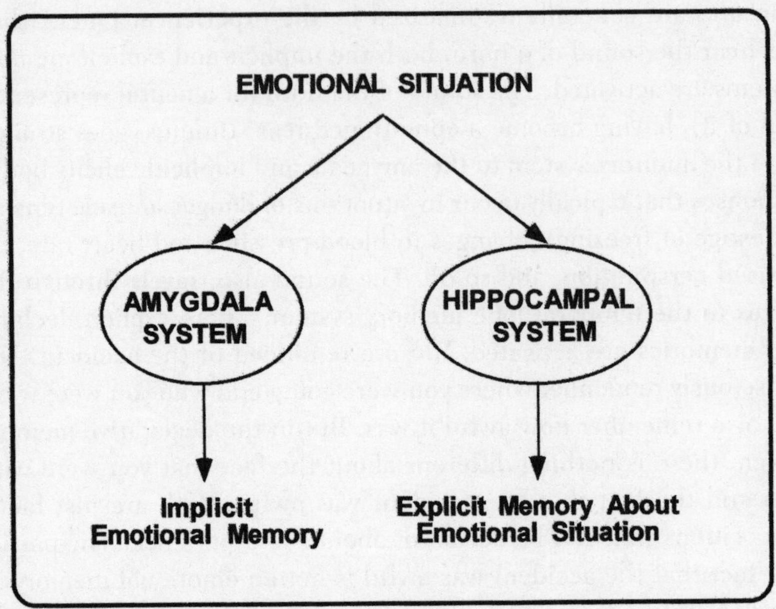

FIGURE 7-5

Brain Systems of Emotional Memory and Memory of Emotion.

It is now common to think of the brain as containing a variety of different memory systems. Conscious, declarative or explicit memory is mediated by the hippocampus and related cortical areas, whereas various unconscious or implicit forms of memory are mediated by different systems. One implicit memory system is an emotional (fear) memory system involving the amygdala and related areas. In traumatic situations, implicit and explicit systems function in parallel. Later, if you are exposed to stimuli that were present during the trauma, both systems will most likely be reactivated. Through the hippocampal system you will remember who you were with and what you were doing during the trauma, and will also remember, as a cold fact, that the situation was awful. Through the amygdala system the stimuli will cause your muscles to tense up, your blood pressure and heart rate to change, and hormones to be released, among other bodily and brain responses. Because these systems are activated by the same stimuli and are functioning at the same time, the two kinds of memories seem to be part of one unified memory function. Only by taking these systems apart, especially through studies of experimental animals but also through important studies of rare human cases, are we able to understand how memory systems are operating in parallel to give rise to independent memory functions.

moment. This unified experience of the past memory and the arousal can then potentially get converted into a new explicit long-term memory, one that will include the fact that you were emotionally aroused last time you remembered the accident. In this case the memory of the accident did not lead to the emotional arousal. The implicit arousal of emotion gave emotional coloration to the explicit memory (see Figure 7-6).

Nevertheless, we know from personal experience that conscious memories can make us tense and anxious, and we need to account for this as well. All that is needed for this to occur is a set of connections from the explicit memory system to the amygdala. There are in fact abundant connections from the hippocampus and the transition regions, as well as many other areas of the cortex, to the amygdala.

It is also possible that implicitly processed stimuli activate the amygdala without activating explicit memories or otherwise being represented in consciousness. As we saw in Chapters 2 and 3, unconscious processing of stimuli can occur either because the stimulus itself is unnoticed or because its implications are unnoticed. For example, suppose the accident described above happened long ago and your explicit memory system has since forgotten about many of the details, such as the fact that the horn had been stuck on. The sound of a horn now, many years later, is ignored by the explicit memory system. But if the emotional memory system has not forgotten, the sound of the horn, when it hits the amygdala, will trigger an emotional reaction. In a situation like this, you may find yourself in the throes of an emotional state that exists for reasons you do not quite understand. This condition of being emotionally aroused and not knowing why is all too common for most of us, and was, in fact, the key condition for which the Schachter-Singer theory of emotion tried to account. But in order for emotion to be aroused in this way, the implicit emotional memory system would have to be less forgetful than the explicit memory system. Two facts suggest that this may be the case. One is that the explicit memory system is notoriously forgetful and inaccurate (as we'll see below). The other is that conditioned fear responses exhibit little diminution with the passage of time. In fact, they often increase in their potency as time wears on, a phenomenon called "the incubation of fear."[57] It is possible to decrease the potency of a conditioned response by presenting the *learned trigger*, the CS,

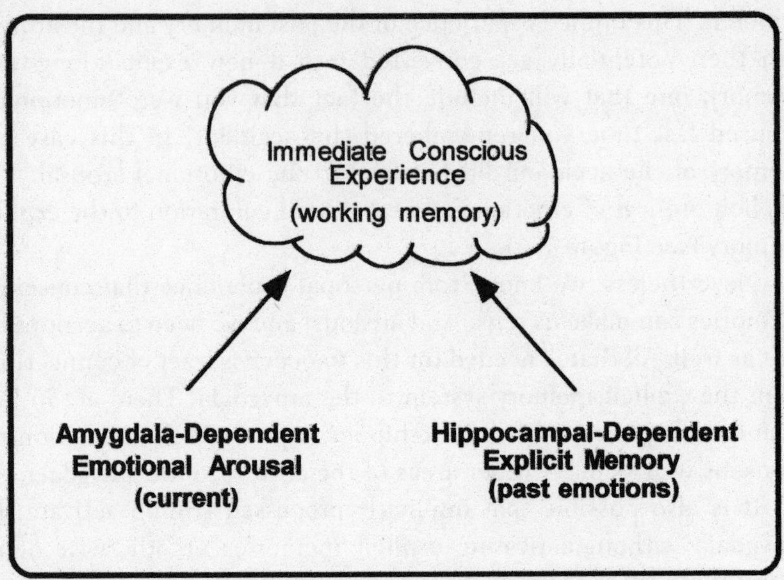

FIGURE 7-6
Intersection of Explicit Memory and Emotional Arousal in Immediate Conscious Experience.

The outcome of activity in the explicit memory system of the hippocampus is a conscious awareness of stored knowledge or personal experiences. The outcome of activity in the amygdala is the expression of emotional (defensive) responses. But we also become conscious of the fact that we are emotionally aroused, allowing us to fuse, in consciousness, explicit memories of past situations with immediate emotional arousal. In this way, new explicit memories that are formed about past memories can be given emotional coloration as well.

over and over again without the US. However, so-called extinguished responses often recur on their own and even when they don't they can be brought back to life by stressful events.[58] Observations like these have led us to conclude that conditioned fear learning is particularly resilient, and in fact may represent an indelible form of learning. This conclusion has extremely important implications for understanding certain psychiatric conditions, as we'll see in the next chapter.

Infantile Amnesia

The idea of separate systems devoted to forming implicit emotional memories and explicit memories of emotions is relevant for understanding infantile amnesia, our inability to remember experiences from early childhood, roughly before age three. Infantile amnesia was first discussed by Freud, who noted that there had not been enough astonishment of the fact that by the time a child is two he can speak well and is at home with complicated mental situations, but if he is later told of some remark made during this time, he will have no memory of it.[59]

Lynn Nadel, together with Jake Jacobs, proposed that the key to infantile amnesia was the relatively prolonged period of maturation that the hippocampus goes through.[60] In order to be fully functional, a brain region has to grow its cells and get them connected with other cells in the various regions with which it communicates. It seems to take the hippocampus a bit longer than most other brain regions to get its act together. So Jacobs and Nadel proposed that we don't have explicit memories of early childhood because the system that forms them is not ready to do its job. Other brain systems, though, must be ready to do their learning and remembering sooner, since children learn lots of things during this amnesic time, even if they don't have conscious memories of the learning.

Jacobs and Nadel were particularly interested in the way that early trauma, though not remembered, might have lasting, detrimental influences on mental life. They proposed that the system that forms unconscious memories of traumatic events might mature before the hippocampus. They did not identify what this unconscious system for traumatic learning and memory was, but we now know, of course, that this system crucially involves the amygdala and its connections.

Although there is a dearth of biological research on the developmental maturation of the amygdala, behavioral studies suggest that the amygdala does indeed mature before the hippocampus. Jerry Rudy and his colleagues at the University of Colorado examined the age at which rats could learn hippocampal-dependent versus amygdala-dependent tasks.[61] They found that the amygdala task was acquired earlier in life than the hippocampal task. The amygdala appears to be functionally mature before the hippocampus.

The separate function and differential maturation of the amygdala has important implications for understanding psychopathological conditions. We'll explore these further in the next chapter.

Flashbulb Memories

Gary Larson, in *The Far Side,* has an illustration with a bunch of animals sitting around in a forest setting.[62] The caption reads something like this: all animals in the forest know where they were and who they were with when they heard the news that Bambi's mother had been shot. You probably recognize this as a takeoff on the phenomenon, characteristic of American baby-boomers and their parents, of being able to remember exactly what they were doing when they heard that President Kennedy had been shot. Psychologists describe this phenomenon as a "flashbulb memory," a memory that is made especially crisp and clear because of its emotional implications.[63] Recent findings, described below, by Jim McGaugh and his colleagues at the University of California at Irvine, together with the idea of separate systems for detecting the emotional implications of a situation and for representing emotional situations in explicit memory, help us understand the biological basis of flashbulb memories.

McGaugh's laboratory has long been concerned with the role of peripheral hormones, like adrenaline, in the solidification of memory processes.[64] His studies show that if rats are given a shot of adrenaline right after learning something, they show an enhanced memory of the learning situation. This suggests that if adrenaline is released naturally (from the adrenal gland) in some situation, that experience will be remembered especially well. Since emotional arousal usually results in the release of adrenaline, it might be expected (as suggested by the flashbulb idea) that the explicit conscious memory of emotional situations would be stronger than the explicit memory of nonemotional situations. It would also be expected that blockade of the effects of adrenaline would neutralize the memory-enhancing effects of emotional arousal.

McGaugh and Larry Cahill tested these hypotheses. They asked human subjects to read a story about a boy riding a bike. For some of the subjects, the boy takes a ride on his bike, goes home, and he and his mom drive to the hospital to pick up his dad, a doctor. For other subjects, the boy takes a ride on his bike, is hit by a car, and rushed to the hospital where his dad, a doctor, works. The words in the two stories are matched as closely as possible, with only the emotional implications manipulated. After reading the stories, and before being tested for recall, half the subjects in each group were given either a shot of placebo or a drug that blocks the effects of adrenaline. For the placebo-treated subjects, those that read the emotional story remembered many more details than those that read the mundane story. However, for the subjects receiving adrenaline blockade, there was no difference in the memory of the emotional and the nonemotional stories—both of these groups performed like the placebo group that read the nonemotional story. Adrenaline blockade, indeed, prevented the memory-enhancing effects of emotional arousal.

McGaugh has suggested some practical applications of this fascinating result. Rescue workers and soldiers in battle are often traumatized by the memories of the horrific scenes they witness. Perhaps it may be possible, immediately after the experience, to block the effects of adrenaline and spare them some anguish later.

But how does an emotional situation lead to the release of adrenaline in the first place? This, not surprisingly, takes us back to the amygdala. As we've repeatedly seen, when the amygdala detects an aversive emotional situation, it turns on all sorts of bodily systems, including the autonomic nervous system. The consequence of autonomic nervous system activation of the adrenal gland is the release of adrenaline into the bloodstream. The adrenaline then appears to influence the brain, although indirectly. This feedback (William James style) then interacts with systems that are also active at the time, such as the hippocampal system that is forming the explicit memory of the situation. Although the manner in which the feedback strengthens the explicit memory is not yet completely understood, it seems that the adrenaline somehow gets back to the brain and influences the functioning of the temporal lobe memory system, strengthening the memories being created there (see Figure 7-7).[65]

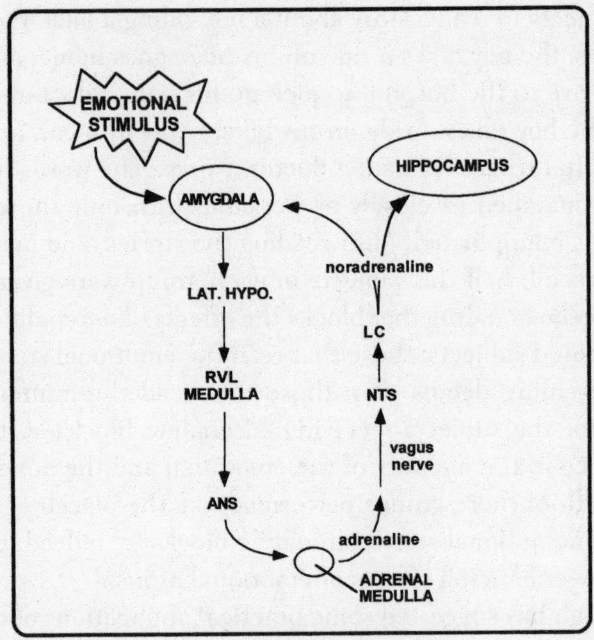

FIGURE 7-7
Modulation of Memory Circuits by Adrenaline.

Studies by McGaugh and colleagues have suggested that the hormone adrenaline, released during stress, stabilizes and strengthens memories. However, since adrenaline cannot normally enter the brain from the blood (adrenaline molecules are too big to cross the blood-brain barrier), the action must be indirect. The diagram shows how adrenaline might have its indirect actions on the brain. Stimuli associated with danger activate the amygdala. By way of pathways through the lateral hypothalamus (LAT. HYPO.) to the rostral ventral lateral (RVL) medulla, the autonomic nervous system (ANS) is aroused. One of the many target organs influenced by ANS arousal is the adrenal medulla. It releases adrenaline, which has widespread actions in the body. Of particular importance in memory modulation appears to be an effect on the vagus nerve, which terminates in the nucleus of the solitary tract (NTS) in the medulla. NTS then sends outputs to the locus coeruleus (LC), which releases noradrenaline in widespread areas of the forebrain, including the amygdala and hippocampus. By influencing amygdala and hippocampal functions, implicit emotional memories and explicit memories of emotion might be modulated.

Some Caveats

The idea that we remember best (or better) those things that are important to us—those things that elicit emotions in us—makes sense intuitively, and is also supported by quite a lot of research. However, there are some important qualifications to this notion that should be kept in mind.

Memory Is Selective: Not all aspects of an experience are remembered equally well, and the memory improvement produced by emotional arousal may affect some aspects more than others.[66] If you are robbed at gunpoint, you will probably later remember the robbery much better than other things that happened to you on the same day. Nevertheless, the vividness and accuracy of your memory of the details about the robbery may vary considerably. In general, the memory of things that are central to the episode (like the appearance of the gun) or that are particularly obvious (like the race and body style of the perpetrator) appear to be remembered better than more peripheral or more subtle details (hair and eye color, the presence or absence of facial hair, the model or color of the getaway car, etc.). Unfortunately, these extra details are often important in tracking down suspects and unequivocally identifying the perpetrator.

The exact details that are remembered probably depend on a variety of individualistic factors, not the least of which is what the victim was focused on at the time. If the victim paid more attention to the gun pointed at his face than to the face of the person pointing the gun, a reasonable thing to do under the circumstances, the appearance of the gun will be remembered better than the face of the perpetrator. An innocent bystander watching all of this may focus on different things, and may have very different explicit memories of the crime than the victim. Explicit memories are very closely related to what gets attended to during the experience.[67]

At the same time, implicit emotional memories may capture aspects of experiences that escape attention and awareness. As we saw in Chapter 3, autonomic responses have been useful in showing the presence of emotional memories that were not consciously encoded. It is conceivable that the physiological measurement of autonomic responses in a victim might be a more sensitive measure of the memory of a perpetrator than the victim's explicit memory. Polygraph tests, which involve measurements of autonomic nervous system function, though unreliable, are sometimes used as a means of getting a suspect's unconscious involuntary reactions to reveal guilt. A reverse polygraph test that probes the involuntary unconscious life of the victim could also be used, though the results of such tests would suffer from uncertainty as well.

Memories Are Imperfect Reconstructions of Experiences: Even though a memory of an emotional experience is strong and vivid, it is not necessarily accurate. Explicit memories, regardless of their emotional implications, are not carbon copies of the experiences that created them. They are reconstructions at the time of recall, and the state of the brain at the time of recall can influence the way in which the withdrawn memory is remembered. As Sir Frederic Bartlett demonstrated long ago, explicit memories involve simplifications, additions, elaborations, and rationalizations of learning experiences, as well as omissions of elements of the initial learning.[68] The memory, in short, occurs in the context of what Bartlett called a cognitive schema, which includes the expectations and biases of the person doing the remembering.[69]

The vulnerability of memory to modification by events that take place after the memory was formed has been documented in many studies and anecdotal reports. The writings of Elizabeth Loftus, a psychologist who specializes in memory malleability, and her colleagues provide lots of examples.[70] One involves Brigadier General Elliot Thorpe, who witnessed the bombing of Pearl Harbor. He described the event one way at the time of his retirement but had given a very different version in an earlier memoir, and both had many inconsistencies with facts established from other sources. Another example comes from the trial of Carl Gustav Christer Pettersson, accused of murdering the prime minister of Sweden, Olof Palme, while Mr. and Mrs. Palme were walking home from a movie.[71] The defense had a memory psychologist evaluate Mrs. Palme's testimony at different times after the murder. The psychologist was called as an expert witness and testified that Mrs. Palme's testimony had become more and more vivid and detailed as more time passed, suggesting that factors other than her experience of the crime influenced the way she remembered the incident. It was argued that she incorporated information from newspaper and television coverage into her memory. A third example comes from an experiment performed by a pioneer cognitive psychologist, Ulric Neisser, who examined people's memories of the explosion of the space shuttle *Challenger* at two times—the day after and several years later.[72] Most of the subjects said their memories of what they were doing when they heard the news were very good. Yet, in many instances, the memory at the later time was very different from the memory reported the day after. These various accounts do not question the vividness of memories established during emotionally arousing experiences, but they encourage a healthy suspicion of the accuracy of explicit memories, even explicit memories of emotional situations, when the details have important consequences.

Memory of Emotional Events May Also Be Poor: It is sometimes said that emotional events, especially traumatic ones, are accompanied by a selective amnesia for the experience, rather than an improved memory of it. Numerous anecdotal reports suggest that combat soldiers or victims of rape, incest, or other violent crimes can have very weak or nonexistent explicit memories of traumatic experiences. These observations are consistent with Freud's theory that unpleasant events are repressed, shunted out of consciousness.[73] The conditions that might lead to the loss of memory as opposed to the facilitation of memory are not understood, but may have something to do with the intensity and duration of the emotional trauma. We will return to this topic, and possible biological mechanisms through which amnesia for traumatic memory might occur, in the next chapter.

Moody Memories

Learning that takes place in one situation or state is generally remembered best when you are in the same situation or state.[74] If you learn a list of words while under the influence of marijuana, your memory of the list may be better when you are again "stoned" than when you are "straight." So-called state-dependent learning applies to lots of situations, not just drug states. Memory for words is better if subjects are tested in the room where the words were learned than if tested in a novel room. And memory for words is also better if the learning and the recall take place while the subject is in the same mood state. A corollary to this is the fact that we are more likely to have unpleasant memories when we are sad, and pleasant ones when happy. The so-called mood congruity of memory is amplified in depressed persons, who seem at times to only be capable of maudlin memories. The existence of separate systems for storing implicit emotional memories and explicit memories of emotions helps us understand how the content of memory is influenced by emotional states.

Many psychologists believe that memories are stored in associative networks, cognitive structures in which the various components of the memory are each separately represented and linked together.[75] In order for a memory to appear in consciousness, the associative network has to reach a certain level of activation, which occurs as a function of the number of components of the memory that are activated and the weight of each activated component. The weight of a component is the contribution that it makes to the overall memory in the network. Things that are essential aspects of a memory will have stronger weights than things that are less essential. The more cues that were present during learning that are also present during remembering, and the stronger the weights of the memory components that are activated by the cues present during remembering, the more likely it is that the memory will occur.

One of the components of an explicit memory of a past emotional experience will be the emotional implications of the experience. The presence of cues that activate this component will facilitate activation of the associative network. The relevant cues, in this case, will be cues from within the brain and body that signal that you are in the same emotional state as during the learning. These cues will occur because the stimuli that act on the explicit system also act on the implicit emotional memory system, causing the return of the emotional state you were in when the explicit memory system did its learning. The match between the current emotional state and the emotional state stored as part of the explicit memory facilitates the activation of the explicit memory. Co-activation of implicit emotional memory may

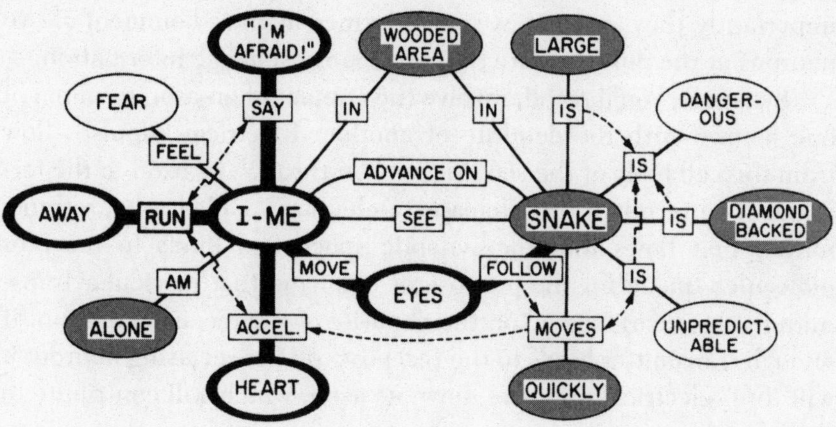

FIGURE 7-8
An Associative Memory Network for Fear.

In associative network models, memory is viewed as being stored as connections between nodes *of knowledge. The more extensively connected the nodes, the more easily retrieved and vivid the memory. The illustration shows a hypothetical associative network that might underlie a snake phobia. In this model, the phobia is maintained by propositional (verbal) information.* (From Figure 7.1 in P. Lang [1984], Cognition in emotion: concept and action. In C.E. Izard, J. Kagan, and R.B. Zajonc, eds., *Emotions, Cognition, and behavior.* New York: Cambridge University Press, © 1984 by Cambridge University Press. Reprinted with the permission of Cambridge University Press.)

thus help the explicit system during remembering as well as during learning.

Synaptic Muscle

So far we've looked at learning and memory from the neural system point of view. Now, it is time to peer deeper into the workings of the brain and take a look at how neurons and their synapses contribute to learning and memory functions.

It is widely believed that learning involves the strengthening of synaptic connections between neurons. From a purely structural point of view, synapses are minuscule spaces between neurons. More importantly, they are the tiny spaces formed by the adjoinment of two neurons at the points where those neurons exchange information.

Synapses, you'll recall, involve the contact of an axon terminal of one neuron with the dendrite of another. Electrical impulses flow from the cell body of the sending neuron through its axon to the terminal. The terminal then releases a chemical, called a neurotransmitter, that flows into the synaptic space and binds to receptor molecules (made for the purpose of receiving that particular transmitter substance) located on the dendrite of the receiving neuron. If enough transmitter binds to the receptors on the receiving neuron, it will "fire" electrical impulses down its axon, which will contribute to the firing of the next neuron, and so on.

In 1949, Donald Hebb, the great Canadian psychologist, proposed a way that learning might take place at the level of synapses.[76] Imagine two neurons, X and Y, that are anatomically interconnected but have a weak synaptic relation. That is, when X fires, Y could potentially fire but does not. However, if on some occasion Y is firing when the impulses from X reach Y, something happens between those two cells—a functional bond is created. As a result, the next time X fires, the likelihood that Y will also fire is increased. A connection between two cells that is strengthened in this way is now referred to a Hebbian synapse.[77] Perhaps nothing captures the essence of the Hebbian idea better than the oft-used slogan "cells that fire together wire together." Hebbian plasticity is shown in Figure 7-9.

For many years, Hebb's hypothesis was considered an interesting but ungrounded idea about how learning might take place. It was a hypothesis in need of a factual basis. In the early 1970s it got just the factual boost it needed to become everyone's favorite idea about how learning surely takes place. The boost came from a series of studies of hippocampal synaptic function carried out by Tim Bliss and Terje Lømo.[78]

It was known at the time that electrical stimulation of the pathway that connects the transition areas with the hippocampus elicits neural activity in the hippocampus. The activity can be measured as a neural response called a field potential, which reflects the overall synaptic response of the various hippocampal cells that are fired by the stimulus. Bliss and Lømo showed that the size of the field potential, and thus the magnitude of the synaptic response, could be in-

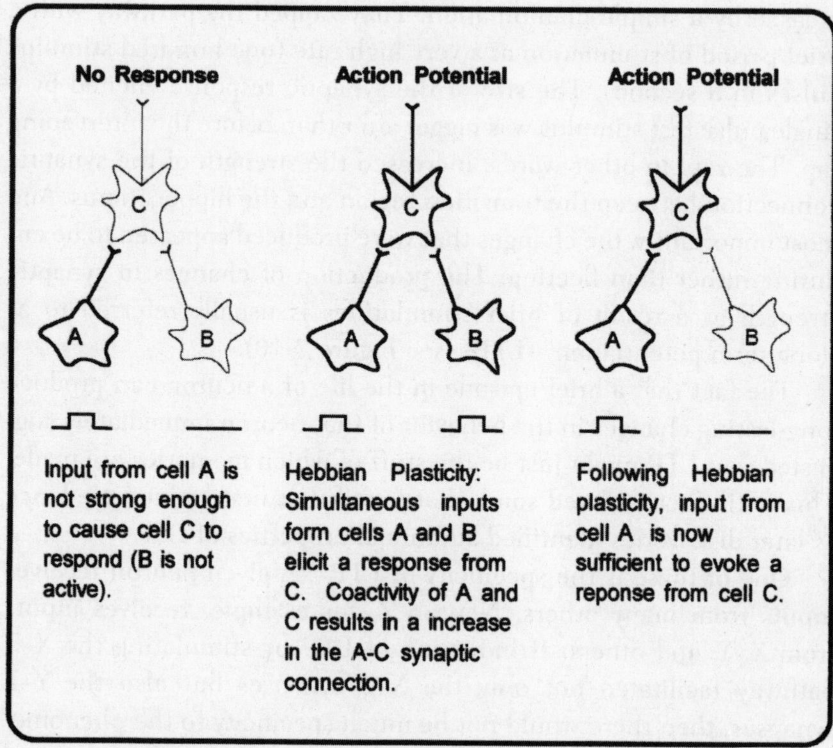

No Response
Action Potential
Action Potential
C
C
C
A
B
A
B
A
B
Input from cell A is not strong enough to cause cell C to respond (B is not active).
Hebbian Plasticity: Simultaneous inputs form cells A and B elicit a response from C. Coactivity of A and C results in a increase in the A-C synaptic connection.
Following Hebbian plasticity, input from cell A is now sufficient to evoke a reponse from cell C.

FIGURE 7-9
Hebbian Plasticity.

In 1949, Donald Hebb proposed that learning might involve changes in neural function brought about when two cells are active at the same time. Today, so-called Hebbian plasticity is everyone's favorite idea about how learning and memory work at the level of individual cells in the brain. As the figure shows, Hebbian plasticity occurs between two cells (A and C) if they fire at the same time. In the illustration, A does not normally fire C but B does. So if B causes C to fire and A happens to fire at the same time, something occurs in the link between A and C such that A acquires the ability to fire C on its own. The exact nature of what occurs between A and C has been a mystery. However, recent work in neuroscience has identified a mechanism that makes Hebbian-like plasticity possible. This mechanism is called long-term potentiation (LTP) and involves glutamate and its receptors. LTP and glutamate receptor function are illustrated in figures 7-10 and 7-l l, respectively.

creased by a simple manipulation. They zapped the pathway with a brief period of stimulation at a very high rate (one hundred stimulus pulses in a second). The size of the synaptic response elicited by a single pulse test stimulus was bigger after than before the intervening zap. The zap, in other words, increased the strength of the synaptic connection between the transition region and the hippocampus. And most importantly, the changes that were produced appeared to be enduring rather than fleeting. The production of changes in synaptic strength as a result of brief stimulations is usually referred to as "long-term potentiation" (LTP) (see Figure 7-10).

The fact that a brief episode in the life of a neuron can produce long-lasting changes in the behavior of that neuron immediately suggested that LTP might just be the stuff of which memories are made. This notion, considered somewhat fanciful at first, gained credence as later discoveries identified additional properties of LTP.

One of these is the specificity of LTP.[79] A given neuron receives inputs from many others. Neuron Z, for example, receives inputs from X, Y, and others. If induction of LTP by stimulating the X–Z pathway facilitated not only the X–Z synapses but also the Y–Z synapses, then there would not be much specificity to the phenomenon and its usefulness as a model of how memories are created through very specific learning experiences would be limited. But zapping the X–Z pathway changes the synaptic strength of this connection and leaves unaltered the strength of the Y–Z connection. LTP does not change the whole post-synaptic neuron, making it more sensitive to any input; it only changes the particular synapses on the post-synaptic neuron that were involved in the experience. Like learning, LTP is experience-specific.

Another important property of LTP is cooperativity.[80] In order for LTP to occur, a certain number of inputs to a cell have to be stimulated so that enough synapses are activated. If too few are stimulated, LTP does not result. Inputs, in other words, have to cooperate for LTP to occur.

A special version of cooperativity that is particularly important for drawing the connection between LTP and learning is associativity.[81] Again consider neuron Z that receives inputs from X and Y. If the X–Z and the Y–Z pathways are zapped at the same time, test stimuli applied to either pathway give a bigger synaptic response than if either

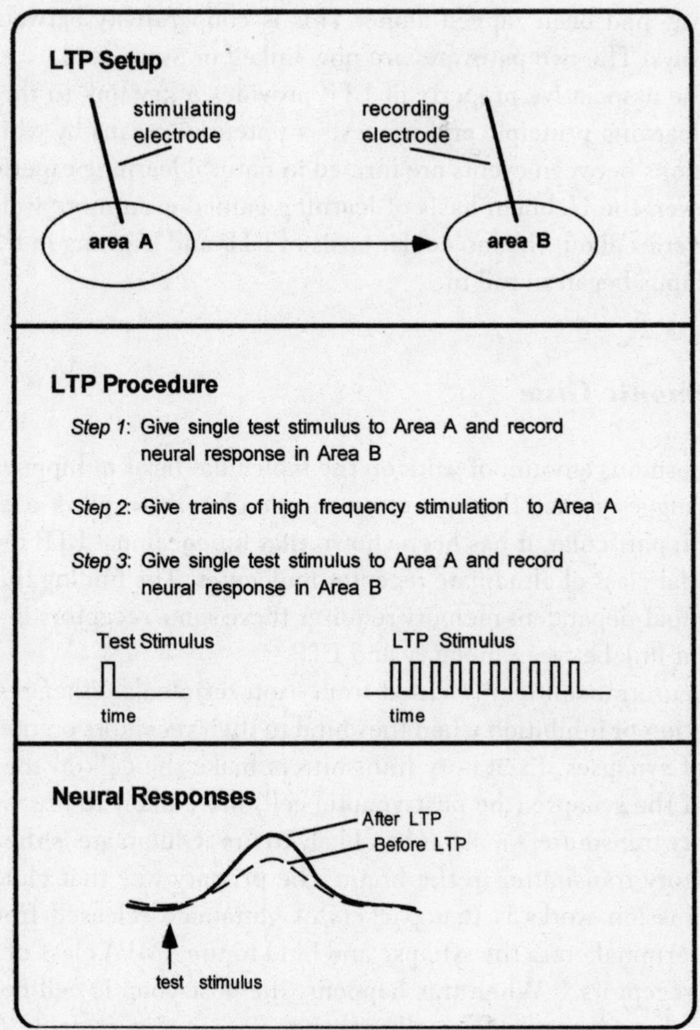

FIGURE 7-10
Long-Term Potentiation (LTP).

LTP involves a strengthening of the functional connection between two brain areas (areas A and B). Because connections between brain areas involve synapses, LTP is believed to involve an enhancement in transmission across synapses. LTP is induced in the laboratory by giving a burst of electrical stimuli to area A. As a result of this treatment, the neural response to a single test stimulus is amplified. Since the same stimulus gives a bigger response after the pathway has been treated with the burst, the burst enhances transmission in the pathway.

pathway had been zapped alone. This is cooperativity between two pathways. The two pathways are now linked or associated.

The associative property of LTP provides a key link to the Hebbian learning principle and suggests a potential means by which associations between events are formed in natural learning experiences. However, the Hebbian basis of learning gained even more weight as discoveries about the molecular basis of LTP and learning in the hippocampus began to roll in.

Mnemonic Glue

An enormous amount of work on the molecular basis of hippocampal LTP suggests that the neurotransmitter glutamate plays a crucial role. In particular, it has been shown that hippocampal LTP requires a special class of glutamate receptor molecules. The finding that hippocampal-dependent memory requires these same receptors is an important link between memory and LTP.

Neurotransmitters released from axon terminals either result in excitation or inhibition when they bind to their receptors on the other side of synapses. Excitatory transmitters make the cell on the other side of the synapse (the postsynaptic cell) more likely to fire, and inhibitory transmitters make it less likely to fire. Glutamate is the major excitatory transmitter in the brain. The primary way that glutamate transmission works is that packets of glutamate released from the axon terminal cross the synapse and bind to the AMPA class of glutamate receptors.[82] When this happens, the postsynaptic cell fires impulses down its axon. Normally, another class of glutamate receptors, NMDA receptors, are cooped up and glutamate reaching them has no effect.[83] But when the postsynaptic cell fires, the NMDA receptors become available to bind glutamate (see Figure 7-11).

The fact that NMDA receptors are only open to the public when the cell that possesses them has just fired allows the NMDA receptor to serve as a means for forming associations between stimuli. The NMDA receptor, in fact, seems to be the way the Hebbian rule (neurons that fire together wire together) is actually realized in the brain.

Imagine that impulses from one input pathway cause the release of glutamate, which binds to the postsynaptic neuron and causes the

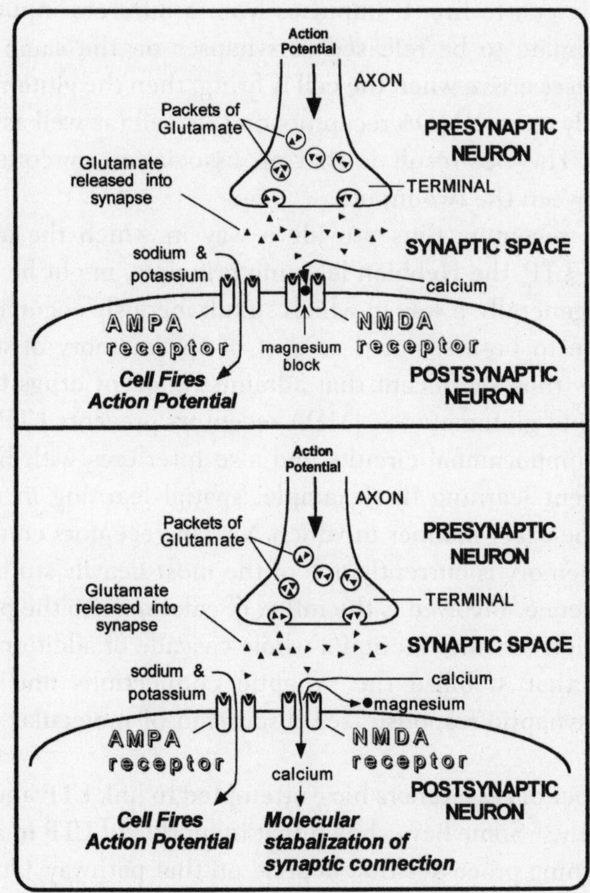

FIGURE 7-11
Glutamate Receptors.

When an action potential comes down the axon to the terminal area, it causes packets of glutamate to be released from the terminal of the presynaptic neuron. The released glutamate diffuses into the synaptic space and binds to AMPA and NMDA receptors on the dendrites of postsynaptic neurons. When glutamate binds to AMPA receptors, sodium and potassium flow into the postsynaptic neuron and help generate an action potential (above). Although NMDA receptors are normally blocked by magnesium, the magnesium block is removed by the action of glutamate at AMPA receptors. Calcium then flows into the cell below, resulting in a host of molecular changes that then strengthen and stabilize the connection between the pre- and postsynaptic neuron. (Illustration based on figure 1 in F.A. Edwards (1992), Potentially right on both sides. *Current Opinion in Neurobiology* 2: 299–401.)

postsynaptic cell to fire. If impulses from a different input pathway cause glutamate to be released at synapses on the same cell, and these impulses arrive when the cell is firing, then the glutamate binds to the briefly open NMDA receptors on this cell (as well as to AMPA receptors). The net result is that an association or connection is formed between the two inputs.

NMDA receptors thus provide a way in which the associative property of LTP, the Hebbian learning principle, might be achieved, and, more generally, a way in which simultaneously occurring events might come to be associated as part of the memory of an experience.[84] It is thus significant that administration of drugs that block the binding of glutamate to NMDA receptors prevents LTP from occurring in hippocampal circuits and also interferes with hippocampal-dependent learning (for example, spatial learning in the water maze).[85] The exact manner in which NMDA receptors contribute to LTP and memory is currently one of the most heavily studied topics in neuroscience. Involved is the influx of calcium into the postsynaptic cell, which sets into motion a whole cascade of additional molecular steps that stabilize the synaptic connections and thus the enhanced synaptic response (see discussion of molecular blindness below).

A number of researchers have attempted to link LTP and memory more directly.[86] Some have shown that induction of LTP in a pathway affects learning processes that depend on that pathway. Others have found that natural learning influences the ease with which LTP occurs. And still others have found that during learning, changes similar to those occurring in LTP take place in pathways that mediate the learning.

While the correspondence between LTP and natural learning is becoming more and more compelling, the case for LTP being the basis of learning remains unproven. No study has actually shown that the changes induced by LTP actually account for learning. Many laboratories are working fast and furiously to convert the correlations between LTP and learning into a causal linkage. Many workers in the field believe that the causal connection is there and that it is just a matter of time until the appropriate way to demonstrate the relation is discovered.

The Molecular Blindness of Memory

Initially, LTP was believed to be mainly a hippocampal phenomenon. This certainly added fuel to the fiery attempts to develop animal models for studying the contribution of the hippocampus to memory. Now, it is known that LTP occurs in many brain regions and in many learning systems. Of special relevance to our concern here is the fact that LTP has been demonstrated in pathways that are involved in fear conditioning,[87] and that blockade of NMDA receptors in the amygdala prevents fear conditioning.[88]

NMDA-dependent synaptic plasticity may be a fairly universal way that the brain learns and stores information at the molecular level. While there are other forms of plasticity that do not depend on NMDA receptors (even in the hippocampus),[89] it nevertheless seems that NMDA-dependent plasticity is one of the major learning devices, and that the brain may in fact have a limited number of learning mechanisms that it uses in a variety of different situations.

If we look more closely at how memories are stabilized, the idea that fairly universal mechanisms are used to form different kinds of memories becomes even more compelling. Studies of species as different as snails, mice, and fruit flies have converged in their conclusions about the kinds of molecular events that convert learning experiences into long-term memories. Protein synthesis, which is controlled by genetic machinery located in the cell nucleus, seems to play a crucial role. If protein synthesis is blocked, long-term memories are not formed.[90] The long-term memory of an experience, in other words, may be maintained by proteins made in cells after learning has taken place. Proteins appear to be important because they make up genes, which control the manufacture of certain chemicals that are required for memory stabilization. Disruption of protein synthesis appears to interfere with the formation of most kinds of long-term memories in most kinds of animals. It also interferes with the long-term maintenance of LTP.[91]

One chemical that appears to be particularly important is cyclic AMP (cAMP). This substance takes over where neurotransmitters leave off. Neurotransmitters allow cells X and Y to communicate with Z; cAMP then helps Z remember that the firing of X and Y at Z occurred at the same time—that X and Y were associated. cAMP is involved in communication between different parts within a cell rather than between cells. The contribution of cAMP to memory was first shown in studies of snails by Eric Kandel, one of the leading researchers of the neurobiology of memory.[92] Kandel also showed that drugs that block the expression of cAMP disrupt hippocampal memory and LTP.[93] New genetic tools have been used to create animals that are incapable of making cAMP. Tim Tully has shown that fruit flies lacking this gene have amnesia for certain long-term memory tasks,[94] and Kandel and Alcino Silva[95] have each shown that genetically engineered mice that are unable to make cAMP have deficient hippocampal LTP and are unable to form new long-term memories in tasks that are dependent on the hippocampus. The mechanisms of memory stabilization appear to be remarkably similar across diverse species and diverse learning procedures. Although there may be more than one such mechanism, the number of them may be relatively small.

The idea that nature might use one or a few molecular mechanisms in many different learning networks in many different kinds of animals has a very important implication. Different forms of learning are not necessarily distinguishable at the level of molecular events, but instead obtain their unique properties by way of the circuits of which they are part. There may well be a universality, or a least a generality, of memory at the molecular level, but there is a multiplicity of memory at the systems level.

Claparede Redux

It should now be clear how it was possible for Claparede's patient to have formed an implicit memory of the pinprick without having an explicit conscious memory of the experience that led to the formation of the implicit memory. Most likely, her temporal lobe memory system was damaged. And, given that the implicit memory she formed involved fear conditioning, it also seems likely that her amygdala was alive and well. Admittedly, these are retrospective guesses since we have no idea where the lesion was in her brain. However, these guesses are based on forty years of research into the neural basis of memory, and even if for some unknown reason they are wrong in her

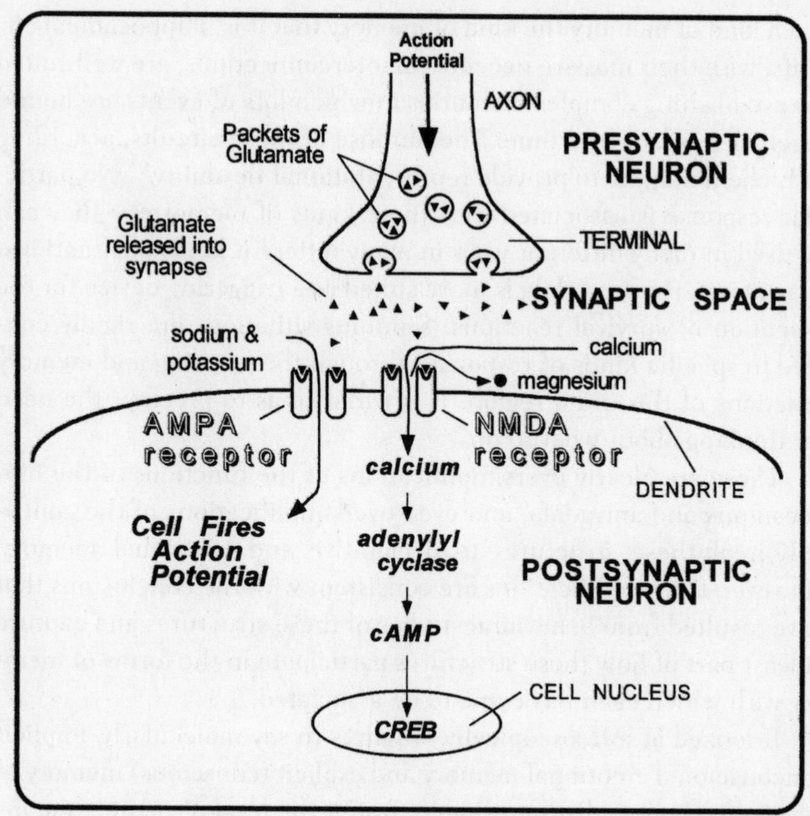

FIGURE 7-12

Molecular Stabilization of Synaptic Plasticity and Memory.

When glutamate binds to the NMDA receptors of a cell that has just fired an action potential, the magnesium block of the NMDA receptor is removed and calcium flows in. Calcium influx, in turn, activates adenylyl cyclase, which leads to an increase in cyclic AMP (cAMP); cAMP elevation then activates cAMP-inducible genes in the cell nucleus by way of the gene transcription factor, CREB. CREB induces proteins, such as synaptic effector proteins, that may contribute to the long-term maintenance of LTP, possibly by stabilizing changes in the structure of the postsynaptic dendrites. (Based on figure 7-12 above and figure 1 in M. Mayford, T. Abel, and E.R. Kandel [1995], Transgenic approaches to cognition. *Current Opinion in Neurobiology* 5:141–48.)

case (something we'll never know), they will surely turn out to often be correct predictions for future cases in which we can verify the locus of brain damage.

The multiplicity of memory at the systems level is what makes a given kind of memory the kind of memory that it is. Hippocampal circuits, with their massive neocortical interconnections, are well suited for establishing complex memories in which lots of events are bound together in space and time. The purpose of these circuits, according to Eichenbaum, is to provide representational flexibility.[96] No particular response is associated with these kinds of memories—they can be used in many different ways in many different kinds of situations. In contrast, the amygdala is more suited as a triggering device for the execution of survival reactions. Stimulus situations are rigidly coupled to specific kinds of responses through the learning and memory functions of this brain region. It is wired so as to preempt the need for thinking about what to do.

These are clearly oversimplifications of the functions of the hippocampus and amygdala, and even oversimplifications of the contribution of these structures to declarative and emotional memory. However, the simplifications are consistent with the conclusions that have resulted from behavioral studies of these structures and capture at least part of how these structures participate in the forms of memory with which each has come to be associated.

If looked at microscopically, which is to say molecularly, implicit (unconscious) emotional memory and explicit (conscious) memory of emotion may be indistinguishable. But at the level of neural systems and their functions, these are clearly unique operations of the brain. Although we know much more at this point about the separate operation of these two systems, we are beginning to also see how they interact. And these interactions are at the core of what gives emotional qualities to memories of emotions past.

8

WHERE THE WILD THINGS ARE

"A phobia, like a psychoanalytic theory, is a story about where the wild things are."

Adam Phillips, *On Kissing, Tickling, and Being Bored: Psychoanalytic Essays on the Unexamined Life*[1]

IN 1793, REVOLUTION WAS in the air in Paris. But the French Revolution about which we are concerned here took place, not in the streets, but in mental asylums. Philippe Pinel had the radical opinion that the mentally ill were not hopeless wild animals that should be incarcerated and tortured, but were people who should be treated with decency and respect. When the Revolution's prison commissioner heard of Pinel's plans to rehabilitate the insane, he asked, "Are you not yourself mad to free these beasts?" Pinel responded, "I am convinced that the *people* are not incurable if they can have air and liberty." Some of the "beasts" recovered under Pinel's guidance. One became his bodyguard.[2]

By 1800 Pinel had become one of the most influential physicians in Paris and he was called upon by the Revolution's Society of Observers of Man to evaluate a truly wild beast, a boy around eleven years old, who, a few months earlier, had been captured in a small village in southern France. As recounted by Roger Shattuck, author of a fascinating book about the Wild Boy of Aveyron, the incident went like this:

> Before dawn on January 9, 1800, a remarkable creature came out of the woods near the village of Saint-Sernin in southern France. No one expected him. No one recognized him. He was human in bodily form and walked erect. Everything else about him suggested an animal. He was naked except for the tatters of a shirt and showed no modesty, no awareness of himself as a human person related in any way to the people who had captured him. He could not speak and made only weird, meaningless cries.[3]

In spite of his prior successes, Pinel felt that rehabilitation was not possible in the case of the Wild Boy. According to Shattuck, Pinel seems not to have seriously pondered whether the boy's condition was due to "organic" or "functional" causes, a distinction that Pinel commonly made in other cases. Such an analysis might have led to a more informed decision about whether the boy was curable. If the problem was organic, say due to brain damage, then his wild state might indeed be untreatable. But if life's circumstances—the lack of nurturing care during his early childhood, the absence of social stimulation, his stressful, traumatic existence in a hostile environment—were the causes, some cure might have been possible. As Shattuck notes, we will never know the answer.

So-called functional disorders are indeed more likely to be treatable than those related to organic causes. However, the distinction between organic and functional maladies needs to be made cautiously, and should in no way imply that some mental disorders are attacks on the brain and others on the mind. As Shakespeare said, the brain is the soul's dwelling place.[4] Mental disorders, like mental order, reflect the workings of the brain.

Actually, Shakespeare's phrase was the soul's *frail* dwelling place, which speaks to the thinness of the line between mental health and illness. We all experience sadness and worry from time to time. But when these become excessive and inappropriate to the circumstances, we slide from normal to pathological emotions.

In this chapter we are going to be especially concerned with the pathological emotions called anxiety disorders. These are among the most common forms of mental illness.[5] I will argue they involve the fear system of the brain, and that the progress we've made in understanding how the fear system normally works also helps us understand what goes wrong in anxiety disorders. I'll propose that anxiety disorders come about when the fear system breaks loose from the cortical controls that usually keep our primitive impulses—the wild things in us—at bay.

A Brief History of Mental Illness

The diagnosis of mental disorders has its roots in the work of Emil Kraepelin in the late nineteenth century. He distinguished schizophrenia from manic-depression by showing that these illnesses take different courses. Freud, Kraepelin's contemporary, was more concerned with neuroses than with psychotic conditions like schizophrenia and emphasized intrapsychic conflict and the resulting anxiety as the cause. According to psychiatrist Peter Kramer, by mid-century American psychiatrists had outdone Freud.[6] They had adopted the spectrum model of mental illness, which assumed that all forms of psychopathology are secondary to anxiety.[7] Neurosis was, in the typical Freudian view, the result of a partially successful defense against anxiety that was accompanied by symptom formation. But under the spectrum model even psychosis came to be viewed as the result of anxiety, such excess anxiety that the ego crumbled and regressed. Mental health and mental illness were distinguished by the degree of anxiety present, and the same treatment, the reduction of inner conflict via psychotherapy, was applicable to all ailments.

The tides have since shifted. Mental health professionals now have a dazzling array of diagnostic categories available to them. All one has to do to see how radically things have changed is to thumb through the diagnostic bible, the American Psychiatric Association's *Diagnostic and Statistical Manual of Mental Disorders* (DSM), first published in 1980, now in its fourth edition.[8] There are a host of phobias, different kinds of panic attacks, a variety of mood and thought disorders, somatization disorders, antisocial personality conditions, numerous forms of substance abuse, and other conditions. In addition, there are overlaps, such as panic with agoraphobia (fear of open or crowded places), or manic-depression with cocaine dependence, and so on.

In spite of this diagnostic diversity, it is clear that some categories of mental illness occur more than others. The U.S. Public Health

Service counts and classifies the prevalence of different forms of mental disorders.[9] In 1994, about 51 million Americans eighteen years and older had some form of diagnosed mental illness, with about 11 million of those involving substance abuse. Of the remaining 40 million, more than half were accounted for by the category of anxiety disorders, and somewhat less than half by mood disorders (especially depression), with schizophrenia and assorted other conditions accounting for the rest.

The high proportion of mental disorders that involve anxiety does not vindicate the spectrum theory, for treating depression or schizophrenia as anxiety is probably not going to get you as far as treating them uniquely. However, it does emphasize the importance of understanding the nature of anxiety and its various manifestations. Fortunately, the understanding of the fear system that has been achieved can help us explain how anxiety disorders arise, and may also help us figure out how to treat them and possibly prevent their occurrence.

Fear and Loathing in Anxiety

Anxiety and fear are closely related. Both are reactions to harmful or potentially harmful situations. Anxiety is usually distinguished from fear by the lack of an external stimulus that elicits the reaction—anxiety comes from within us, fear from the outside world. The sight of a snake elicits fear, but the remembrance of some unpleasant experience with a snake or the anticipation that you may encounter a snake are conditions of anxiety. Anxiety has also been described as unresolved fear.[10] Fear, according to this view, is related to the behavioral acts of escape and avoidance in threatening situations, and when these actions are thwarted, fear becomes anxiety.

Fear and anxiety are normal reactions to dangers (real or imagined) and are not themselves pathological conditions. When fear and anxiety are more recurrent and persistent than what is reasonable under the circumstances, and when they impede normal life, then a fear/anxiety disorder exists.[11]

Conditions that reflect anxiety and its defenses (conversion, repression, displacement)[12] were called neuroses by Freud. Today, the field of psychiatry is less devoutly Freudian than it once was and the term "neurosis" is deemphasized in DSM to avoid the implication that symptoms of anxiety necessarily reflect Freudian defense mechanisms.[13] Consequently, while DSM "anxiety disorders" include conditions that Freud called anxiety neuroses, more contemporary diagnoses appear as well.[14] The full complement of DSM anxiety disorders are: panic, phobias, post-traumatic stress disorder, obsessive-compulsive disorder, and generalized anxiety.

The characteristic features of these disorders are intense feelings of anxiety and avoidance of situations that are likely to bring on these feelings.[15] *Phobias* are fears of specific stimuli or situations that are in excess of the actual threat posed. Exposure to the phobic object or situation reliably elicits a profound state of anxiety. The person will go to great lengths to avoid the object or situation. *Panic* attacks involve discrete periods of intense anxiety and discomfort. The afflicted person often feels like he or she is suffocating. Unlike phobias, the attacks are often unpredictable and frequently not related to any particular external stimulus or situation. Sometimes panic is accompanied by agoraphobia. In severe cases, avoidance of such situations can lead to a sheltered existence. *Post-traumatic stress disorder* (PTSD) involves severe anxiety elicited by stimuli that were present during some extreme trauma or that are somehow related to stimuli that occurred during the trauma. It is common in war veterans but also occurs in victims of severe physical or sexual abuse or natural disasters. Situations or even thoughts that are likely to remind the person of the trauma are avoided. *Obsessive-compulsive disorder* involves intrusive, repetitive, and persistent thoughts and/or repetitive behaviors that are performed in a very precise way in response to obsessive thoughts. The compulsive behaviors are meant to neutralize anxiety, but the behaviors are either not well connected to the situation or are excessive responses to the situation that they are intended to neutralize. *Generalized anxiety,* also known as free-floating anxiety, involves excessive worry about unrelated things for a long period of time.

DSM outlines symptoms and situational factors that allow skilled clinicians to distinguish the various anxiety disorders. However, Arne Öhman, a leader in the study of human fear and anxiety, has recently argued that, "when comparing the physiological responses seen in phobics exposed to their feared objects with those seen in PTSD patients exposed to relevant traumatic scenes for the disorder, and with physiological responses during panic attacks, one is much more struck by the similarities than by the differences."[16] He goes on to argue that panic, phobic fear, and PTSD reflect the "activation of one and the same underlying anxiety response." This is essentially the case that I will make. However, I state the idea in terms of brain systems rather than symptoms: anxiety disorders reflect the operation of the fear system of the brain. Öhman leaves generalized anxiety out of his grouping because it involves a stable personality trait rather than discrete episodes of anxiety, a distinction that is often referred to as one between trait and state anxiety. However, generalized anxiety most likely involves the same underlying brain system (at least partly) as the other anxiety disorders.

Little Albert Meets Little Hans

Anxiety disorders can arise at any time, but most often appear in early adult life. Why does this happen? How does the brain go from a state in which it is not especially anxious to one in which it is pathologically worried or exhibiting neurotic behaviors that keep the worry in check?

Most theorists from Freud onward have assumed that clinically debilitating anxiety is the result of traumatic learning experiences that create unpleasant memories. Breuer and Freud,[17] in the famous case of Anna O., for example, argued that "hysterics suffer mainly from reminiscences," or as Matthew Erdelyi puts it, "traumatic memories which they have expunged from consciousness."[18] Since fear conditioning is the *sine qua non* of traumatic learning, it should come as no surprise that fear conditioning has been proposed to be involved in the genesis of pathogenic anxiety. Though long considered controversial and incomplete, as we will see, new findings have made it seem more likely, and even quite plausible, that fear conditioning contributes significantly to anxiety disorders.[19]

The conditioning theory of anxiety arose in the 1920s, a time when psychologists were beginning to explain most aspects of behavior in terms of learning experiences, and particularly in terms of Pavlov's conditioned reflexes.[20] John Watson, the father of behaviorism, claimed to have conditioned an animal phobia in an eleven-month-old boy, Little Albert, by making a loud clanging sound while the boy was happily playing with a rat.[21] Thereafter, the boy avoided playing with the rat and cried when he was near it. To explain this finding, Watson proposed that certain stimuli (loud noises, painful stimuli, sudden loss of physical support) are innately capable of eliciting fear reactions. When these unconditioned stimuli occur, other stimuli that happen to be present acquire the capacity to elicit conditioned fear. According to Watson, neuroses arise as a result of these traumatic learning situations and then persist and influence behavior throughout life.[22]

Watson's theory of anxiety, as well as his behaviorist view of psychology, was based on Pavlovian conditioned reflex learning. But by the 1930s, another form of learning, called instrumental conditioning, had come to be of equal importance to behaviorists.[23] In instrumental conditioning, an arbitrary response (like pressing a bar or making a turn in a maze) is learned if it is reinforced, which means it is either followed by the presentation of a reward or the omission of a punishment. The response is learned because it is reinforced, and thereafter is performed in order to get the reward or avoid the punishment. While Pavlovian conditioning involves the transfer of meaning from an emotionally arousing to a neutral stimulus, in instrumental conditioning the association is between an emotionally arousing stimulus and neutral response.

Behaviorism and psychoanalysis were radically different approaches, but both sought to understand why we act the way we do. O. Hobart Mowrer, a leading behaviorist, saw value in both approaches and set out in the 1940s to translate Freud's theory of anxiety neurosis into the language of learning theory.[24] Using the principles of Pavlovian and instrumental conditioning, Mowrer hoped to solve what he called the "neurotic paradox": "a normal sensible man, or even a beast to the limits of his intelligence, will weigh and balance the consequences of his acts. . . . If the net effect is unfavorable, the action producing it will be inhibited, abandoned. In neurosis, however, one sees actions which have predominantly unfavorable consequences, yet they persist over a period of months, years, or a lifetime."[25]

Anxiety, according to Mowrer, motivates us to deal with traumatic events in advance of their occurrence. And because anxiety reduction brings about relief or security, it is a powerful reinforcer of instrumental behaviors (arbitrary responses that are learned because they satisfy some need or accomplish some goal). Responses that reduce anxiety are thus learned and maintained.

Mowrer felt that anxiety is initially learned much like Watson had suggested—stimuli that are present during painful or traumatic stimulation acquire the capacity to elicit anxiety. Because anxiety is uncomfortable, when the stimuli that elicit it are present the anxious person will be motivated to change the circumstances, to remove himself from where the anxiety-causing stimuli are, and to avoid such situations in the future. The reduction in anxiety that these responses produce then reinforces the behaviors and perpetuate their performance. This is often useful, but sometimes it leads to neurotic symptoms.

Consider a real-life example. A man is mugged in an elevator. From that day on, he becomes afraid of riding in elevators. He avoids them as much as possible. He consults a therapist, who tries to reassure him that it is highly unlikely that he will be mugged again in an elevator, especially if he rides at busy times. But the reassurance is not helpful. The man must get to his office on the thirteenth floor. This makes him anxious. In spite of the inconvenience that it causes him, each day he takes the stairs. The reduction in anxiety that results from taking the stairs, according to Mowrer's theory, maintains the neurotic behavior of taking the stairs.

Mowrer, like existentialist philosophers, saw anxiety as an important part of human existence, as fundamental to what is special about humans, but also as a clue to our frailty:

> By and large, behavior that reduces anxiety also operates to lessen the danger that it presages. An antelope that scents a panther is likely not only to feel less uneasy (anxious) if it moves out of the range of the odor of the panther but also likely to be in fact somewhat safer. A primitive village that is threatened by marauding men or beasts sleeps better after it has surrounded itself with a deep moat or a sturdy stockade. And a modern mother is made emotionally more comfortable after her child has been properly vaccinated against a dreaded disease. This capacity to be made uncomfortable

> by the mere prospect of traumatic experiences, in advance of their actual occurrence (or reoccurrence), and to be motivated thereby to take realistic precautions against them, is unquestionably a tremendously important and useful psychological mechanism, and the fact that the forward-looking, anxiety-arousing propensity of the human mind is more highly developed than it is in lower animals probably accounts for many of man's unique accomplishments. But it also accounts for some of his most conspicuous failures.[26]

Mowrer paved the way for a behavioral interpretation of Freud, but this pursuit was most successfully implemented by another behavioral psychologist, Neal Miller.[27] Miller had been attempting to work out in detail how fear might serve as a drive, like hunger or sex, an internal signal that motivates one to act in a way that reduces the drive. Just as a hungry animal looks for food, a fearful one tries to get away from the stimuli that arouse fear. He trained rats to avoid being shocked by jumping over a hurdle that separated two compartments whenever a buzzer sounded.[28] The first phase involved fear conditioning: the buzzer came on and the rats were shocked. Then, through random actions, they learned that if they jumped over the hurdle during the buzzer, they could avoid getting shocked. Once the rat figured this out, it would jump every time it heard the buzzer, even if the shock was turned off. The shock was no longer present and was thus no longer the motivator. The avoidance response seemed, as Mowrer had suggested, to be maintained by the anticipation of shock, by the fear elicited by the warning signal. But to prove that fear was the motivator, Miller changed the rules on the rat. Previously, when the rat jumped over the hurdle, the buzzer went off, and turning the buzzer off seemed to be sufficient reinforcement to keep the rat jumping. But now the buzzer stayed on when the rat jumped and would only go off if the rat pressed a lever. And once this was learned Miller changed the game again, forcing the rat to learn still another response to turn the buzzer off. While the initial response was learned because it allowed the rat to avoid the shock, the subsequent ones were never associated with the shock. They were reinforced by the fact that they turned off the sound. According to Miller, the findings showed that fear is a drive, an internal energizer of behavior, and that behaviors that reduce fear are reinforced and thereby become habitual ways of acting (note, however, that "fear" is an internal bodily signal, like hunger, and does not necessarily refer to subjective, consciously experienced fear in this theory).

Miller felt that this new view of fear as a drive was the key to a truly scientific approach to psychoanalytic principles. Together with John Dollard, a trained analyst, Miller attempted to account for unconscious neurotic conflict and its expression as symptoms in terms of the principles of animal learning.[29] Just as a rat could learn any response that allowed it to escape from or avoid an anxiety-provoking situation, humans learn all sorts of instrumental responses that allow them to escape or avoid anxiety and guilt caused by neurotic conflict.[30] As Dollard and Miller put it:

> the symptoms of the neurotic are the most obvious aspects of his problem. These are what the patient is familiar with and feels he should be rid of. The phobias, inhibitions, avoidances, compulsions, rationalizations, and psychosomatic symptoms of the neurotic are experienced as a nuisance by him and by all who have to deal with him. . . . When a successful symptom occurs it is reinforced because it reduces neurotic misery. The symptom is thus learned as a habit.[31]

Conditioned fear theories of anxiety took a different turn in the early 1960s. In contrast to the tradition of Mowrer and Miller, who saw Freud as scientifically imprecise but on the right track, the new theorists had little patience with the psychoanalytic view of anxiety and its emphasis on unresolved and unconscious conflict. Joseph Wolpe was one of these. He reinterpreted Freud's famous phobic case, Little Hans,[32] in terms of simple Pavlovian conditioning.[33] Hans, a five-year-old boy, became afraid of horses one day while witnessing a frightening event in which a horse fell down. Freud's view was that the horse phobia was an unresolved Oedipal conflict—Hans' fear of being castrated by his father for desiring his mother was displaced to horses. The trauma of witnessing the horse falling was the occasion that allowed the phobia to cover for the underlying conflict. But Wolpe saw it differently. Like all good conditioning theorists, he argued that a neutral stimulus, like a horse, that occurs in the presence of a trauma will acquire the capacity to elicit fear reactions, and that phobias are nothing more than fear (anxiety) that has been conditioned to some otherwise meaningless event. In making his case,

Wolpe severely criticized Freud's selective use of information that confirmed his theory and his selective disregard for information that went against it. For example, Hans himself supposedly said that he "got the nonsense" when he saw the horse fall down, and his father, in support of this view, said the anxiety broke out immediately after the incident. Freud dismissed these surface explanations, but Wolpe took them at face value. For Wolpe, Little Hans was just like Little Albert. The conditioning theory had come full circle.

The distinction between Watson's and Wolpe's purely Pavlovian approach and Mowrer's and Miller's psychoanalytic translations is more than just one of the language used to describe how anxiety arises. It also impacts importantly on the issue of how anxiety should be treated. Freudians, and their behavioral protégés, saw the goal of therapy as the resolution of unconscious conflict. The other school, typified by Wolpe, had no use for unconscious explanations and saw neurotic symptoms as nothing more and nothing less than conditioned responses. In the words of Stanley Rachman and Hans Eysenck, two other leaders in this movement, "Get rid of the symptom . . . and you have eliminated the neurosis."[34]

In spite of many important differences, there is a common theme that runs through psychoanalytic and the various conditioning theories—anxiety is the result of traumatic learning experiences. Since traumatic learning involves (at least in part) fear conditioning, it is possible that similar brain mechanisms contribute to pathogenic anxiety in humans and conditioned fear in animals. If so, findings from easily performed animal experiments could be used to understand how anxiety is learned, unlearned, and controlled in humans. However, before we can accept this rather strong, and some would say controversial, conclusion, we need to consider some additional ideas about the relation of fear conditioning to anxiety disorders, and some additional facts about the organization and function of the fear system of the brain.

Ready to Fear

In the early 1970s, Martin Seligman, an experimental psychologist who had been studying conditioned fear in animals, pointed out some striking differences between human anxiety and laboratory conditioned fear.[35] Especially important to Seligman was the fact that avoidance conditioning extinguishes quickly if the animal is prevented from making the avoidance response and alternative solutions for escape or avoidance are not provided. Recall that Miller's rats kept jumping over the hurdle when the buzzer sounded even when the shock was turned off. They never had the chance to find out that the shock was off because they kept jumping. But Seligman's point is that if the hurdle is replaced with a wall, thus preventing the avoidance response, the rat soon learns that the buzzer is no longer followed by a shock and begins to ignore the buzzer. If the wall is now removed and the hurdle returned, jumping no longer occurs in response to the buzzer. Forcing the rat to see that the buzzer doesn't lead to danger extinguishes the fear and this leads to the extinction of the neurotic avoidance response. In contrast, telling an acrophobic that no one has ever accidentally fallen off the Empire State Building and that he will be just fine if he goes to the top, or forcing him to go up there to prove the point, does not help, and can even make the fear of heights worse rather than better. Human phobias seem more resistant to extinction, and more irrational, than conditioned fears in animals.

The key to this difference, in Seligman's view, is the fact that while laboratory experiments use arbitrary, meaningless stimuli (flashing lights or buzzers), phobias tend to involve specific classes of highly meaningful objects or situations (insects, snakes, heights). He argued that perhaps we are prepared by evolution to learn about certain things more easily than others, and that these biologically driven instances of learning are especially potent and long lasting. Phobias, in this light, reflect our evolutionary preparation to learn about danger and to retain the learned information especially strongly.

In a relatively stable environment, it is generally a good bet that the dangers a species faces will change slowly. As a result, having a ready-made means of rapidly learning about things that were dangerous to one's ancestors, and theirs, is in general useful. But since our environment is very different from the one in which early humans lived, our genetic preparation to learn about ancestral dangers can get us into trouble, as when it causes us to develop fears of things that are not particularly dangerous in our world.

With the notion of preparedness, Seligman injected a dose of biological realism into the plain vanilla conditioning theory that Watson and later behaviorists popularized. Ironically, the phenomenon of preparedness may have played a seminal role in Watson's conditioning of Little Albert. Several later studies failed to reproduce Watson's findings[36] and these results have often been used as ammunition against fear conditioning theories of anxiety. But Seligman notes that in choosing a furry animal as the conditioned stimulus, Watson may have unwittingly used a prepared stimulus, and the failure of the later studies may well be because they used inanimate, meaningless stimuli.

Preparedness theory quickly received strong support from studies by Susan Mineka.[37] It had long been thought that monkeys have an inherited fear of snakes, so that the first time a monkey saw a snake it would act afraid and protect itself. However, Mineka showed that laboratory-reared monkeys are in fact not afraid on the first exposure to a snake. Most of the earlier work had involved testing of the young monkeys in the presence of their mothers. If the young monkey is shown the snake when separated from its mother, it doesn't act afraid. It appears that the infant learns to be afraid of the snakes by seeing its mother acting afraid. The young monkeys did not learn about nonfrightening things in this way, suggesting that there is something special about biologically relevant stimuli that makes them susceptible to rapid and potent observational learning. Humans learn many things by observing others in social situations and it has been proposed that anxiety, especially pathological anxiety, is sometimes or even often learned by social observation.[38]

In recent years, preparedness theory has been championed by Öhman.[39] Öhman believes that evolution has equipped contemporary humans with a propensity to associate fear with situations that threatened the survival of our ancestors. To the extent that this propensity evolved, it must be based in our genes, and genetic variation must therefore exist. As a result, although humans are in general prepared to acquire fears of ancestral dangers easily, some individuals must be more prepared than others to acquire specific fears. These super-prepared humans are, he proposes, vulnerable to phobias.

Öhman has subjected preparedness theory to stringent tests. He started with the assumption that snakes and insects are common objects of phobias and are likely to be prime examples of prepared stimuli, whereas flowers are not common phobic objects. He then used these fear-relevant (prepared) and fear-irrelevant stimuli in conditioning studies in humans. In support of preparedness theory, he found that conditioned fear (measured by autonomic nervous system responses) was more resistant to extinction with fear-relevant than with fear-irrelevant stimuli. Further, when modern fear-relevant stimuli (guns and knives) were used, no evidence for resistance to extinction was found, suggesting that evolution has not yet had enough time to build these dangers in. He also showed that phobics respond to a greater degree when they see stimuli relevant to their own phobia than when they see other fear-relevant stimuli—snake phobics gave bigger conditioned responses to snake pictures than to spider pictures and spider phobics did the reverse. This is consistent with his contention that phobics are super-prepared genetically to respond to the objects of their phobia. Finally, using special procedures to prevent conditioned stimuli from being consciously perceived, he was able to produce the prepared conditioning in the absence of awareness of the conditioned stimuli. This shows that phobias can be learned and expressed independently of consciousness, which may be related to their seemingly irrational nature.

Preparedness theory goes a long way toward dealing with some of the shortcomings of the traditional fear conditioning theories of anxiety, particularly the fact that in anxiety disorders fear doesn't extinguish easily and is especially irrational. Nevertheless, important aspects of phobias and other anxiety disorders remained unexplained. People become anxious about objects and situations that are not evolutionarily prepared—like fear of cars or elevators. Anxiety disorders can and often do exist in the absence of a memory of a traumatic experience, suggesting that maybe traumatic conditioning is not so important. And sometimes a clear trauma precedes the onset of an anxiety disorder, but the trauma is unrelated to the disorder (for example, the death of one's mother preceding the development of a fear of heights)—this doesn't make sense if the anxiety was conditioned by the trauma. However, our understanding of the brain mechanisms of conditioned fear, together with new observations about the effects of stress on the brain, give us additional clues that help fill these gaps.

New Twists on Anxiety: Clues from the Brain

In further pursuing the nature of anxiety disorders, we'll draw upon the notion, developed in the previous chapter, of multiple memory systems. In particular, we'll examine some of the implications of the idea that during a traumatic learning situation, conscious memories are laid down by a system involving the hippocampus and related cortical areas, and unconscious memories established by fear conditioning mechanisms operating through an amygdala-based system. These two systems operate in parallel and store different kinds of information relevant to the experience. And when stimuli that were present during the initial trauma are later encountered, each system can potentially retrieve its memories. In the case of the amygdala system, retrieval results in expression of bodily responses that prepare for danger, and in the case of the hippocampal system, conscious remembrances occur.

It is very helpful to keep the workings of the declarative system separate from other memory systems when considering how anxiety disorders might arise and be maintained. This point was made by Jake Jacobs and Lynn Nadel in a 1985 article that greatly influenced my thinking about the effects of stress on the fear system.[40]

Stress-Induced Loss and Recovery of Traumatic Memories: The fact that some clinically anxious persons do not recall any particular traumatic event that might be the cause of their anxiety has been an especially sharp thorn in the side of conditioning theories. In contrast, the main competition, Freud's psychoanalytic theory, assumes that anxiety will only result when traumatic memories are dispatched to the unconscious corners of the mind. Not wanting to call upon anything so mysterious and scientifically unfounded as repression, conditioning theorists have struggled with instances where there is no memory of an instigating trauma. Either no trauma, and thus no conditioning, occurred, or the trauma occurred but is not remembered. Both possibilities leave conditioning theorists with something to explain.

A possible solution to this puzzle has emerged from recent work showing that stressful events can cause malfunctions in the hippocampus. This suggests that at least in some instances the failure to recall an instigating trauma may be due to a stress-induced breakdown in hippocampal memory function.[41] In order to understand how and why this occurs, we need to explore the biological effects of stress.

When people or other animals are exposed to a stressful situation, the adrenal gland secretes a steroid hormone into the bloodstream.[42] Adrenal steroids play an important role in helping the body mobilize its energy resources to deal with the stressful situation. As we saw in Chapter 6, the amygdala is critically involved in the control of the release of adrenal steroids. When the amygdala detects danger, it sends messages to the hypothalamus, which in turn sends messages to the pituitary gland, and the result is the release of a hormone called ACTH. ACTH flows through the blood to the adrenal gland to cause the release of steroid hormone. In addition to reaching target sites in the body, the steroid hormone flows through the blood into the brain, where it binds to receptors in the hippocampus, amygdala, prefrontal cortex, and other regions. Because the adrenal and pituitary secretions are reliably elicited by stressful events, they are called stress hormones.

It has been recognized for some time that the hippocampal steroid receptors are part of a control system that helps regulate how much adrenal steroid hormone is released.[43] When the hormone binds to receptors in the hippocampus, messages are sent to the hypothalamus to tell it to tell the pituitary and adrenal glands to slow down the release. In the face of stress, the amygdala keeps saying "release" and the hippocampus keeps saying "slow down." Through multiple cycles through these loops the concentration of the stress hormones in the blood is delicately matched to the demands of the stressful situation.

If stress persists too long, the hippocampus begins to falter in its ability to control the release of the stress hormones, and to perform its routine functions. Stressed rats are unable to learn and remember how to perform behavioral tasks that depend on the hippocampus.[44] For example, they fail to learn the location of the safe platform in the water maze task described in the last chapter. Stress also interferes with the ability to induce long-term potentiation in the hippocampus,[45] which probably explains why the memory failure occurs. Im-

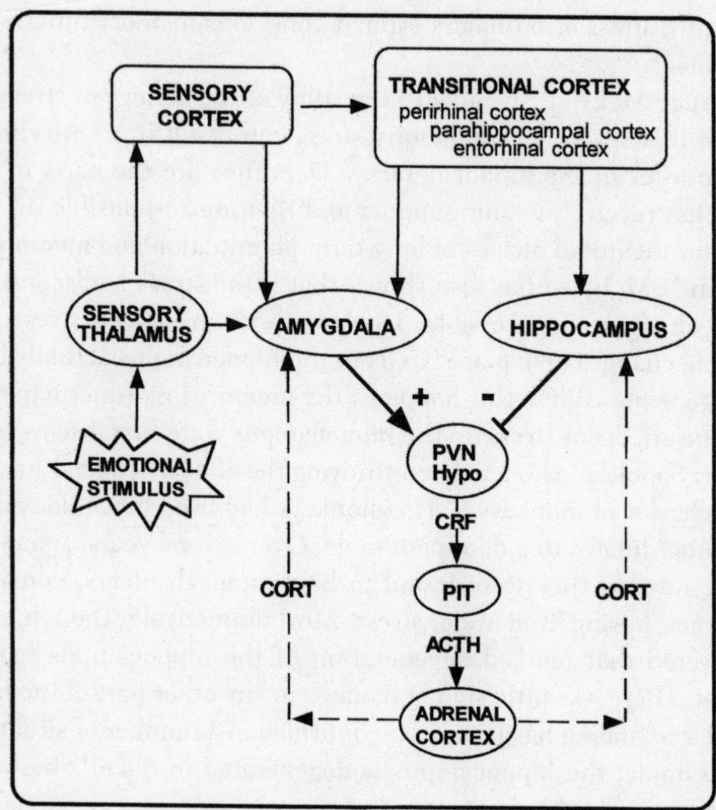

FIGURE 8-1
Stress Pathways.

Stimuli associated with danger activate the amygdala. By way of pathways from the amygdala to the paraventricular nucleus of the hypothalamus (PVN Hypo), corticotrophin-releasing factor (CRF) is sent to the pituitary gland, which, in turn, releases adrenocorticotropic hormone (ACTH) into the bloodstream. ACTH then acts on the adrenal cortex, causing it to release steroid hormones (CORT) into the bloodstream. CORT freely travels from the blood into the brain, where it binds to specialized receptors on neurons in regions of the hippocampus and amygdala, as well as other regions. Through the hippocampus, CORT inhibits the further release of CRF from the PVN. However, as long as the emotional stimulus is present, the amygdala will attempt to cause PVN to release CRF. The balance between the excitatory inputs (+) from the amygdala and the inhibitory inputs (-) from the hippocampus to PVN determines how much CRF, ACTH, and ultimately CORT will be released.

portantly, stress also impairs explicit conscious memory functions in humans.[46]

Bruce McEwen, a leader in the study of the biology of stress, has shown that severe but temporary stress can result in a shriveling up of dendrites in the hippocampus.[47] Dendrites are the parts of neurons that receive incoming inputs and that are responsible, in large part, for the initial phases of long-term potentiation and memory formation.[48] McEwen has also shown that if the stress is discontinued these changes are reversible. However, with prolonged stress, irreversible changes take place. Cells in the hippocampus actually begin to degenerate. When this happens, the memory loss is permanent.

The effects of stress on the hippocampus were first discovered by Robert Sapolsky, who had been studying the effects of social stress on the behavior of monkeys.[49] The monkeys had lived in a colony as social subordinates to a dominant male. Over several years, some died. Upon autopsy, they were found to have stomach ulcers, consistent with their having lived under stress. Most dramatically, though, it was discovered that marked degeneration of the hippocampus had occurred. There was little sign of damage to any other part of the brain. This basic finding has now been confirmed in a number of situations. For example, the hippocampus is degenerated in mice living under social stress.[50]

Recent studies have shown that the human hippocampus too is vulnerable to stress.[51] In survivors of trauma, like victims of repeated childhood abuse or Vietnam veterans with post-traumatic stress disorder, the hippocampus is shrunken. These same persons exhibit significant deficits in memory ability, without any loss in IQ or other cognitive functions. Stressful life events can alter the human hippocampus and its memory functions.

It seems clear that adrenal steroids account for these physical changes in the hippocampus and in the memory problems that result.[52] For example, there is a condition called Cushing's disease in which tumors develop in the adrenal gland and excess steroid hormone is secreted. These persons have long been known to have memory problems. Recent studies have also shown that the hippocampus is shrunken in this disease. Also, if rats or humans are injected with high levels of steroids, mimicking the effects of severe stress, hippocampal cell death and memory problems result. And if rats are

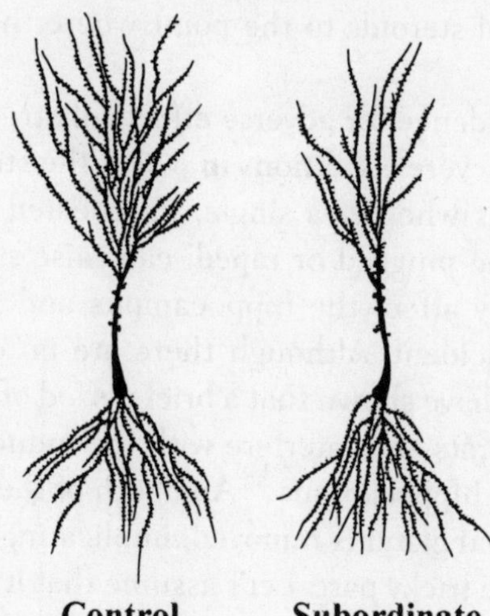

FIGURE 8-2
Dendrites Shriveled by Social Stress.

Neurons are shown from unstressed (control) and stressed (subordinate) tree shrews, a mammalian species related to early primate evolution. The stress in this experiment involved exposing subordinate males to a dominant male. Repeated social stress of this type reduced the branching and length of dendrites. Compare the top half of the cell from the unstressed control and from the stressed subordinate. (Reprinted from A.M. Magarinos, B.S. McEwen, G. Flugge, and E. Fuchs [1996], Chronic psychosocial stress causes apical dendritic atophy of hippocampal CA3 pyramidal neurons in subordinate tree shrews. *Journal of Neuroscience* 16 (3534–40.)

given drugs that block the effects of steroids, they are made immune to the effects of stress on the hippocampus and on memory.

There's one more relationship between stress and memory that's worth pointing out. One of the consequences of excess life stress is depression, and depressed persons sometimes have poor memory. It is quite possible that the memory disturbances that occur in depression are closely tied up with the effects of stress on the hippocampus.

Sometimes stress helps in the formation of explicit memories, making them stronger (recall the flashbulb hypothesis), but it can also devastate explicit memory. We now have a plausible explanation for this paradox. Memory is likely to be enhanced by mild stress, due to the facilitatory effects of adrenaline (Chapter 7), but may be interfered with if the stress is sufficiently intense and prolonged to raise the level of adrenal steroids to the point where the hippocampus is adversely affected.

Most of the evidence for adverse effects of stress on memory has come from rather severe conditions in which the stress continued for days. A key issue is whether a single, unrepeated traumatic experience, such as being mugged or raped, can raise steroid levels sufficiently to adversely affect the hippocampus and produce a loss of memory for the incident. Although there are no definitive answers yet, recent studies have shown that a brief period of stress can disrupt spatial memory in rats and interfere with the induction of long-term potentiation in the hippocampus.[53] And both of these effects are prevented if the adrenal gland is removed, implicating adrenal steroids.

Now comes the tricky part. Let's assume that it is indeed possible for a temporary period of trauma to lead to an amnesia for the experience. Can one then later recover a memory of these events? Although we can identify in a general sense the kinds of conditions under which recovery is possible or impossible, we can't say whether it occurred in a particular instance. For example, if the hippocampus was completely shut down by the stress to the point where it had no capacity to form a memory during the event,[54] then it will be impossible through any means to dredge up a conscious memory of the event. If no such memory was formed, then no such memory can be retrieved or recovered. On the other hand, if the hippocampus was only partially affected by the trauma, it may have participated in the formation of a weak and fragmented memory. In such a situation, it may be possible to mentally reconstruct aspects of the experience. Such memories will by necessity involve "filling in the blanks," and the accuracy of the memory will be a function of how much filling in was done and how critical the filled-in parts were to the essence of the memory.

Explicit, conscious memories, as I emphasized in the last chapter, are reconstructions that blend information stored in long-term memory with one's current frame of mind. Even memories that are formed with a perfectly well-functioning hippocampus are easily distorted by experiences that occur between the formation of the memory and its retrieval. This has been demonstrated in numerous experiments by Elizabeth Loftus and her colleagues.[55] Particularly important are their studies showing how easy it is to induce a false memory by controlling events that happen after the memory is established, or to create from scratch a memory of an experience that never happened. The subjects in these studies fully believe their memories, but because they have occurred in controlled laboratory experiments it is possible to show that the memory is fabricated. At the same time, there are also carefully controlled laboratory studies showing that information that was initially processed consciously and stored, but later forgotten, can be brought back, a phenomenon called hypermesia that we looked at in Chapter 3.[56]

The only thing that is clear about memory recovery in real life is that there is no way for outsiders to definitely determine whether a particular memory is real or fabricated in the absence of solid corroborating evidence (fabrication does not imply that the person is lying, only that the memory is false). There are surely victims of horrible incidents who have lost their memory of the event, and there may be some who can later piece together a memory of what happened. However, distinguishing between fabricated and real memories simply on the basis of self-knowledge can be tricky. Salvador Dali once said, "The difference between false memories and true ones is the same as for jewels: it is always the false ones that look the most real, the most brilliant."[57] Whether he was right might be debated, but as we saw earlier (Chapters 2 and 3), introspective knowledge of thought processes provides a highly inaccurate window into the mind, even in mundane (nontraumatic) situations. Things are likely to be even worse when confusion abounds, as it must during and following trauma. The waters of memory recovery are treacherous and should be walked through very carefully.

As far as is known, stress does not interfere with the workings of the amygdala, and, as we'll see below, stress may even enhance amygdala functions. It is thus completely possible that one might have poor conscious memory of a traumatic experience, but at the same time form very powerful implicit, unconscious emotional memories through amygdala-mediated fear conditioning. And because of other effects of stress to be described below, these potent unconscious fears can become very resistant to extinction. They can, in other words, become unconscious sources of intense anxiety that potentially exert their opaque and perverse influences throughout life. However, there is no way for these powerful implicit memories to then be converted into explicit memories. Again, if a conscious memory wasn't formed, it can't be recovered.

That Freud was correct in his belief that aspects of traumatic experiences are sometimes stored in memory systems that are not directly accessible from consciousness seems clear. Less certain is whether repression (in the Freudian sense) is involved. The failure to remember traumatic events may sometimes be due to a stress-induced shutdown of the hippocampus, although this remains to be proven. In light of this, though, there is nothing particularly devastating to the conditioning theory of anxiety about the fact that the traumatic origin of the anxiety is not always remembered. Of course, repression of unpleasant experience may well be a real phenomenon, one that we still don't understand scientifically. And some anxiety disorders may develop without an initial trauma. Nevertheless, we at least have a possible mechanism that might account for some aspects of these disorders in easily understood biological terms.

Amplification of Emotional Memory by Irrelevant Stressors: There is a flip side to the debilitating effects of intense stress on explicit conscious memory of trauma. The same amount of stress that can lead to an amnesia for a trauma may amplify implicit or unconscious memories that are formed during the traumatic event.

For example, recent studies have shown that if rats are given injections of adrenal steroids at levels that mimic very severe stress, there is a dramatic decrease in the amount of a certain chemical, called corticotropin-releasing factor (CRF), in the part of the hypothalamus that controls the release of the stress hormone, ACTH, from the pituitary gland.[58] CRF is in fact the neurotransmitter that stimulates ACTH release. The decrease in CRF in this pathway reflects the negative feedback control over stress hormones by the hippocampus—once the blood level of adrenal steroids reaches a certain level, the hippocampus tells the hypothalamus to slow down the secretions. And when the steroid level reaches a critical point, the hippocampal circuits begin to falter. In stark contrast, there is a dramatic increase in CRF in the central nucleus of the amygdala under the same conditions—as blood levels of steroids increase, the amygdala may keep getting more and more active. The bottom line is that the effects of stress on the amygdala seem very different from the effects on the hippocampal-hypothalamic circuit.

On the basis of these observations, Keith Corodimas, Jay Schulkin, and I predicted that during intense stress the learning and memory processes mediated by the amygdala might be facilitated and we examined the effects of stress hormone overload on conditioned fear behavior.[59] In line with the prediction, we found that the strength of learned fear was increased in the steroid-treated rats relative to other rats that didn't have the steroids. Although this result is somewhat preliminary, studies using other forms of Pavlovian conditioning have also found that stress enhances conditioned responses.[60]

If indeed the hippocampus is impaired and the amygdala facilitated by stress, it would suggest the possibility that stress shifts us into a mode of operation in which we react to danger rather than think about it. It's not clear whether this is a specific adaptation or whether we're just lucky that when the higher functions break down our fallback position is one in which we can let evolution do the thinking for us.

The finding that stress hormones can amplify conditioned fear responses has an important implication for our understanding of anxiety disorders, and in particular for understanding why these sometimes seem to occur or get worse after unrelated stressful events.[61] During stress, weak conditioned fear responses may become stronger. The responses could be weak either because they were weakly conditioned, or because they were previously extinguished or were otherwise treated into remission. Either way, their strength might be increased by stress. For example, a snake phobic might be in remission for years but upon the death of his spouse the phobia returns. Alternatively, a mild fear of heights, one that causes few problems in everyday life, might be converted into a pathological fear under the amplifying influences of stress. The stress is unrelated to the disorder that develops and is instead a condition that lowers the threshold for an anxiety disorder, making the individual vulnerable to anxiety, but not dictating the nature of the disorder that will emerge. The latter is probably determined by the kinds of fears and other vulnerabilities that the person has lurking inside.

Brain Malfunctions Can Make Unprepared Learning Resistant to Extinction: Neurotic fears are notoriously difficult to shake. This is the bane of a therapist's professional existence, but also his or her bread and butter. While preparedness provides one way out of this dilemma, there is another. Fear responses conditioned to arbitrary tones or lights in rats can be made highly resistant to extinction if certain cortical areas that project to the amygdala are damaged. This suggests that these areas of the cortex may be malfunctioning in some cases of pathogenic anxiety, allowing ordinary stimuli to be conditioned by the amygdala in a way that resists extinction.

Several years ago we were examining the effects of damage to visual areas of the cortex on the ability of rats to be conditioned to visual stimuli.[62] The lesioned rats learned just fine, supporting our contention that there are subcortical pathways that take sensory information to the amygdala during conditioning. But when we tried to extinguish the fear responses in these animals, something unusual happened. We couldn't do it. Normal rats, after several days of seeing the light without the shock, stopped acting afraid in the presence of the light. But the rats with lesions of the visual cortex were like Energizer batteries—they just kept going and going and going.

We never thought that the visual cortex was the seat of extinction. Instead, we proposed that the visual cortex might be a necessary link between the visual world and other higher order cortical areas that are necessary for extinction. One area that seemed like a possible regulator of extinction was the medial prefrontal cortex. This area receives signals from the sensory regions of the cortex and from the amygdala, and sends connections back to the amygdala, as well as to many of the areas to which the amygdala projects.[63] The medial prefrontal cortex is thus nicely situated to be able to regulate the outputs of the amygdala on the basis of events in the outside world as well as on the basis of the amygdala's interpretation of those events. When Maria Morgan made lesions of this region, rats continued to act fearful in the presence of a conditioned fear stimulus long after rats without lesions of this area had stopped acting afraid.[64]

The amygdala of the cortically lesioned rat, like the neurotic human, stubbornly expresses its fear memories in the face of information showing that the stimulus is no longer associated with danger. Extinction appears to involve the cortical regulation over the amygdala, and even unprepared conditioned fear can be resistant to extinction when the amygdala is freed from these cortical controls.

One of the hallmarks of frontal lobe damage in humans is perseveration, the inability to stop doing something once it is no longer appropriate.[65] For example, when frontal lobe patients are performing a task in which a rule must be followed, they have great difficulty in changing their behavior when the rule is switched. In a standard version of this test, the patient is given a stack of cards, each with one or more colored symbols on it. The patient's job is to figure out, on the basis of feedback about whether each response is correct, which kind of cue (color, shape, or number) is the current solution. Once they get going on a principle (like shape) they can do the task fine. But if all of a sudden the principle shifts (say, to color), they keep following the old rule. Sometimes they even know what they should do, but can't make their behavior match their knowledge. They are rigid and inflexible, and perseverate in their ways, even when it is obvious that the behavior is not appropriate to the situation. This seems to characterize their behavior in real life as well.

Although perseveration is usually thought of as a cognitive or thought disorder, it seems that our findings about fear extinction in rats with prefrontal lesions might reflect the same kind of difficulty, but in the domain of emotion. In fact, we used the expression "emotional perseveration" to describe the failure of our rats to extinguish conditioned fear responses.[66] While cognitive perseveration is produced by damage to the lateral areas of the prefrontal cortex, emotional perseveration resulted from damage to a small part of the medial prefrontal region.[67] The lateral and medial prefrontal areas may perform the same operation, adapting behavior to changing conditions, with the involvement in cognitive or emotional functions determined by the areas with which the prefrontal region works in conjunction. The medial cortex, in other words, engages in response switching behavior because it is part of the prefrontal cortex, and it engages in response switching guided by emotional information because it is connected with the amygdala. Edmund Rolls has proposed a similar role for the medial prefrontal cortex in emotion on the basis of studies in which he has recorded from neurons in this region while monkeys performed tasks where the reinforcer (reward or punishment) associated with certain responses changed frequently.[68] Other ideas about the contribution of prefrontal cortex to emotion have been proposed as well, and the work of Antonio Damasio is particularly notable.[69] Some of these ideas will be considered in the next chapter on emotional consciousness.

The prefrontal cortex, like the hippocampus, may be altered by stress. Recent research has shown that the prefrontal cortex, like the hippocampus, offers a counterforce that keeps too much of the stress hormones from being released.[70] Since prolonged stress results in a breakdown in this negative feedback control function, it may be the case that both the prefrontal cortex and hippocampus are adversely affected. A stress-induced shutdown of the prefrontal cortex might release the brakes on the amygdala, making new learning stronger and more resistant to extinction, and possibly allowing previously extinguished conditioned fears to be expressed anew.

Just because clinical fear is difficult to extinguish does not mean that it involves a different brain system from the one that mediates extinguishable conditioned fears in animals. Differences in the ease of extinction of conditioned fear in laboratory experiments and in anxious persons are more likely to reflect differences in the way the fear system works in normal and anxious brains rather than differences in the system used by the brain to learn conditioned fear and clinical anxiety. This doesn't mean that anxious persons, like our rats, are walking around with holes in their prefrontal cortex. There are many subtle ways in which disruptions in electrical and chemical functions can adversely affect a brain region, with lesions being just an extreme example of this.

Gone but Not Forgotten—The Indelibility of Emotional Memory: Our finding that when the medial prefrontal cortex is damaged routine fear conditioning becomes resistant to extinction has another important implication. It also suggests that extinction prevents the expression of conditioned fear responses but does not erase the implicit memories that underlie these responses.[71] Extinction, in other words, involves the cortical control over the amygdala's output rather than a wiping clean of the amygdala's memory slate.

The idea that extinction does not involve the erasure of emotional memories but instead prevents their expression is consistent with a number of findings about conditioned responses.[72] Pavlov, for example, found that extinguished responses would, with simply the passage of time, *spontaneously recover.* It is also known that if a rat is conditioned by pairing a tone and shock in one box, and the fear response elicited by the tone is completely extinguished in another box, the conditioned response elicited by the tone will be *renewed* if the rat is returned to the original training box. An extinguished response can also be *reinstated* by giving the rat an exposure to the US or, importantly, to other forms of stressful stimulation. Stress, in other words, can bring back extinguished, or perhaps weakly established but unextinguished, conditioned responses.[73] Each of these examples, like our lesion study, demonstrates that emotional memories are not erased by extinction but are simply held in check. Extinguished memories, like Lazarus, can be called back to life.

I recently had a scientific "ah ha" experience, one of those rare, wonderful moments when a new set of findings from the lab suddenly makes you see something puzzling in a new, crystal clear way. The studies involved recordings of electrical activity of the amygdala before and after fear conditioning by Greg Quirk, Chris Repa, and me.[74] We found dramatic increases in electrical responses elicited by the tone CS after conditioning, and these increases were reversed by extinction. However, because we were recording from multiple individual neurons at the same time, we were also able to look at the activity relationships between the cells. Conditioning increased the functional interactions between neurons so that the likelihood that two cells would fire at the same time dramatically increased. These interactions were seen both in the response to the stimulus and in the spontaneous firing of the cells when nothing in particular was going on. What was most interesting was that in some of the cells, these functional interactions were not reversed by extinction. Conditioning appears to have created what Donald Hebb called "cell assemblies,"[75] and some of these seemed to be resistant to extinction. Although the tone was no longer causing the cells to fire (they had extinguished), the functional interactions between the cells, as seen in their spontaneous firings, remained. It is as if these functional couplings are holding the memory even at a time when the external triggers of the memory (for example, phobic stimuli) are no longer effective in activating the memory and its associated behaviors (for example, phobic responses). Although highly speculative at this point, the observations suggest clues as to how memories can live in the brain at a time when they are not accessible by external stimuli (Figure 8-3). All that it would take to reactivate those memories would be a change in the strength of the input to the cell assembly. This may be something that stress can accomplish.

Unconscious fear memories established through the amygdala appear to be indelibly burned into the brain. They are probably with us for life. This is often very useful, especially in a stable, unchanging world, since we don't want to have to learn about the same kinds of dangers over and over again. But the downside is that sometimes the things that are imprinted in the amygdala's circuits are maladaptive. In these instances, we pay dearly for the incredible efficiencies of the fear system.

Psychiatrist Roger Pitman has astutely noted that findings from studies of fear conditioning in rats have important implications for how anxiety is treated.[76] The classic treatment, based on Mowrer's and Miller's theory, was to force the patient to be exposed to the anxiety-causing stimuli without allowing any avoidance or escape behavior and thereby try to extinguish the anxiety that the stimuli elicit. But in light of the indelibility of the amygdala's hold on traumatic memories, he suggests a bleaker, though perhaps more realistic, assessment. We may not be able to get rid of the implicit memories that underlie anxiety disorders. If this is the case, the best we can hope for is to exercise control over them.

The Fear System and Specific Anxiety Disorders

Until fairly recently, the various anxiety disorders were not distinguished and were not treated differently.[77] Panic and PTSD, for example, did not appear in the DSM until 1980. And although phobias have long been associated with neuroses, they were typically thought of as neurotic symptoms rather than a particular kind of anxiety disorder. With the emergence of clear diagnostic distinctions between different anxiety disorders, disorder-specific fear conditioning theories have been proposed. Below, I'll attempt to buttress disorder-specific theories of phobias, PTSD, and panic with findings about the brain mechanisms of fear conditioning.[78]

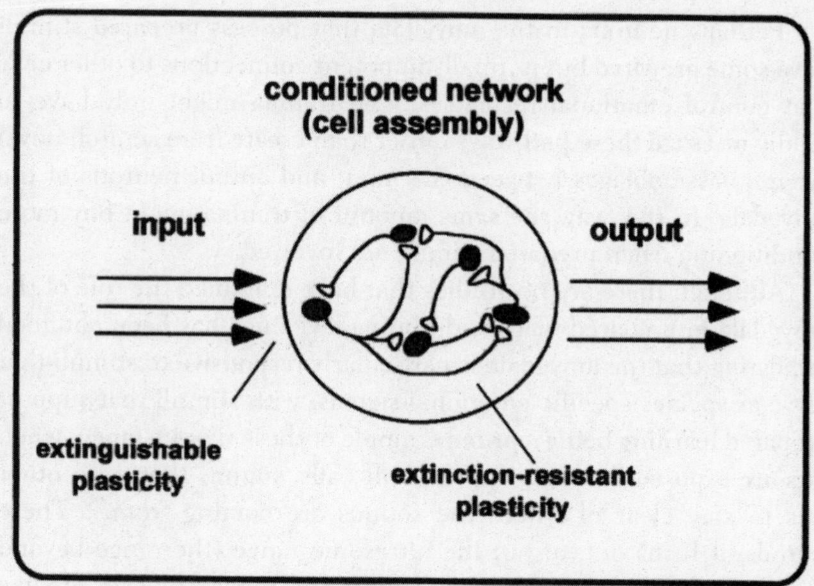

FIGURE 8-3

Creation of Extinction-Resistant Learning in the Brain.

Recent studies have recorded neural activity in the amygdala during conditioning and extinction. After conditioning, the response of individual cells to the conditioned stimulus is increased (the same input produces a bigger output). In addition, individual cells develop stronger interconnections so that when one fires the others also fire. These interconnected neurons are called a cell assembly. While the response of individual cells to the conditioned stimulus diminishes during extinction, in some cases the conditioned interconnections persist. These cell assemblies within the amygdala, or between the amygdala and cortical areas, may constitute an important aspect of the long-term, extinction-resistant, implicit memory created by fear conditioning.

Phobic Fears: Contemporary ideas about phobias continue to be centered around the notion of preparedness. Normally, the strength of conditioning is determined mainly (though not exclusively) by how traumatic the unconditioned stimulus is. But in prepared fear conditioning, the CS also contributes some of the emotional impact. As a result, given two conditioned stimuli, one biologically prepared to be conditioned to danger and the other not, the same unconditioned stimulus should support the establishment of a stronger conditioned response for the prepared stimulus. How might that work in the brain?

Perhaps neurons in the amygdala that process prepared stimuli have some prewired but normally impotent connections to other cells that control emotional responses. The trauma might only have to mildly massage these pathways rather than create from scratch novel synaptic assemblages between the input and output neurons of the amygdala. In this way, the same amount of trauma might buy more conditioning when prepared stimuli are involved.

Although there are no studies that have examined the role of the amygdala in prepared fear conditioning, evidence has been obtained indicating that the amygdala is particularly responsive to stimuli that serve as species-specific emotional signals, with stimuli that support prepared learning being a prime example of these. For instance, when rats are exposed to a cat, they give off calls, sounds that warn other rats to stay clear of where the sounds are coming from.[79] These sounds, it turns out, are in the ultrasonic range (the range beyond human hearing). Since cats can't hear in this range, the calls are like secret encrypted messages that pass undetected through enemy lines. In recent experiments, Fabio Bordi and I found some neurons in the rat amygdala that responded especially briskly to ultrasounds similar to the warning calls.[80] The rat amygdala may be evolutionarily prepared to respond to these sounds and to learn about them. In fact, the amygdala of all creatures may be prepared to respond to species-relevant cues.[81] For example, faces are important emotional signals in the lives of primates, and neurons in the monkey amygdala respond briskly to the sight of monkey faces.[82]

As we saw in Chapter 6, information about external stimuli reaches the amygdala from two pathways, one subcortical, the other cortical. The subcortical pathway is shorter and faster but imprecise, and the cortical pathway has the opposite attributes. And as we saw in Chapter 7, learning and memory appear to involve the potentiation of synaptic transmission in these pathways. In the normal brain the potentiation probably occurs in both pathways, which work together in the conditioning and expression of fear responses to external stimuli. But suppose, because of genetic predisposition or past experiences, phobic learning were to involve the subcortical pathway to a greater extent than the cortical pathway, especially for prepared stimuli. This might explain why phobias generalize broadly—as Öhman has pointed out, phobics can sometimes lose track of what they are afraid of when fear generalizes.[83] The subcortical pathway, not being very capable of making fine distinctions, may produce learning that more freely spreads to other stimuli. And this pathway, being subcortical, would also presumably be particularly difficult to gain conscious, cortical control over. Interestingly, the high-frequency sounds that turned on amygdala cells so effectively did so through the quick and dirty subcortical pathways.

Although amygdala-mediated fear conditioning is a form of implicit learning (regardless of the input pathways involved), phobics are consciously afraid of their phobic stimuli. This means that they also have an explicit conscious memory, formed through their temporal lobe memory system, that reminds them that they are afraid of snakes, heights, or whatever. This memory might be established during the initial traumatic learning situation, but some phobics do not recall such a learning experience, possibly because of a stress-induced memory loss. In such instances, the conscious memory of being phobically afraid could be established in later experiences with the phobic object. When the object is encountered, the amygdala will unconsciously detect the stimulus and produce the bodily expression of fear. Upon becoming aware of this bodily response, the person attributes (à la Schachter and Singer) the arousal to the most likely object and forms the memory that they are afraid of objects of that type. In the case of standard phobic objects (snakes, spiders, heights), these phobic attributions are probably facilitated by the fact that the person knows that people are often afraid of these things. Once this explicit memory is created, its retrieval into consciousness becomes a potent stimulus that is itself capable of activating the amygdala and producing anxiety by way of connections from cortical areas (including the hippocampus) to the amygdala. Even if one does not have a conscious memory of the initial learning, there is likely to be an awareness of the phobic condition stored in explicit memory.

Not everyone exposed to a traumatic event develops a phobia. Some people's brains, because of their genetic makeup or past experiences, must be predisposed to react to traumatic learning experiences in this particular way. In these people, the amygdala may be supersensitive to some class of prepared stimuli or the amygdala may have other alterations that make fear conditioning especially potent. On the other hand, as we've seen, changes in the frontal lobe may predispose some people to develop fears that resist extinction, even when unprepared stimuli are involved.

Traumatic Stress: PTSD was once known as shell shock, battle fatigue, or war neurosis, because it was most commonly diagnosed in war veterans.[84] Although it occurs in victims of many kinds of trauma, the following quotation from a Vietnam veteran illustrates the phenomenon:

> I can't get the memories out of my mind! The images come flooding back in vivid detail, triggered by the most inconsequential things, like a door slamming or the smell of stir-fried pork. Last night I went to bed, was having a good sleep for a change. Then . . . there was a bolt of crackling thunder. I awoke instantly, frozen in fear. I am right back in Vietnam. . . . My hands are freezing, yet sweat pours from my entire body. I feel each hair on the back of my neck standing on end. I can't catch my breath and my heart is pounding. . . . The next clap of thunder makes me jump so much that I fall to the floor. . . .[85]

The similarity between disorders of this type and laboratory-conditioned fear has not escaped the notice of psychiatrists. Conditioned fear was in fact proposed as the explanation of war neuroses in veterans of World War I.[86] Two of the most noted contemporary psychiatrists who study PTSD are Dennis Charney of Yale and Roger Pitman of Harvard, both of whom champion the notion that fear conditioning is involved in the disorder.[87]

The difference between a fear conditioning theory of phobia and PTSD is one of where the conditioning process gets its strength. In the case of prepared phobic learning, the conditioned stimulus makes the learning especially strong. The unconditioned stimulus is typically unpleasant and may even be painful, but is not necessarily extraordinary. However, in the case of PTSD, the conditioned stimulus events are less notable than the unconditioned stimulus. PTSD, in fact, is defined in DMS-III-R as involving a trauma that is far outside the realm of experiences in ordinary life.

Once we assume that the trauma in PTSD is an extraordinary event, an especially potent US, a fairly standard view of the way the amygdala mediates conditioned fear provides a plausible account of this disorder. Admittedly, we don't know exactly what combination of factors come together to make up the horrendous US at the neuronal level, but we can easily imagine that such a neural condition exists, one that bombards the amygdala with electrical and chemical signals that are particularly potent as reinforcers of Pavlovian conditioning. These powerful reinforcing stimuli are then linked synaptically with the sounds, sights, and smells of the battle, which also reach the amygdala. Later, the occurrence of these same conditioned stimuli, or stimuli related to them, elicit profound fear responses by reactivating these powerfully potentiated amygdala circuits.

Conditioned stimuli activate the amygdala unconsciously, but at the same time reach the temporal lobe memory system and can lead to the recall of the initial trauma or to the recall of recent episodes in which the initial trauma is relieved. These conscious memories, together with the awareness of now being in a state of strong emotional arousal (due to the unconscious activation of fear responses through the amygdala), then gives rise to conscious anxiety and worry. These cognitions about the emotional arousal, in turn, flow from the neocortex and hippocampus to further arouse the amygdala. And the bodily expression of the amygdala's responses keeps the cortex aware that emotional arousal is ongoing, and further facilitates the anxious thoughts and memories. The brain enters into a vicious cycle of emotional and cognitive excitement and, like a runaway train, just keeps picking up speed.

It is possible that in PTSD, as proposed for phobic learning, the direct projections to the amygdala from subcortical sensory processing regions are involved. If this were so, it would explain why the attacks are so impulsive and uncontrollable, and tend to generalize so readily (from gunshots to lightning to slamming doors). As we've seen, the subcortical pathways are quick and dirty transmission routes. They turn the amygdala on and start emotional reactions before the cortex has a chance to figure out what it is that is being reacted to. And since these pathways are not very capable of distinguishing between stimuli, generalization readily occurs (a slamming door may indeed not sound very different from a gunshot to this circuit). Perhaps trauma, for some reasons (genetic or experiential) in some persons, biases the brain in such a way that the thalamic pathways to the amygdala predominate over the cortical ones, allowing these low-level processing networks to take the lead in the learning and storage of information. Later exposure to stimuli that even remotely resemble those occurring during the trauma would then pass, like greased lightning, over the potentiated pathways to the amygdala, unleashing the fear reaction. Quite possibly, it is harder for one to gain conscious willful control over these subcortical pathways. At the same time, because conscious memories are formed during anxiety attacks, the bodily sensations associated with those attacks, when recognized consciously, become potent elicitors or at least facilitators of anxiety. Next, we'll see just how bodily sensations can drive anxiety in panic disorders, which often occur in conjunction with PTSD.

Panic: Panic attacks are the most commonly diagnosed anxiety disorder.[88] They are similar to phobic and PTSD reactions in the sense that the patient suffers from strong emotional arousal, including intense activation of the sympathetic nervous system. However, while phobic and PTSD responses occur in the presence of external stimuli, panic attack appears to be more related to internal stimuli.[89] And because panic involves internal events, it is especially difficult for the person to avoid the stimuli that bring it on. Panic patients thus differ in this respect from patients with PTSD and phobia, who engage in extensive avoidance behavior.[90]

A panic attack can be induced by having the patient hyperventilate or inhale a gaseous mixture rich in carbon dioxide, or giving the patient an intravenous injection of sodium lactate.[91] These procedures give rise to internal signals (bodily sensations) similar to those that are typically present during a naturally occurring attack. Panic can also be induced by the provision of false feedback about the rate at which the heart is beating, making the patient believe that heightened bodily arousal is occurring when it is not.[92] The belief that panic is occurring may be an important link in the chain of events that tie together the occurrence of bodily sensations and full-blown panic.

There are a number of theories of why panic occurs, including biological explanations (e.g., supersensitivity to carbon dioxide) and psychological ones (e.g., a history of childhood separation anxiety).[93] I will make no attempt to review or evaluate the various theories here. My aim instead is to discuss one theory, the conditioning theory, and to consider how it might be implemented in the brains of panic patients.

One common view is that artificial panic induction procedures lead to bodily sensations that then serve as conditioned stimuli.[94] Having experienced panic before, the patient learns the warning signs. When these internal signals occur (even when artificially induced) the patient feels that panic is starting.[95] This cognitive appraisal of bodily sensations then drives the system into panic. Induced panic, and presumably natural panic, by this way of thinking, is a conditioned response to internal stimuli that occurred during past panic attacks. It has even been argued that these internal sensations might be prepared stimuli, thus further linking panic and phobia and their underlying mechanism.[96] Support for the preparedness of such internal stimuli comes from Donald Klein's theory that panic represents the activation of an evolutionarily old suffocation alarm system.[97]

The most complete conditioning theory of panic has been developed by Wolpe.[98] He has argued that the first panic attack is the result of experiencing the consequences of hyperventilation, which increases the carbon dioxide in the lungs and blood and results in a variety of unpleasant bodily sensations (dizziness, racing heart, the feeling of suffocation). The hyperventilation can arise for a variety of reasons. Certain drugs like cocaine, amphetamine, or LSD, or exposure to toxic chemicals in the workplace, can be the cause. However, according to Wolpe, most often panic occurs in persons who are particularly anxious and worried and who have been under a lot of stress. One study cited by Wolpe found that severe marital conflict occurred during the year before the first panic attack in 84 percent of the patients surveyed, emphasizing again that cognitive factors can lift anxiety over the threshold.

According to Wolpe, the cause of the first panic is not important. It can be organic or psychological. Regardless, once panic occurs, the stimuli that happen to be present at the time will become conditioned fear stimuli. But unlike typical fear conditioning situations, the critical stimuli are internal rather than external. For example, an elevation of blood pressure that occurs in response to hyperventilation might become a conditioned fear stimulus. If blood pressure happens to increase for some other reason, such as talking to a superior or being in some other socially tense situation, the noxious sensations previously elicited by hyperventilation, having been conditioned to increases in blood pressure levels, are now brought on. These sensations are then noticed and interpreted as indicative of the onset of a panic attack. In contrast, the CS (elevation of blood pressure) is not easily noticed (high blood pressure is in fact sometimes called the "silent killer"), and the panic appears to be spontaneous. External stimuli can also become conditioned panic stimuli. If the first panic occurred in a car, then being in cars may make it more likely that panic will occur there. Nevertheless, in Wolpe's model, the internal stimuli play the leading role.

Let's now consider the sequence of events by which the amygdala might participate in conditioned panic. There are neurons in the lower brain stem that are very sensitive to changes in blood level of carbon dioxide.[99] The amygdala, it turns out, receives inputs from the neurons in this region.[100] The amygdala also receives information about the status of the internal organs—the rate at which the heart is beating, the level of blood pressure, and other vital statistics from the inner core of the body.[101] By integrating these internal signals about the state of bodily organs (the conditioned stimuli) with information about the level of carbon dioxide in the blood (the unconditioned stimulus), the amygdala could form synaptic linkages between the co-occurring events, allowing the internal signals to substitute for the carbon dioxide effects in producing a profound activation of the sympathetic nervous system through the outputs of the amygdala. Once the sympathetic nervous system is activated in this way, the person becomes aware of the bodily arousal and is reminded, through explicit memory, that the symptoms being experienced tend to occur in panic attacks, suggesting that one might be starting. These conscious memories and thoughts about the possibility of panic might, then, by way of projections to the amygdala from the hippocampus and neocortex, lead to further and continued activation of the sympathetic nervous system, and to the build-up of a full-blown panic attack. Alternatively, in the case of false feedback about the status of heart rate or other bodily functions, the chain of events probably starts with cortical cognitions (for example, the belief that the heart is beating fast), which then serve as retrieval cues for explicit memories of past experiences in which fast heart beating occurred (past panic attacks).

These conscious thoughts and explicit memories, again by way of connections from neocortical areas and the hippocampus to the amygdala, then trigger the amygdala and its sympathetic outflow as before.

These neuro-scenarios, of course, are hypothetical, as there has not been any research on the role of the amygdala in panic. However, while the contribution of these circuits to human panic disorder is hypothetical, the circuits and their functions are real and it is quite conceivable that they might contribute to panic in the way described.

Bad Habits and Anxious Thoughts

The avoidance responses that so typify anxiety disorders fall somewhere between what I described earlier as innate emotional *reactions* and voluntary emotional *actions*. Avoidance responses are instrumental responses that are learned because they are reinforced. They are then performed habitually, which is to say automatically, when the appropriate stimuli occur. But unlike innate responses, avoidance responses are more or less arbitrarily related to danger. Innate emotional reactions occur when the amygdala is turned on (by innate or learned triggers) because the response is hardwired to the amygdala. In contrast, for avoidance, the brain has learned some response that can be performed in the presence of a learned trigger that short-circuits the innate response. For example, initially rats freeze when they hear a sound that predicts a shock. With time, they may learn to jump up at just the right moment during the sound to avoid the shock, or to jump over a barrier during the sound, or to turn a wheel to inactivate the shock. These responses, once learned, prevent emotional arousal. They are performed automatically, without conscious decision. They become habits, ways of automatically responding to stimuli that routinely warn of danger. Like conditioned fear responses, they are performed automatically, but they are learned rather than innate responses.

Emotional habits can be very useful. If you find out that going to a certain water hole is likely to put you face to face with a bloodthirsty predator, then the best thing to do is to avoid going there. But if you stop going to water holes because you become anxious whenever you begin to look for water, or you start drinking less water than you need to maintain your health whenever you do get around to drinking, then your avoidance response has become detrimental to routine life. You have an anxiety disorder.

The automatic nature of emotional habits can be extremely useful, allowing you to avoid routine dangers without having to give them much thought. However, when emotional habits become anxiety disorders, then the rigid unextinguishable learning that typifies avoidance behavior becomes a liability.

Many of the leading drugs for treating anxiety have been developed because of their efficiency in reducing avoidance behavior in animals. For example, if a rat is shocked when it steps off a platform in a test chamber, it will remain on the platform when it is placed in the chamber the next day. However, if the rat gets a shot of Valium just before being placed on the platform on the second day, it will be much more likely to step off the platform to figure out if the danger still exists. In other words, the rat is less fearful, less anxious, about the situation when it receives the drug.

As Mowrer and Miller proposed, avoidance learning is usually thought of as taking place in two stages. First, fear conditioning occurs. Then, a response is learned because it supposedly reduces the learned fear. We know that the amygdala is required for the fear conditioning part, but the brain mechanisms involved in the instrumental avoidance response are less clearly understood. It seems that structures like the basal ganglia, frontal cortex, and hippocampus may be involved.[102] There is controversy as to just where in the brain drugs like Valium have their anxiety-reducing effects.[103] In fact, however, they probably act in a number of places.

Let's consider how a drug like Valium might work in the amygdala. Valium belongs to the class of drugs known as benzodiazepines. These drugs have natural receptors in the brain. When you take Valium, it binds to the benzodiazepine receptors all over the brain. These receptors do a very specific thing. They facilitate the effects of the inhibitory neurotransmitter, GABA. So you basically increase inhibition in a variety of brain areas. In some brain regions, this will not have any consequence for anxiety because that region is not involved in that function. Basically, if a brain region is involved in anxiety, whatever it does during anxiety-provoking situations, it will probably do less of it in the presence of Valium. For example, the lateral nucleus is the sensory-input region of the amygdala. The increase of inhibition in this region will raise the threshold for anxiety. Stimuli that would normally elicit fearful responses through the amygdala no longer do so (see Figure 8-4). Jeffrey Gray has proposed that the antianxiety drugs work through the hippocampus (albeit indirectly).[104] This may be true as well, reducing the ability of explicit memories to make us anxious and afraid.

The brain circuits of avoidance are far less clear than the circuits of fear conditioning. Avoidance is more complex: it involves fear conditioning plus instrumental learning. Also, there are many ways in which avoidance conditioning studies can be performed and a great variety of responses can be conditioned this way. Avoidance responses are arbitrary. Anything that reduces the exposure to fear-eliciting events can be an avoidance response. These factors make the brain systems of avoidance more difficult to track down. However, now that we have a good handle on the brain mechanisms involved in the first phase of avoidance learning (the fear conditioning phase), we can more wisely approach the second phase.

Psychotherapy: Just Another Way to Rewire the Brain

Freud's psychoanalytic theory and the various conditioning theories all assume that anxiety is the result of traumatic learning experiences that foster the establishment of anxiety-producing long-term memories. In this sense, psychoanalytic and conditioning theories have drawn similar conclusions about the origins of anxiety. However, the two kinds of theories lead to different therapeutic approaches. Psychoanalysis seeks to help make the patient conscious of the origins of inner conflict, whereas behavior therapy, the name given to therapies inspired by conditioning theories, tries to rid the person of the symptoms of anxiety, often through various forms of extinction therapy. There is a good deal of debate about the best treatment strategy: psychoanalysis, behavioral therapy, or most recently cognitive therapy.[105] However, extinction therapies, either alone or in combination with other approaches, are commonly recommended for many anxiety disorders.[106]

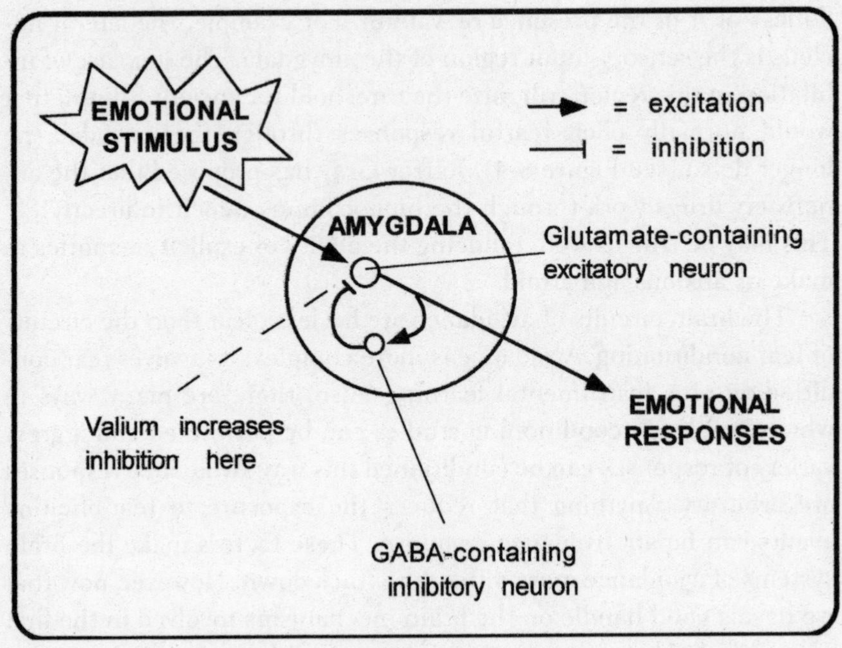

FIGURE 8-4
One Way Valium Might Reduce Fear and Anxiety.

Valium and some other antianxiety drugs act by increasing the ability of inhibitory neurons to prevent excitatory transmission. When we are under the influence of Valium, external emotional stimuli (as well as thoughts) are less capable of producing emotional responses, in part (perhaps) because of an action on GABA inhibitory neurons in the amygdala.

The prototypical extinction therapy pioneered by Wolpe starts off with relaxation training.[107] Once the patient learns to feel comfortable in the therapeutic setting, he or she is asked to produce emotional images, starting with less frightening images and working toward more frightening ones. This is called systematic desensitization. The desensitization can then move from images to real objects and situations that cause anxiety, again starting with the least and moving toward the more frightening. Erdelyi interpreted systematic desensitization in the language of conditioning: present the CS in degrees until the conditioned emotional responses drop out.[108] The CS comes to be associated with a new US, safety, and the new conditioned response is no response. Erdelyi suggests that the standard techniques of psychoanalytic cathartic therapy (hypnotic induction, lying on a couch, trust in the therapist, image production) may accomplish the same thing as Wolpian therapy: extinction of the learned emotional reaction.

Figuring out the brain mechanisms of extinction is obviously going to be an important part of understanding how therapy works. As we've seen, extinction appears to involve interactions between the medial prefrontal cortex and the amygdala. And work by Michael Davis has shown that extinction occurs through the same kind of synaptic mechanism that conditioning does: NMDA-dependent synaptic plasticity in the amygdala.[109] When NMDA receptors are blocked, it may be that the amygdala can't learn what the prefrontal cortex is trying to teach it—to inhibit a particular emotional memory.

These observations give us a different kind of understanding of therapy. Therapy is just another way of creating synaptic potentiation in brain pathways that control the amygdala. The amygdala's emotional memories, as we've seen, are indelibly burned into its circuits. The best we can hope to do is to regulate their expression. And the way we do this is by getting the cortex to control the amygdala.

Behavior (extinction) therapy and psychoanalysis have the same goal—help the person with their problem. In both cases, the effects may be achieved by helping the cortex gain control over the amygdala. However, the neural roads taken may be different. Extinction therapy may take place through a form of implicit learning involving the prefrontal-amygdala circuit, whereas psychoanalysis, with emphasis on conscious insight and conscious appraisals, may involve the control of the amygdala by explicit knowledge through the temporal lobe memory system and other cortical areas involved in conscious awareness (see Chapter 9). Interestingly, it is well known that the connections from the cortical areas to the amygdala are far weaker than the connections from the amygdala to the cortex.[110] This may explain why it is so easy for emotional information to invade our conscious thoughts, but so hard for us to gain conscious control over our emotions. Psychoanalysis may be such a prolonged process because of this asymmetry in connections between the cortex and amygdala.

(No) Thanks for the Memories

The ability to rapidly form memories of stimuli associated with danger, to hold on to them for long periods of time (perhaps eternally), and use them automatically when similar situations occur in the future is one of the brain's most powerful and efficient learning and memory functions. But this incredible luxury is costly. We sometimes, perhaps all too often, develop fears and anxieties about things that we would just as well not have. What is so useful about being afraid of heights or elevators or certain foods or means of travel? While there are risks associated with each of these things, the chances of them causing harm are usually relatively small. We have more fears than we need, and it seems that our utterly efficient fear conditioning system, combined with an extremely powerful ability to think about our fears and an inability to control them, is probably at fault. As we'll see in the next chapter, though, there is some hope that the future evolution of the human brain will take care of this imbalance.

9

ONCE MORE, WITH FEELINGS

"Men believe themselves to be free, simply because they are conscious of their actions, and unconscious of the causes whereby those actions are determined."

Baruch Spinoza, *Ethics*[1]

"How small the cosmos . . . how paltry and puny in comparison to human consciousness, to a single individual recollection. . . ."

Vladimir Nabokov, *Speak, Memory*[2]

THE PICTURE OF EMOTION I've painted so far is largely one of automaticity. I've shown how our brains are programmed by evolution to respond in certain ways to significant situations. Significance can be signaled by information built into the brain by evolution or by memories established through past experiences. In either case, though, the initial responses elicited by significant stimuli are automatic and require neither conscious awareness of the stimulus nor conscious control of the responses.

This scenario, you may say, is fine for the control of the bodily responses. But these are not the essence of an emotion. They occur during an emotion, but an emotion is something else, something more. An emotion is a subjective experience, a passionate invasion of consciousness, a feeling.

I've spent most of this book trying to show that much of what the brain does during an emotion occurs outside of conscious awareness.

It's now time to give consciousness its due. It's time to see what role consciousness has in emotion, and what role emotion has in consciousness. It's time to look at emotion once again, this time with feelings as part of the picture.[3]

A Simple Idea

My idea about the nature of conscious emotional experiences, emotional feelings, is incredibly simple. It is that a subjective emotional experience, like the feeling of being afraid, results when we become consciously aware that an emotion system of the brain, like the defense system, is active. In order for this to occur, we need at least two things. We need a defense system and we need to have the capacity to be consciously aware of its activity. The upside of this line of thought is that once we understand consciousness we will also understand subjective emotional experiences. The downside is that in order to understand subjective emotional experiences, we've got to figure out consciousness.

To my way of thinking, then, emotional experience is not really a problem about emotion. It is, instead, a problem about how conscious experiences occur. Because the scientific study of emotions has mostly been about conscious emotional experiences,[4] scientists who study emotions have set things up so that they will not understand emotions until they've understood the mind-body problem, the problem of how consciousness comes out of brains, arguably the most difficult problem there is and ever was.[5]

The field got this way at the beginning, when William James brought up the business with the bear. He started with a question about why the sight of a bear makes us run away (the stimulus-to-response problem in emotion) but ended up with a question about why we feel afraid when we see the bear (the stimulus-to-feeling problem in emotion). Ever since, the study of emotion has been focused on where conscious feelings come from.[6]

All areas of psychology have had to deal with consciousness. Perception and memory, for example, also involve conscious experiences. To perceive an apple is to be aware that an apple is there, and to remember something about an apple is to be aware of that particular thing about an apple. The difficulty of scientifically understanding the conscious content that occurs during perception, memory, or emotion is what led to the behaviorist movement in psychology.[7] And the success of the cognitive movement as an alternative to behaviorism was largely due to the fact that it could deal with the mind in terms of processes that occur unconsciously, and thus without having to first solve the problem of how conscious content is created. But because emotion was left out of the cognitive revolution,[8] it somehow did not reap the benefits that come from thinking of minds in terms of unconscious processes rather than in terms of conscious content. The study of emotion is, as a result, still focused on where subjective feelings come from rather than on the unconscious processes that sometimes do and sometimes do not give rise to those conscious states.

By treating emotions as unconscious processes that can sometimes give rise to conscious content, we lift the burden of the mind-body problem from the shoulders of emotion researchers and allow them to get on with the problem of figuring out how the brain does its unconscious emotional business. But we also see how conscious emotional experiences are probably created. They are probably created the same way that other conscious experiences are—by the establishment of a conscious representation of the workings of underlying processing systems.[9] Although much remains unknown about how conscious representations come about, recent studies have begun to provide important clues.

Short Stuff

There have been many ideas about what consciousness is and isn't.[10] While you could hardly say that there is a consensus on this topic, many of the theories that have been proposed in recent years are built around the concept of working memory.[11]

Remember this number: 783445. Now close your eyes and repeat it, and then count backward from 99 to 91 by 2s and try repeating the number again. Chances are you can't. The reason for this is that thinking occurs in a mental workspace that has a limited capacity. When you started using the workspace to do the subtraction problem you bumped the stored number out. This workspace is called working memory, a temporary storage mechanism that allows several pieces of information to be held in mind at the same time and compared, contrasted, and otherwise interrelated.[12]

Working memory is pretty much what used to just be called short-term memory. However, the term working memory implies not just a temporary storage system but an active processing mechanism used in thinking and reasoning.

Much of our understanding of working memory is owed to the pioneering work of Alan Baddeley in the early 1970s.[13] It was known from a famous study performed by one of the pioneers of cognitive psychology, George Miller, that short-term memory has a capacity limit of about seven pieces of information.[14] Baddeley reasoned that if he had subjects actively remember six things, like six digits, they should have trouble performing, at the same time, other tasks that require temporary storage since the mental workspace would be mostly used up. To test this, he had his subjects rehearse the digits out loud while, at the same time, reading sentences and pressing buttons to verify whether the sentence referred to something true or false. Baddeley found that sentence comprehension was greatly slowed down, but to his surprise the subjects could still do it to some extent.

Baddeley's experiment led him to reformulate the notion of short-term memory. He replaced the generic notion of short-term memory with the concept of working memory, which, he suggested, consists of a general-purpose temporary storage system utilized in all active thinking processes and several specialized temporary storage systems that are only called into play when specific kinds of information have to be held on to.

Borrowing a term from computer technology, memory researchers sometimes refer to temporary storage mechanisms as buffers. It is now believed that a number of specialized buffers exist. For example, each sensory system has one or more temporary buffers. These aid in perception, allowing the system to compare what it is seeing or hearing now to what it saw or heard a moment ago. There are also temporary buffers associated with aspects of language use (these help you keep the first part of a sentence in mind until you've heard the last part so that the whole thing can be understood). The specialized memory buffers work in parallel, independent of one another.

The general-purpose system consists of a *workspace*, where information from the specialized buffers can be held on to temporarily, and a set of so-called *executive* functions that control operations performed on this information. The executive functions take care of the overall coordination of the activities of working memory, such as determining which specialized systems should be attended to at the moment and shuffling information in and out of the workspace from these and other systems.

Although only a limited amount of information can be held in the general workspace at any one time, any kind of information can be held on to. As a result, different kinds of information can be interrelated in working memory (the way something looks, sounds, and smells can be associated with its name in working memory). And thanks to "chunking," another of George Miller's many insights into the cognitive mind, the capacity limit of working memory (about seven pieces of information) can be overcome to some degree: we can remember seven of just about anything (letters, words, or ideas), so that the amount of information actually represented by the seven pieces of information can be enormous (think of all that is implied by the names of seven countries).[15]

The stuff in working memory is the stuff we are currently thinking about or paying attention to. But working memory is not a pure product of the here and now. It also depends on what we know and what kinds of experiences we've had in the past. In other words, it depends on long-term memory. In order to be aware that you are looking at a basketball, it is not enough for the basketball to be represented as a purely visual pattern (a round, orange object with thin black lines around it) by your visual system. The pattern has to also have grabbed the attention of the working memory executive. This means that the pattern is what is being held in the visual short-term memory buffer and that the visual buffer, as opposed to the auditory or other buffers, is the one with which the executive is working. But neither is this enough. Only when the visual pattern is matched with information in long-term memory (stored facts about and stored past experiences with similar objects) does the visual stimulus become recognized as a basketball. But in addition to being important in figuring out the meaning of information being picked up by lower level specialized systems, stored knowledge also influences the workings of the lower level systems. For example, once memories having to do with basketballs are activated and made available to working memory, the operation of the specialized processors becomes biased toward detecting and picking up on external information that is relevant to basketballs. This influence of memory on perception is an example of what cognitive scientists sometimes call top-down processing, which contrasts with the build-up of perceptions from sensory processing, known as bottom-up processing.

Working memory, in short, sits at the crossroads of bottom-up and top-down processing systems and makes high-level thinking and reasoning possible. Stephen Kosslyn, a leading cognitive scientist, puts it this way:

> Working memory . . . corresponds to the activated information in long-term memories, the information in short-term memories, and the decision processes that manage which information is activated in the long-term memories and retained in the short-term memories. . . . This kind of working memory system is necessary for a wide range of tasks, such as performing mental arithmetic, reading, problem solving and . . . reasoning in general. All of these tasks require not only some form of temporary storage, but also an interplay between information that is stored temporarily and a larger body of stored knowledge.[16]

The Here and the Now in the Brain

How, then, does working memory work in the brain? Studies conducted in the 1930s by C.F. Jacobsen provide the foundation for our understanding of this problem.[17] He trained monkeys using something called the delayed response task. The monkey sat in a chair and watched the experimenter put a raisin under one of two objects that were side by side. A curtain was then lowered for a certain amount of time (the delay) and then the monkey was allowed to choose. In order to get the raisin, the monkey had to remember not which object the raisin was under but whether the raisin was under the left or the right object. Correct performance, in other words, required that the monkey hold in mind the spatial location of the raisin during the delay period (during which the playing field was hidden from view). At very short delays (a few seconds), normal monkeys did quite well, and performance got predictably worse as the delay increased (from seconds to minutes). However, monkeys with damage to the prefrontal cortex performed poorly, even at the short delays. On the basis of this and research that followed, the prefrontal cortex has come to be thought of as playing a role in temporary memory processes, processes that we now refer to as working memory.

In the last chapter, we examined the role of the *medial* prefrontal cortex in the extinction of emotional memory. In contrast, it is the *lat-*

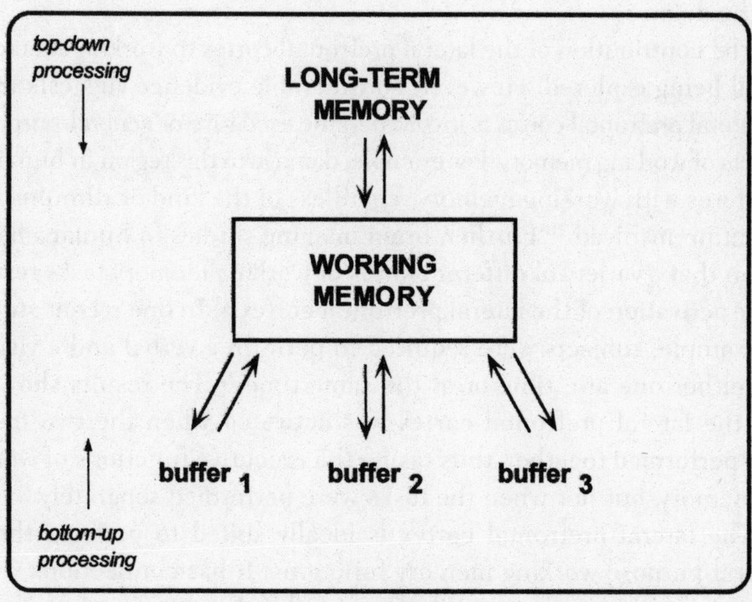

FIGURE 9-1

Relation of Specialized Short-Term Buffers, Long-Term Explicit Memory, and Working Memory.

Stimuli processed in different specialized systems (such as sensory, spatial, or language systems) can be held simultaneously in short-term buffers. The various short-term buffers provide potential inputs to working memory, which can deal most effectively with only one of the buffers at a time. Working memory integrates information received from short-term buffers with long-term memories that are also activated.

eral prefrontal cortex that has most often been implicated in working memory. The lateral prefrontal cortex is believed to exist only in primates and is considerably larger in humans than in other primates.[18] It is not surprising that one of the most sophisticated cognitive functions of the brain should involve this region.

In recent years, the role of the lateral prefrontal cortex in working memory has been studied extensively by the laboratories of Joaquin Fuster at UCLA and Pat Goldman-Rakic at Yale.[19] Both researchers have recorded the electrical activity of lateral prefrontal neurons while monkeys performed delayed response tasks and other tests requiring short-term storage. They have shown that cells in this region become particularly active during the delay periods. It is likely that these cells are actively involved in holding on to the information during the delay.

The contribution of the lateral prefrontal cortex to working memory is still being explored. However, considerable evidence suggests that the lateral prefrontal cortex is involved in the executive or general-purpose aspects of working memory. For example, damage to this region in humans interferes with working memory regardless of the kind of stimulus information involved.[20] Further, brain imaging studies in humans have shown that a variety of different kinds of working memory tasks result in the activation of the lateral prefrontal cortex.[21] In one recent study, for example, subjects were required to perform a verbal and a visual task either one at a time or at the same time.[22] The results showed that the lateral prefrontal cortex was activated when the two tasks were performed together, thus taxing the executive functions of working memory, but not when the tasks were performed separately.

The lateral prefrontal cortex is ideally suited to perform these general-purpose working memory functions. It has connections with the various sensory systems (like the visual and auditory systems) and other neocortical systems that perform specialized temporary storage functions (like spatial and verbal storage) and is also connected with the hippocampus and other cortical areas involved in long-term memory.[23] In addition, it has connections with areas of the cortex involved in movement control, allowing decisions made by the executive to be turned into voluntarily performed actions.[24] Recent studies have begun to show how the lateral prefrontal cortex interacts with some of these areas. Best understood are interactions with temporary storage buffers in the visual cortex.

Cortical visual processing begins in the primary visual area located in the occipital lobe (the rear-most part of the cortex). This area receives visual information from the visual thalamus, processes it, and then distributes its outputs to a variety of other cortical regions. Although the cortical visual system is enormously complex,[25] the neural pathways responsible for two aspects of visual processing are fairly well understood. These involve the determination of "what" a stimulus is and "where" it is located.[26] The "what" pathway involves a processing stream that travels from the primary visual cortex to the temporal lobe and the "where" pathway goes from the primary cortex to the parietal lobe.

Goldman-Rakic and colleagues recorded from cells in the parietal lobe "where" pathway during short-term memory tests requiring the temporary remembrance of the spatial location of visual stimuli. They found that cells there, like cells in the lateral prefrontal cortex, were active, suggesting that they were keeping track of the location, during the delay.[27] The parietal and frontal regions in question are anatomically interconnected—the parietal area sends axons to the prefrontal region and the prefrontal region sends axons back to the parietal area. These findings suggest that the parietal lobe visual area works with the lateral prefrontal cortex to maintain information about the spatial location of visual stimuli in working memory. Similarly, Robert Desimone found evidence for reciprocal interactions between the visual areas of the temporal lobe (the "what" pathway) and the lateral prefrontal cortex in studies involving the recognition of whether a particular object had been seen recently.[28]

The maintenance of visual information in working memory thus appears to crucially depend on interactions between the lateral prefrontal region and specialized areas of the visual cortex.[29] The pathway from the specialized visual areas tell the prefrontal cortex "what" is out there and "where" it is located (bottom-up processing). The prefrontal cortex, by way of pathways back to the visual areas, primes the visual system to attend to those objects and spatial locations that are being processed in working memory (top-down processing). As we've seen, these kinds of top-down influences on sensory processing

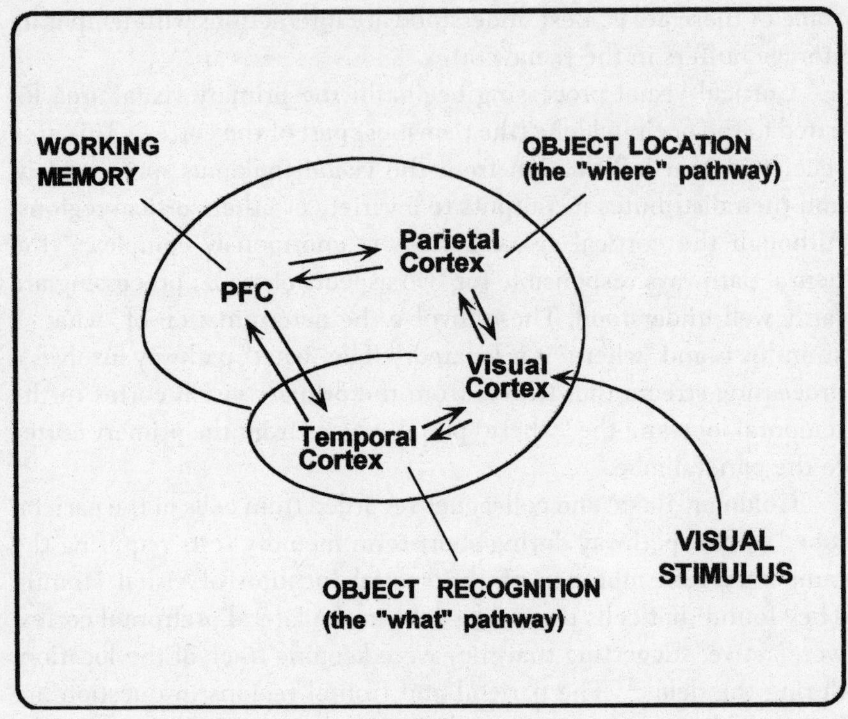

FIGURE 9-2
Relation of the "What" and "Where" Visual Pathways to Working Memory.

Visual information, received by the visual cortex, is distributed to cortical areas that perform specialized visual processing functions. Two well-studied specialized functions are those involved in object recognition (mediated by the "what" pathway) and object location (mediated by the "where" pathway). These specialized visual pathways provide inputs to the prefrontal cortex (PFC), which plays a crucial role in working memory. The specialized systems also receive inputs back from the prefrontal cortex, allowing the information content of working memory to influence further processing of incoming information. Leftward-going arrows represent bottom-up processing and rightward-going ones top-down processing.

are believed to be important aspects of the executive control functions of working memory.

Recent studies, especially by Goldman-Rakic and associates, have raised questions about the role of the prefrontal cortex as a general-purpose working memory processor.[30] For example, they have found that different parts of the lateral prefrontal cortex participate in working memory when animals have to determine "what" a visual stimulus is as opposed to "where" it is located, suggesting that different parts of the prefrontal cortex are specialized for different kinds of working memory tasks. While these findings show that parts of the prefrontal cortex participate uniquely in different short-term memory tasks, they do not rule out the existence of a general-purpose workspace and a set of executive functions that coordinate the activity of the specialized systems, especially since the tasks studied did not tax the capacity of working memory in a way that would reveal a limited-capacity system.[31] Studies that have taxed the system, like the imaging studies in humans described above, suggest that neurons in the lateral prefrontal cortex are part of a general-purpose working memory network. At the same time, it is possible, given Goldman-Rakic's findings, that the general-purpose aspects of working memory are not localized to a single place in the lateral prefrontal cortex but instead are distributed over the region. That this may occur is suggested by the fact that some cells in the specialized areas of the lateral prefrontal cortex participate in multiple working memory tasks.[32]

There is also evidence that the general-purpose functions of working memory involve areas other than the lateral prefrontal cortex. For example, imaging studies in humans have shown that another area of the frontal lobe, the anterior cingulate cortex, is also activated by working memory and related cognitive tasks.[33] Like the lateral prefrontal cortex, the anterior cingulate region receives inputs from the various specialized sensory buffers, and the anterior cingulate and the lateral prefrontal cortex are anatomically interconnected.[34] Moreover, both regions are part of what has been called the frontal lobe attentional network, a cognitive system involved in selective attention, mental resource allocation, decision making processes, and voluntary movement control.[35] It is tempting to think of the general-purpose aspects of working memory as involving neurons in the lateral prefrontal and anterior cingulate regions working together. Earlier (Chapter 4) we saw that the cingulate cortex was once considered the seat of the soul (consciousness). Given the new work implicating the cingulate region in working memory, the older idea may not be so far off the mark.

One other area of the prefrontal cortex, the orbital region, located on the underneath side of the frontal lobe, has emerged as important as well. Damage to this region in animals interferes with short-term memory about reward information, about what is good and bad at the moment,[36] and cells in this region are sensitive to whether a stimulus has just led to a reward or punishment.[37] Humans with orbital frontal damage become oblivious to social and emotional cues and some exhibit sociopathic behavior.[38] This area receives inputs from sensory processing systems (including their temporary buffers) and is also intimately connected with the amygdala and the anterior cingulate region. The orbital cortex provides a link through which emotional processing by the amygdala might be related in working memory to information being processed in sensory or other regions of the neocortex. We'll have more to say about this later.

There is still much to be learned about working memory and its neural basis. It is not clear, for example, whether both the temporary workspace and the executive functions are actually located in the frontal cortex. It is possible that the prefrontal areas do not store anything but instead just control the activity of other regions, allowing the activity in some areas to rise above the threshold for consciousness and inhibiting the activity of the others.[39] In spite of the fact that we still have much to learn, the researchers in this area have made considerable progress on this very tough, and very important, problem.

The Platform of Awareness

Tennessee Williams said, "Life is all memory except for the one present moment that goes by you so quick you hardly catch it going."[40] What Williams didn't realize is that even the immediate present involves memory—what we know about the one present moment is basically what is in our working memory. Working memory allows us to know that the "here and now" is "here" and is happening "now." This insight underlies the notion, adopted by a number of contemporary cognitive scientists, that consciousness is the awareness of what is in working memory.

For example, Stephen Kosslyn argues that to be aware of something, that something must be in working memory.[41] John Kihlstrom

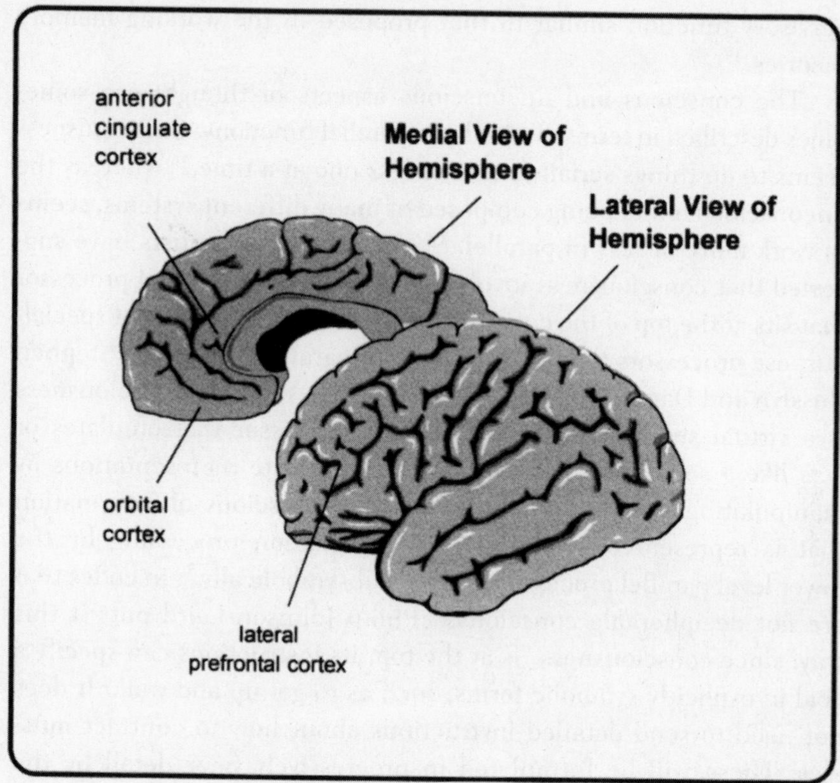

FIGURE 9-3
Frontal Cortex Areas Involved in Aspects of Working Memory.
Some areas of the frontal lobe that have been implicated in working memory functions include the lateral prefrontal cortex and the orbital and anterior cingulate cortex.

proposes that a link must be made between the mental representation of an event and a mental representation of the "self" as the agent or experiencer in order for us to be conscious of that event. These integrated episodic representations, according to Kihlstrom, reside in working memory.[42] Philip Johnson-Laird notes that the contents of working memory are what we are conscious of at the moment.[43] Bernard Baars, in an influential book called *A Cognitive Theory of Consciousness,* treats "consciousness as a kind of momentary working memory."[44] And several contemporary theories equate consciousness with focused attention, which is achieved through an executive or supervisory function similar to that proposed in the working memory theories.[45]

The conscious and unconscious aspects of thought are sometimes described in terms of serial and parallel functions. Consciousness seems to do things serially, more or less one at a time,[46] whereas the unconscious mind, being composed of many different systems, seems to work more or less in parallel. Some cognitive scientists have suggested that consciousness involves a limited-capacity serial processor that sits at the top of the cognitive hierarchy above a variety of special-purpose processors that are organized in parallel (some, like Stephen Kosslyn and Daniel Dennett, have even suggested that consciousness is a virtual serial processor—a parallel processor that emulates or acts like a serial one).[47] Serial processors create representations by manipulating symbols,[48] and we are only conscious of information that is represented symbolically.[49] Information processing by the lower level parallel processors occurs subsymbolically,[50] in codes that are not decipherable consciously. Philip Johnson-Laird puts it this way: since consciousness "is at the top, its instructions can specify a goal in explicitly symbolic terms, such as to get up and walk. It does not need to send detailed instructions about how to contract muscles. These will be formulated in progressively finer detail by the processors at lower levels. . . . It [consciousness] receives the results of computations from the lower processors, but again in a high-level and explicitly symbolic form."[51] This reasoning yields an explanation for why we are conscious of the outcome of mental computations but not of the computations themselves and how we can produce behaviors without knowing how individual muscles are controlled. In other words, the consciousness processor works at the symbolic level, which yields introspectively accessible content, but the parallel processors work subsymbolically and their operations are not directly accessible from consciousness.[52] And since not all subsymbolic processors necessarily feed into the consciousness processor, some subsymbolic processing remains inacessible.

Working memory is the limited-capacity serial processor that creates and manipulates symbolic representations. It is where the integrated monitoring and control of various lower level specialized processors takes place. Working memory is, in other words, a crucial part of the system that gives rise to consciousness.

The advantage of a working memory concept of consciousness over many other formulations is that it allows the problem to be posed in a concrete fashion. Concreteness for the sake of concreteness would not be so good, but in this case it seems to buy us something. As working memory, consciousness can be thought of in terms of a computational system, a system that creates representations by performing computations, by processing information. Viewed in computational terms, consciousness can be explored both psychologically and neurologically, and its underlying processes can even be modeled using computer simulations.

However, it is not clear that consciousness is computable. Johnson-Laird reminds us that a computer simulation of the weather is not the same thing as rain or sunshine.[53] Working memory theories, in dealing with consciousness in terms of processes rather than as content, try to explain what kinds of computational functions might be responsible for and underlie conscious experiences but they do not explain what it is like to have those experiences.[54] These theories provide an account of the way human minds work, in a general sense, rather than an account of what a particular experience is like in a particular mind. They can suggest how a representation might be created in working memory but not what it is like to be aware of that representation. They suggest how decision processes in working memory might lead to movement but not what it is like to actually decide to move. In other words, working memory is likely to be an important, and possibly an essential, aspect of consciousness. It is in fact likely to be the platform on which a conscious experience stands. But consciousness, especially its phenomenal or subjective nature, is not completely explained by the computational processes that underlie working memory, at least not in a way that anyone presently comprehends.[55]

Figuring out the exact nature of consciousness and the mechanisms by which it emerges out of collections of neurons is truly an important problem. Many questions remain to be answered about how working memory is mediated by the brain and how consciousness relates to the working memory system and/or other brain systems. However, it is not necessary for emotions researchers to solve these problems, nor is it necessary for us to wait for the solutions before studying how emotions work. Emotion researchers need to figure out how emotional information gets represented in working memory. The rest of the problem, figuring out how the contents of working memory become consciously experienced and how these subjective phenomena emerge from the brain, belongs on the shoulders of all mind scientists. Emotions researchers certainly have a lot to contribute to the study of consciousness, but figuring out consciousness is not their job, or at least theirs alone. Although this may seem obvious, the study of emotion has been so focused on the problem of emotional consciousness that the basic underlying emotional mechanisms have often been given short shrift.

The Emotional Present

I admit that I've passed the emotional consciousness buck. I've redefined the problem of emotional feelings as the problem of how emotional information comes to be represented in working memory. This won't make you happy if you want to know exactly what a feeling is or if you want to know how something as intangible as a feeling could be part of something so tangible as a brain. It won't, in other words, solve the mind-body problem. However, as important as solving the mind-body problem would be, it's not the only problem worth solving. And figuring out the mind-body problem wouldn't tell us what's unique about those states of mind we call emotions, nor would it explain why different emotions feel the way they do. Neither would it tell us what goes wrong in emotional disorders or suggest ways of treating or curing them. In order to understand what an emotion is and how particular emotional feelings come about we've got to understand the way the specialized emotion systems operate and determine how their activity gets represented in working memory.

Some might say I'm taking a big chance. I'm resting our understanding of our feelings, our most private and intimate states of mind, on the possibility that working memory is the key to consciousness. But really what I'm doing is using working memory as an "in principle" way of explaining feelings. I'm saying that feelings come about when the activity of specialized emotion systems gets represented in the system that gives rise to consciousness, and I'm using working memory as a fairly widely accepted version of how the latter might come about.

We've gone into great detail as to how one specialized emotion system, the defense system, works. So let's now see how the activity of this system might come to be represented in working memory and thereby give rise to the feeling we know as fear.

From Conscious Appraisals to Emotions: You encounter a rabbit while walking along a path in the woods. Light reflected from the rabbit is picked up by your eyes. The signals are then transmitted through the visual system to your visual thalamus, and then to your visual cortex, where a sensory representation of the rabbit is created and held in a short-term visual object buffer. Connections from the visual cortex to the cortical long-term memory networks activate relevant memories (facts about rabbits stored in memory as well as memories about past experiences you may have had with rabbits). By way of connections between the long-term memory networks and the working memory system, activated long-term memories are integrated with the sensory representation of the stimulus in working memory, allowing you to be consciously aware that the object you are looking at is a rabbit.

A few strides later down the path, there is a snake coiled up next to a log. Your eyes also pick up on this stimulus. Conscious representations are created in the same way as for the rabbit—by the integration in working memory of short-term visual representations with information from long-term memory. However, in the case of the snake, in addition to being aware of the kind of animal you are looking at, long-term memory also informs you that this kind of animal can be dangerous and that you might be in danger.

According to cognitive appraisal theories, the processes described so far would constitute your assessment of the situation and should be enough to account for the "fear" that you are feeling as a result of encountering the snake. The difference between the working memory representation of the rabbit and the snake is that the latter includes information about the snake being dangerous. But these cognitive representations and appraisals in working memory are not enough to turn the experience into a full-blown emotional experience. Davy Crockett, you may remember, said his love for his wife was so hot that it mighty nigh burst his boilers. There is nothing equivalent to boiler bursting going on here. Something else is needed to turn cognitive appraisals into emotions, to turn experiences into emotional experiences. That something, of course, is the activation of the system built by evolution to deal with dangers. That system, as we've seen, crucially involves the amygdala.

Many but not all people who encounter a snake in a situation such as the one described will have a full-blown emotional reaction that includes bodily responses and emotional feelings.[56] This will only occur if the visual representation of the snake triggers the amygdala. A whole host of output pathways will then be activated. Activation of these outputs is what makes the encounter with the snake an emotional experience, and the absence of activation is what prevents the encounter with the rabbit from being one.[57]

What is it about the activation of amygdala outputs that converts an experience into an emotional experience? To understand this we need to consider some of the various consequences of turning on amygdala outputs. These outputs provide the basic ingredients that, when mixed together in working memory with short-term sensory representations and the long-term memories activated by these sensory representations, create an emotional experience.

Ingredient 1: Direct Amygdala Influences on the Cortex: The amygdala has projections to many cortical areas.[58] In fact, as we've alredy seen, the projections of the amygdala to the cortex are considerably greater than the projections from the cortex to the amygdala (see Figure 9-4). In addition to projecting back to cortical sensory areas from which it receives inputs, the amygdala also projects to some sensory processing areas from which it does not receive inputs. For example, in order for a visual stimulus to reach the amygdala by way of the cortex, the stimulus has to go through the primary cortex, to a secondary region, and then to a third cortical area in the temporal lobe (which does the short-term buffering of visual object information). This third area then projects to the amygdala. The amygdala projects back to this area, but also to the other two earlier visual processing regions. As a result, once the amygdala is activated, it is able to influence the cortical areas that are processing the stimuli that are activating it (see Figure 9-4). This might be very important in directing attention to emotionally relevant stimuli by keeping the short-term object buffer focused on the stimuli to which the amygdala is assigning significance. The amygdala also has an impressive set of connections with long-term memory networks involving the hippocampal system and areas of the cortex that interact with the hippocampus in long-lasting information storage. These pathways may contribute to the activation of long-term memories relevant to the emotional implications of immediately present stimuli. Although the amygdala has relatively meager connections with the lateral prefrontal cortex, it sends rather strong connections to the anterior cingulate cortex, one of the other partners in the frontal lobe working memory executive circuitry. It also sends connections to the orbital cortex, another player in working memory that may be especially involved in working memories about rewards and punishments. By way of these connections with specialized short-term buffers, long-term memory networks, and the networks of the frontal lobe, the amygdala can influence the information content of working memory (Figure 9-5). There is obviously a good deal of redundancy built into this system, making it possible for the conscious awareness of amygdala activity to come about in several ways.

In sum, connections from the amygdala to the cortex allow the defense networks of the amygdala to influence attention, perception, and memory in situations where we are facing danger. At the same time, though, these kinds of connections would seem to be inadequate in completely explaining why a perception, memory, or thought about an emotional event should "feel" different from one about a nonemotional event. They provide working memory with information about whether something good or bad is present, but are insufficient for producing the feelings that come from the awareness that something good or bad is present. For this we need other connections as well.

Ingredient 2: Amygdala-Triggered Arousal: In addition to the direct influences of the amygdala on the cortex, there are a number of indirect channels through which the effects of amygdala activation can impact on cortical processing. An extremely important set of such connections involves the arousal systems of the brain.

It has long been believed that the difference between being

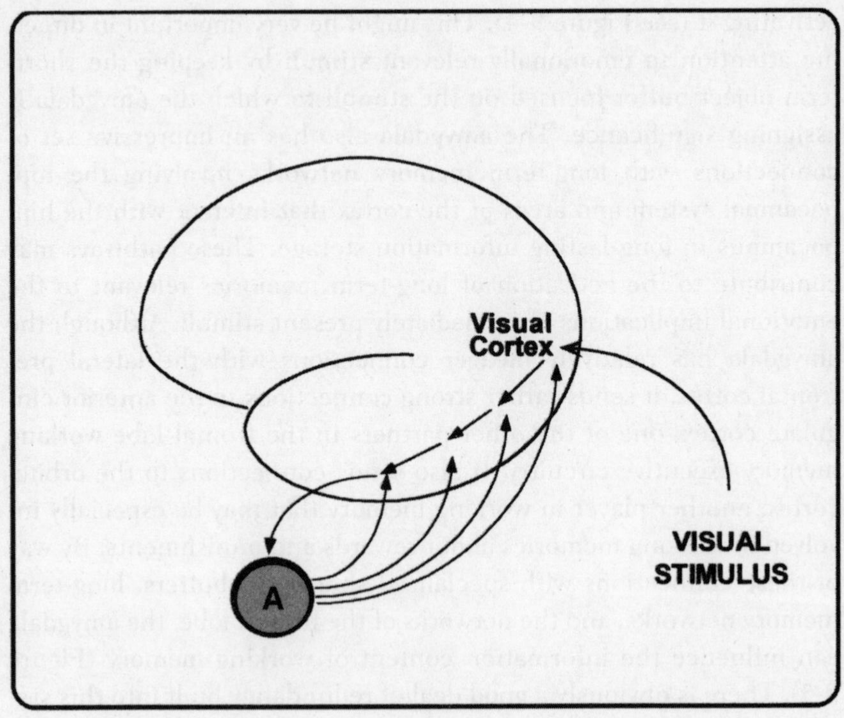

FIGURE 9-4

The Amygdala's Influence on Sensory Areas of Cortex Is Greater Than the Influence of the Same Areas on the Amygdala.

The amygdala receives inputs from the latest stages of cortical processing within the sensory systems, but projects back to all stages of cortical processing, even the earliest. An example from the visual system is shown.

awake and alert, on the one hand, and drowsy or asleep on the other is related to the arousal level of the cortex.[59] When you are alert and paying attention to something important, your cortex is aroused. When you are drowsy and not focusing on anything, the cortex is in the unaroused state. During sleep, the cortex is in the unaroused state, except during dream sleep when it is highly aroused. In dream sleep, in fact, the cortex is in a state of arousal that is very similar to the alert waking state, except that it has no access to external stimuli and only processes internal events.[60]

Cortical arousal can be easily detected by putting electrodes on the scalp of a human. These electrodes pick up the electrical activity

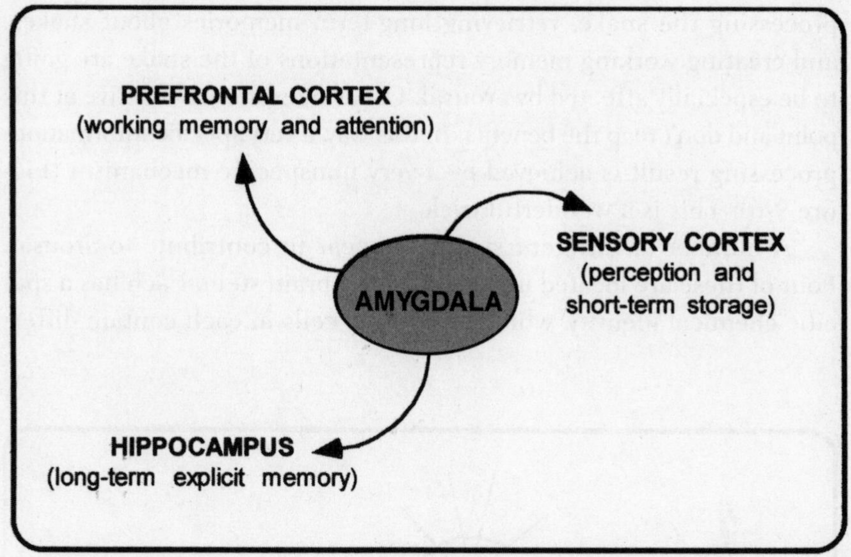

FIGURE 9-5

Some Cortical Outputs of the Amygdala and Their Function.

Areas of the amygdala project to a wide variety of cortical areas. Included are projections to all stages of cortical sensory processing (see figure 9-4), to prefrontal cortex, and to the hippocampus and related cortical areas. Through these projections, the amygdala can influence ongoing perceptions, mental imagery, attention, short-term memory, working memory, and long-term memory, as well as the various higher-order thought processes that these make possible.

of cortical cells through the skull. This electroencephalogram or EEG is slow and rhythmic when the cortex is not aroused and fast and out of sync (desynchronized) during arousal.

When arousal occurs, cells in the cortex, and in the thalamic regions that supply the cortex with its major inputs, become more sensitive.[61] They go from a state in which they tend to fire action potentials at a very slow rate and more or less in synchrony to a state in which they are generally out of sync but with some cells being driven especially strongly by incoming stimuli.

While much of the cortex is potentially hypersensitive to inputs during arousal, the systems that are processing information are able to make the most use of this effect. For example, if arousal is triggered by the sight of a snake, the neurons that are actively involved in processing the snake, retrieving long-term memories about snakes, and creating working memory representations of the snake are going to be especially affected by arousal. Other neurons are inactive at this point and don't reap the benefits. In this way, a very specific information-processing result is achieved by a very nonspecific mechanism (Figure 9-6). This is a wonderful trick.

A number of different systems appear to contribute to arousal. Four of these are located in regions of the brain stem. Each has a specific chemical identity, which means the cells in each contain differ-

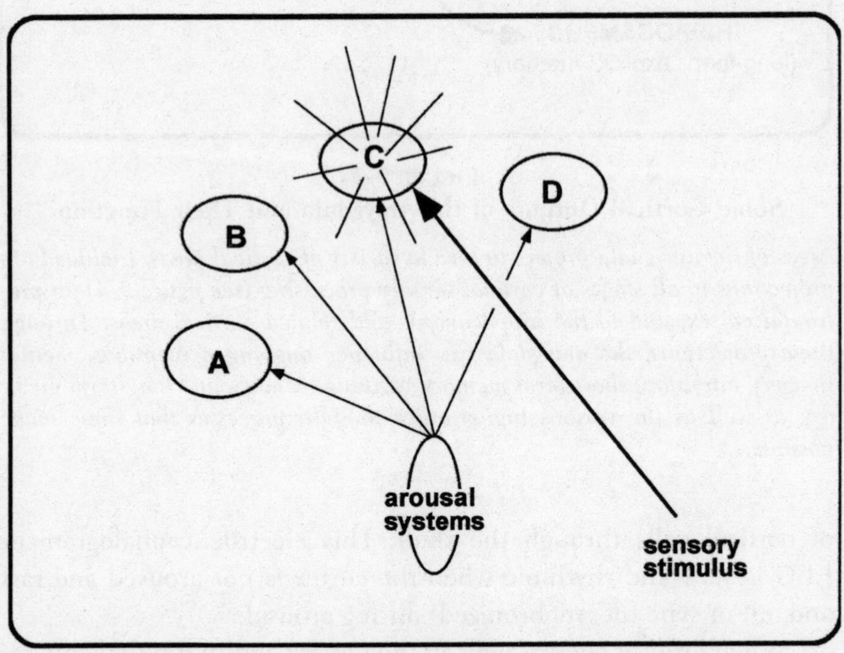

FIGURE 9-6

How Nonspecific Arousal Achieves a Specific Effect.

Arousal systems act in nonspecific ways throughout the forebrain. One of their main contributions is to make cells more sensitive to incoming signals. Those cells that are processing stimuli during arousal will be especially affected. In this way, very specific effects are achieved by nonspecific arousal. In the example shown, arousal systems potentially influence areas A, B, C, and D. However, the effects of arousal are greatest in area C, which is processing a stimulus. Other areas are inactive during arousal and thus do not reap the benefits of arousal.

ent neurotransmitters that are released by their axon terminals when the cells are activated. One group makes acetylcholine (ACh), another noradrenaline, another dopamine, and another serotonin. A fifth group, also containing ACh, is located in the forebrain, near the amygdala. The axons of each of these cell groups terminate in widespread areas of the forebrain. In the presence of novel or otherwise significant stimuli the axon terminals release their neurotransmitters and "arouse" cortical cells, making them especially receptive to incoming signals.

Arousal is important in all mental functions. It contributes significantly to attention, perception, memory, emotion, and problem solving. Without arousal, we fail to notice what is going on—we don't attend to the details. But too much arousal is not good either. If you are overaroused you become tense and anxious and unproductive. You need to have just the right level of activation to perform optimally.[62]

Emotional reactions are typically accompanied by intense cortical arousal. Certain emotion theories around mid-century proposed that emotions represent one end of an arousal continuum that spans from being completely unconscious (in a coma), to asleep, to awake but drowsy, to alert, to emotionally aroused. This high level of arousal is, in part, the explanation for why it is hard to concentrate on other things and work efficiently when you are in an emotional state. Arousal helps lock you into the emotional state you are in. This can be very useful (you don't want to get distracted when you are in danger), but can also be an annoyance (once the fear system is turned on, it's hard to turn it off—this is the nature of anxiety).

Although each of the arousal systems probably contributes to arousal in the presence of stimuli that are dangerous or that warn of danger, it appears that interactions between the amygdala and the nearby ACh-containing system in the forebrain are particularly important.[63] This ACh-containing system is called the nucleus basalis. Damage to the amygdala or to the nucleus basalis prevents stimuli that warn of danger, like conditioned fear stimuli, from eliciting arousal. Moreover, stimulation of the amygdala or the nucleus basalis elicits cortical arousal artificially. And administration of drugs that block the actions of ACh in the cortex prevents these effects on arousal of conditioned stimuli, amygdala stimulation, or nucleus basalis stimulation from occurring. Together, these and other findings suggest that when the amygdala detects danger it activates the nucleus basalis, which then releases ACh throughout the cortex. The amygdala also interacts with the other arousal systems located in the brain stem, and the overall effect of amygdala activation on arousal certainly involves these as well.[64]

Although there are a number of different ways that the nucleus basalis cells can be turned on, the way they are turned on by a dangerous stimulus is through the activity of the amygdala.[65] Other kinds of emotional networks most likely have their own ways of interacting with the arousal systems and altering cortical processing.

Arousal occurs to any novel stimulus that we encounter and not just to emotional stimuli. The difference is that a novel but insignificant stimulus will elicit a temporary state of arousal that dissipates almost immediately but arousal is prolonged in the presence of emotional stimuli. If you are face-to-face with a predator it is crucial that you not lose interest in what is going on or be distracted by some other event. While this seems so obvious as to be silly, it is only so because the brain does it so effortlessly.

Why is arousal perpetuated to emotional but not to other stimuli? Again, the answer probably has to do with the involvement of the amygdala. The arousal elicited by a novel stimulus does not require the amygdala. Instead, it is mediated by direct inputs from sensory systems to arousal networks.[66] These kinds of arousal effects quickly habituate. If the stimulus is meaningful, say dangerous, then the amygdala is brought into the act and it also activates arousal systems. This adds impetus to keep arousal going. The continued presence of the stimulus and its continued interpretation by the amygdala as dangerous continues to drive arousal systems, and these systems, in turn, keep cortical networks that are processing the stimulus in a state of hypersensitivity. The amygdala, it should be noted, is also the recipient of arousal system axons, so that amygdala activation of arousal systems also helps keep the amygdala aroused. These are self-perpetuating, vicious cycles of emotional reactivity. Arousal locks you into whatever emotional state you are in when arousal occurs, unless something else occurs that is significant enough and arousing enough to shift the focus of arousal.

The information content provided by arousal systems is weak. The cortex is unable to discern that danger (as opposed to some other emotional condition) exists from the pattern of neural messages it receives from arousal systems. Arousal systems simply say that something important is going on. The combination of nonspecific cortical arousal and specific information provided by direct projections from the amygdala to the cortex allows the establishment of a working memory that says that something important is going on and that it involves the fear system of the brain. These representations converge in working memory with the representations from specialized short-term memory buffers and with representations from long-term memory triggered by current stimuli and by amygdala processing. The continued driving of the amygdala by the dangerous stimulus keeps the arousal systems active, which keeps the amygdala and cortical networks actively engaged in the situation as well. Cognitive inference and decision making processes controlled by the working memory executive become actively focused on the emotionally arousing situation, trying to figure out what is going on and what should be done about it. All other inputs that are vying for the attention of working memory are blocked out.

We now have many of the basic ingredients for a complete emotional experience. But one more is needed.

Ingredient 3: Bodily Feedback: As we've seen in earlier chapters, activation of the amygdala results in the automatic activation of networks that control the expression of a variety of responses: species-specific behaviors (freezing, fleeing, fighting, facial expressions), autonomic nervous system (ANS) responses (changes in blood pressure and heart rate, piloerection, sweating), and hormonal responses (release of stress hormones, like adrenaline and adrenal steroids, as well as a host of peptides, into the bloodstream). The ANS and hormonal responses can be considered together as visceral responses—responses of the internal organs and glands (the viscera). When these behavioral and visceral responses are expressed, they create signals in the body that return to the brain.

The opportunities for bodily feedback during emotional reactions to influence information processing by the brain and the way we consciously feel are enormous. Nevertheless, much debate has occurred over whether feedback has any effect on emotional experience and if so how much (see Chapter 3). William James, you'll recall, is the father of the feedback theory. He argued that we do not cry because we are sad or run from danger because we are afraid, but instead we are sad because we cry and are afraid because we run. James was attacked by Cannon, who argued that feedback, especially from the viscera, would be too slow and undifferentiated to determine what emotion you are feeling at the moment. Let's ignore the fact that James included somatic as well as visceral feedback in his theory for now and just consider the validity of Cannon's claims about the viscera.

In Cannon's day, the visceral systems were indeed thought to respond uniformly in all situations. However, we now know that the ANS, which controls the viscera, has the ability to respond selectively, so that visceral organs can be activated in different ways in different situations. Recent studies show, for example, that different emotions (anger, fear, disgust, sadness, happiness, surprise) can be distinguished to some extent on the basis of different autonomic nervous system responses (like skin temperature and heart rate).[67]

The main hormone that was thought to be important for emotional experience in Cannon's time was adrenaline, which is under the control of the ANS and thus was thought to respond uniformly in different situations. However, we now know that there are steroid and peptide hormones that are released by body organs during emotional arousal and that travel in the blood to the brain. It is conceivable that activation of different emotional systems in the brain results in different patterns of hormone release from body organs, which in turn would produce different patterns of chemical feedback to the brain that could have unique effects in different emotions.

Regardless of their specificity, though, visceral responses have relatively slow actions, too slow in fact to be the factor that determines what emotion you experience in a given moment. At a minimum, it takes a second or two for signals to travel from the brain to the viscera and then for the viscera to respond and for the signals created by these responses to return to the brain. For some systems the delay is even longer. It's not so much the travel time from the brain to the organs by way of nerve pathways that's slow, it's the response time of the organs themselves. Visceral organs are made up of what is called "smooth muscle," which responds much more slowly than the striated muscles that move our skeleton during behavioral acts. Also, for hormonal responses the travel time in the blood to the brain can be slow, and for some hormones (like adrenal steroids) the effects on the brain can require the synthesis of new proteins and can take hours to be achieved.

On the other hand, emotional states are dynamic. For example, fear can turn into anger or disgust or relief as an emotional episode unfolds, and it is possible that visceral feedback contributes to these emotional changes over time. While arousal is nonspecific and tends to lock you into the state you are in when the arousal occurs, unique patterns of visceral, especially chemical, feedback have the potential for altering which brain systems are active and thus may contribute to transitions from one emotion to another within a given emotional event.

So Cannon was on target about the inability of visceral responses to determine emotional feelings, but more because of their slow time course than their lack of specificity. At the same time, though, Cannon's critique was somewhat inappropriate given that James had argued for the importance of somatic as well as visceral feedback. And the somatic system clearly has the requisite speed and specificity to contribute to emotional experiences (it takes much less than a second for your striated muscles to respond to a stimulus and for the sensations from these responses to reach your cortex). This point was noted many years ago by Sylvan Tomkins and was the basis of his facial feedback theory of emotion,[68] which has been taken up and pursued in recent years by Carroll Izard.[69]

While most contemporary ideas about somatic feedback and emotional experience have been about feedback from facial expressions, a recent theory by Antonio Damasio, the somatic marker hypothesis, calls upon the entire pattern of somatic and visceral feedback from the body.[70] Damasio proposes that such information underlies "gut feelings" and plays a crucial role in our emotional experiences and decision making processes.

When all the interactions between the various systems are taken together, the possibilities for the generation of emotion-specific patterns of feedback are staggering. This is especially true when considered from the point of view of what would be necessary to scientifically document the existence of these patterns, or, even more difficult, to prove that feedback is not important.

One approach to this problem has been to study emotional feelings in persons who have spinal cord injuries, in whom the flow of information from the brain to the body and from the body back to the brain is, to a great degree, interrupted. An early study claimed that patients with the most severe damage had a dulling of the intensity of emotional feelings and a reduction in the range of emotions experienced, lending support to the idea that feedback plays an important role.[71] Later studies suggest that the first study was flawed and that when the experiment is done properly no deficits in emotional feelings result.[72] However, spinal cord injury does not completely interrupt information flow between the brain and body. For example, spinal cord injury can spare the vagus nerve, which transmits much information from the visceral organs to the brain, and it also fails to interfere with the flow of hormones and peptides from the brain to the body and from the body to the brain. And, of course, the nerves controlling facial movements and sending sensations from facial movements back to the brain are intact, since these go directly between the brain and face without going through the spinal cord. Failure to find a dulling of emotional experiences, or a restriction of the range of emotional experiences, in these patients does not really prove anything.[73]

There is one remaining argument against a contribution of feedback to emotion that needs to be considered. Although somatic responses, like facial or somatic muscle movements, have the requisite speed and specificity to contribute to emotional feelings, it has been argued that these cannot do the trick either. The same response (like running) can occur during different emotions (running to obtain food or to escape from danger) and diametrically opposed responses can occur during the same emotion (we can run or freeze in fear). While these comments are obviously true, it is important to remember that bodily feedback occurs in a biological context. Bodily feedback, when detected by the brain, is recorded by the systems that produced the responses in the first place. Although we may run both to get food and to escape from danger, the feedback from the somatic and visceral responses that return to the brain will interact with different systems in these two instances. The feedback from running from danger will find the food-seeking system idle but the defense system active. The same pattern of feedback can have unique contributions when it interacts with specific brain systems.

William James said that he found it impossible to imagine an emotional experience occurring in the absence of the bodily responses that accompany it—he didn't believe in disembodied emotions.[74] I have to agree for several reasons. First of all, it seems from personal experience that emotions work this way. Most of us feel our emotions in our body, which is why we have such expressions as "an aching heart" and a "gut-wrenching" experience. While personal experience is not a good way to prove anything (we've seen the perils of introspection as scientific data), there's nothing wrong with using it as a takeoff point for a more penetrating analysis. Second, the evidence against feedback playing a role is weak—the spinal cord studies are inconclusive at best. Third, there is plenty of feedback available during emotional responses, and quite a bit of it is fast enough and specific enough to play a role in subjective experiences. Fourth, studies by Paul Ekman and by Robert Zajonc have shown that feedback is indeed used.[75] For example, Ekman had subjects move certain facial muscles. Unbeknownst to the subjects, they were being made to exhibit the facial expressions characteristic of different emotions. They then had to answer some questions about their mood. It turns out that the way the subjects felt was significantly influenced by whether they had been wearing positive or negative emotion expressions. Putting on a happy face may not be such a bad idea when you are feeling blue.

It's hard to believe that after all these years we actually still don't have a clear and definitive understanding of the role of body states in emotions. However, I'm placing my bets in favor of feedback playing a role. Emotional systems evolved as ways of matching bodily responses with the demands being made by the environment, and I can't see many ways that a full-blooded emotional feeling could exist without a body attached to the brain that is trying to have the feeling.

There is one way, though, that should be mentioned. It involves what Damasio calls "as if" loops. In certain situations, it may be possible to imagine what bodily feedback would feel like if it occurred. This "as if" feedback then becomes cognitively represented in working memory and can influence feelings and decisions. We saw evidence that this could occur in Chapter 3 when we considered a study by Valins. He gave subjects false feedback about their heart rate. Their belief that their heart rate was changing was enough to make them feel as though they were emotionally aroused and that they liked certain pictures more than others. This kind of situation, obviously, could only exist in a brain that had experienced real feedback many times so that the way feedback feels and works could be imagined and used to actually influence subjective experiences. This reinforces, rather than questions, the role of body states in emotional feelings.

Feelings: The Bare Essentials

We now have all the ingredients of an emotional feeling, all the things needed to turn an emotional reaction into a conscious emotional experience. We've got a specialized emotion system that receives sensory inputs and produces behavioral, autonomic, and hormonal responses. We've got cortical sensory buffers that hold on to information about the currently present stimuli. We've got a working memory executive that keeps track of the short-term buffers, retrieves information from long-term memory, and interprets the contents of the short-term buffers in terms of activated long-term memories. We also have cortical arousal. And finally, we have bodily feedback—somatic and visceral information that returns to the brain during an act of emotional responding. When all of these systems function together a conscious emotional experience is inevitable. When some components are present and others lacking, emotional experiences may still occur, depending on what's there and what's not. Let's see what's dispensable and indispensable for the emotion fear.

- You *can't* have a conscious emotional feeling of being afraid without aspects of the emotional experience being represented in working memory. Working memory is the gateway to subjective experiences, emotional and nonemotional ones, and is indispensable in the creation of a conscious emotional feeling.

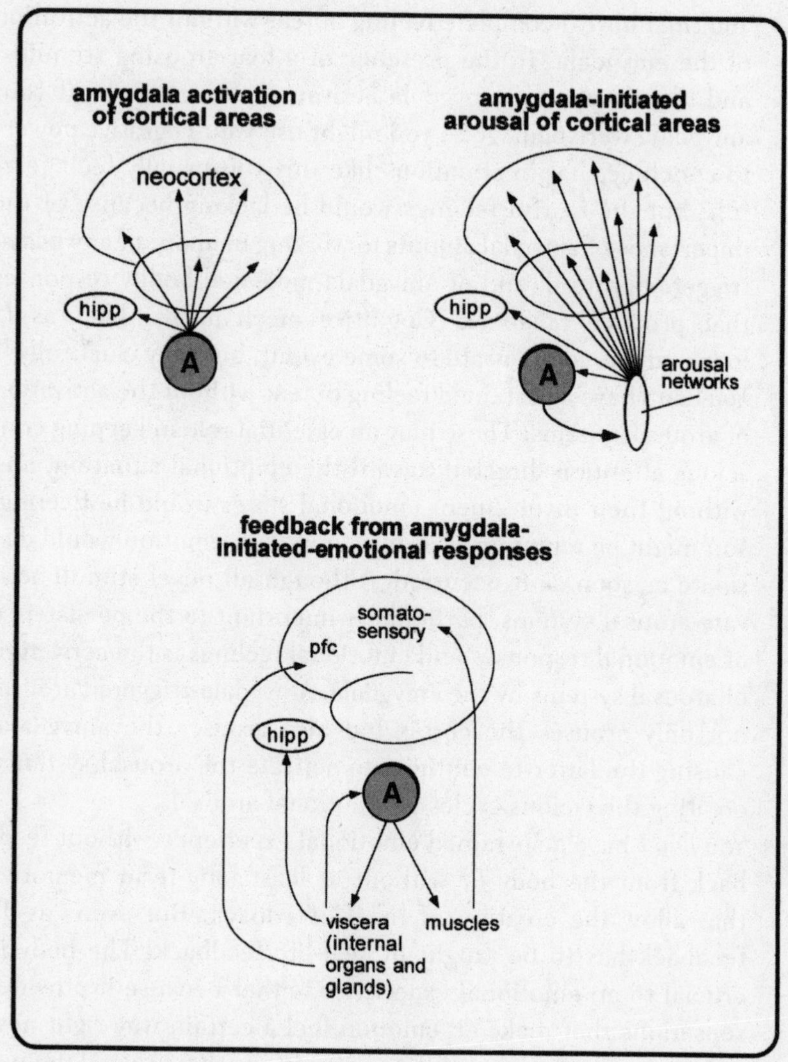

FIGURE 9-7

Some *Neural* Ingredients of a Conscious Emotional Experience.

Conscious emotional experiences are made up of a number of ingredients. Some of the factors that contribute are illustrated, including direct inputs from the amygdala to cortical areas (sensory and higher-order processing regions), inputs from the amygdala to nonspecific arousal systems and from these to widespread areas of the forebrain (cortical and subcortical areas), and feedback to the amygdala and cortical areas from the bodily expression of emotion. Note that the bodily expressions (visceral and muscular) are themselves controlled by the amygdala. A, amygdala; hipp, hippocampus; pfc, prefrontal cortex.

- You *can't* have a complete feeling of fear without the activation of the amygdala. In the presence of a fear-arousing stimulus, and the absence of amygdala activation (for example, if your amygdala were damaged), you might use your cognitive powers to conclude that in situations like this you usually feel "fearful," but the fearful feelings would be lacking because of the importance of amygdala inputs to working memory, of amygdala-triggered arousal, and of amygdala-mediated bodily responses that produce feedback. Cognitive mechanisms, like "as-if" loops, might compensate to some extent, but they can't fully.[76]
- You *can't* have a sustained feeling of fear without the activation of arousal systems. These play an essential role in keeping conscious attention directed toward the emotional situation, and without their involvement emotional states would be fleeting. You might be temporarily aroused but your emotion would dissipate as soon as it occurred. Although all novel stimuli activate arousal systems, particularly important to the persistence of emotional responses and emotional feelings is the activation of arousal systems by the amygdala. Amygdala-triggered arousal not only arouses the cortex but also arouses the amygdala, causing the latter to continue to activate the arousal systems, creating the vicious cycles of emotional arousal.
- You *can't* have a sustained emotional experience without feedback from the body or without at least long-term memories that allow the creation of "as-if" feedback. But even "as-if" feedback has to be taught by real-life feedback. The body is crucial to an emotional experience, either because it provides sensations that make an emotion feel a certain way right now or because it once provided the sensations that created memories of what specific emotions felt like in the past.
- You probably *can* have an emotional feeling without the direct projections to the cortex from the amygdala. These help working memory know which specialized emotion system is active, but this can be figured out indirectly. Nevertheless, the emotion will be different in the absence of this input than in its presence.
- You *can* have an emotional feeling without being conscious of the eliciting stimulus—without the actual eliciting stimulus

being represented in a short-term cortical buffer and held in working memory. As we saw in Chapter 3, stimuli that are not noticed, or that are noticed but their implications aren't, can unconsciously trigger emotional behaviors and visceral responses. In such situations, the stimulus content of working memory will be amplified by the arousal and feedback that result, causing you to attribute the arousal and bodily feelings to the stimuli that are present in working memory. However, because the stimuli in working memory did not trigger the amygdala, the situation will be misdiagnosed (recall Schachter and Singer's subjects who were artificially aroused and who misattributed their arousal to their surroundings). And if there is nothing particular occupying working memory, you will be in a situation where your feelings are not understood. If emotions are triggered by stimuli that are processed unconsciously, you will not be able to later reflect back on those experiences and explain why they occurred with any degree of accuracy. Contrary to the primary supposition of cognitive appraisal theories, the core of an emotion is not an introspectively accessible conscious representation. Feelings do involve conscious content, but we don't necessarily have conscious access to the processes that produce the content. And even when we do have introspective access, the conscious content is not likely to be what triggered the emotional responses in the first place. The emotional responses and the conscious content are both products of specialized emotion systems that operate unconsciously.

What's Different About Thoughts and Feelings?

Conscious emotional feelings and conscious thoughts are in some sense very similar. They both involve the symbolic representation in working memory of subsymbolic processes carried out by systems that work unconsciously. The difference between them is not due to the system that does the consciousness part but instead is due to two other factors. One is that emotional feelings and mere thoughts are generated by different subsymbolic systems. The other is that emotional feelings involve many more brain systems than thoughts.

When we are in the throes of emotion, it is because something important, perhaps life threatening, is occurring, and much of the brain's resources are brought to bear on the problem. Emotions create a flurry of activity all devoted to one goal. Thoughts, unless they trigger emotional systems, don't do this. We can daydream while doing other things, like reading or eating, and go back and forth between the daydream and the other activities. But when faced with danger or other challenging emotional situations, we don't have time to kill nor do we have spare mental resources. The whole self gets absorbed in the emotion. As Klaus Scherer has argued, emotions cause a mobilization and synchronization of the brain's activities.[77]

Do Fish Have Feelings Too?

Philosophers have something called "the problem of other minds." In simple terms, it is the difficulty, if not the impossibility, of proving that anyone, other than oneself, is conscious. This is an onerous problem that applies both to the minds of other humans and other animals. We are somewhat better off in the case of other humans than other animals though. Depending on how strict we are (philosophically), we can usually convince ourselves that most other humans have emotional feelings and other conscious states of mind because we can talk to them and compare notes about our mental experiences with them—this is one of the beauties of having natural language. We may not be completely justified philosophically in our conclusion that other people are conscious, but from a practical point of view it is useful to live our lives in violation of philosophical certainty and treat others as if they are conscious. Fortunately, though, there is another reason to adopt the belief that other humans are conscious. Since all humans have pretty much the same kind of brain architecture we can assume that, barring pathological conditions, the same general kinds of functions come out of all human brains—if I'm conscious and you have the same kind of brain that I do, then you are probably conscious as well. This kind of reasoning holds for brain functions that we know something about (like perception and memory), so we might reasonably expect it to also hold for conscious awareness.

But no matter how firm or flimsy the arguments about consciousness in other humans are, when it comes to making the leap to the minds of other animals, we are on considerably shakier ground. Our ability to hold conversations with other animals is somewhere between not at all and not much.[78] And while our brain is, in many ways, incredibly similar to the brains of other creatures (this is what makes much of brain research possible), it also differs in some important ways. The human brain, most especially the cerebral cortex, is much larger than it should be, given our body size.[79] This alone would give us reason to be cautious about attributing consciousness to other animals. However, there are other facts to take into account. First, as we've seen, the part of the human cortex that has increased in size the most is the prefrontal cortex,[80] which is the part of the brain that has been implicated in working memory, the gateway to consciousness. Some brain scientists believe that this part of the cortex doesn't even exist except in primates.[81] And there is behavioral evidence that only the higher primates, in whom the prefrontal cortex is especially well developed, are self-aware, as determined by their ability to recognize themselves in a mirror.[82] Second, natural language only exists in the human brain.[83] Although the exact nature of the brain specialization involved in making language possible is not fully understood, something changed with the evolution of the human brain to make language happen. Not surprisingly, the development of language has often been said to be the key to human consciousness.[84] Clearly, the human brain is sufficiently different from the brains of other animals to give us reasons for being very cautious about attributing consciousness beyond our species. As a result, the arguments that allow us to say with some degree of confidence that other humans have conscious states do not allow us to insert consciousness into the mental life of most other animals.

My idea about consciousness in other animals is this. Consciousness is something that happened after the cortex expanded in mammals. It requires the capacity to relate several things at once (for example, the way a stimulus looks, memories of past experiences with that stimulus or related stimuli, a conception of the self as the experiencer).[85] A brain that cannot form these relations, due to the absence of a cortical system that can put all of the information together at the same time, cannot be conscious. Consciousness, so defined, is undoubtedly present in humans. To the extent that other animals have the capacity to hold and manipulate information in a generalized mental workspace, they probably also have the potential capacity to be conscious. This formulation allows the possibility that some other mammals, especially (but not exclusively) some other primates, are conscious. However, in humans, the presence of natural language alters the brain significantly. Often we categorize and label our experiences in linguistic terms, and store the experiences in ways that can be accessed linguistically. Whatever consciousness exists outside of humans is likely to be very different from the kind of consciousness that we have.

The bottom line is this. Human consciousness is the way it is because of the way our brain is. Other animals may also be conscious in their own special way due to the way their brains are. And still others are probably not conscious at all, again due to the kinds of brains they have. At the same time, though, consciousness is neither the prerequisite to nor the same thing as the capacity to think and reason. An animal can solve lots of problems without being overtly conscious of what it is doing and why it is doing it. Obviously, consciousness elevates thinking to a new level, but it isn't the same thing as thinking.

Emotional feelings result when we become consciously aware that an emotion system of the brain is active. Any organism that has consciousness also has feelings. However, feelings will be different in a brain that can classify the world linguistically and categorize experiences in words than in a brain that cannot. The difference between fear, anxiety, terror, apprehension, and the like would not be possible without language. At the same time, none of these words would have any point if it were not for the existence of an underlying emotion system that generates the brain states and bodily expressions to which these words apply. Emotions evolved not as conscious feelings, linguistically differentiated or otherwise, but as brain states and bodily responses. The brain states and bodily responses are the fundamental facts of an emotion, and the conscious feelings are the frills that have added icing to the emotional cake.

Qué Será Será

Where is evolution taking our brain? While it is true that whatever will be will be, we have the opportunity to take a peek at what evolution is up to. It's not that evolution is forward thinking. It only has hindsight.[86] However, we *are* evolution in progress and we can see what sorts of changes might be happening in our brain by looking at trends in brain evolution across related species.

As things now stand, the amygdala has a greater influence on the cortex than the cortex has on the amygdala, allowing emotional arousal to dominate and control thinking. Throughout the mammals, pathways from the amygdala to the cortex overshadow the pathways from the cortex to the amygdala. Although thoughts can easily trigger emotions (by activating the amygdala), we are not very effective at willfully turning off emotions (by deactivating the amygdala). Telling yourself that you should not be anxious or depressed does not help much.

At the same time, it is apparent that the cortical connections with the amygdala are far greater in primates than in other mammals. This suggests the possibility that as these connections continue to expand, the cortex might gain more and more control over the amygdala, possibly allowing future humans to be better able to control their emotions.

Yet, there is another possibility. The increased connectivity between the amygdala and cortex involves fibers going from the cortex to the amygdala as well as from the amygdala to the cortex. If these nerve pathways strike a balance, it is possible that the struggle between thought and emotion may ultimately be resolved not by the dominance of emotional centers by cortical cognitions, but by a more harmonious integration of reason and passion. With increased connectivity between the cortex and amygdala, cognition and emotion might begin to work together rather than separately.

Oscar Wilde once said, "It is because Humanity has never known where it was going that it has been able to find its way."[87] But wouldn't it be wonderful if we did understand where our emotions were taking us from moment to moment, day to day, and year to year, and why? If the trends toward cognitive-emotional connectivity in the brain are any indication, our brains may, in fact, be moving in this direction.

NOTES

Chapter 1: What's Love Got to Do with It?

1. Dreiser (1900).
2. The study of emotions in brain science has gone through cycles. We'll review this history in detail in Chapter 4. For now, I'll just note that neuroscientists have, in recent decades, been much more interested in the intellectual or cognitive aspects of the mind than in emotions. However, this is starting to change. Although there are still relatively few neuroscientists who claim *emotions* as their main interest, emotional functions of the brain are becoming more popular as research topics. There are several reasons for this shift. One has to do with the recognition that the mind is more than cognition, so that the focus on cognitive processes by neuroscientists only reveals part of the mind. Another is the realization that the subjective states of awareness that accompany emotions are just part of the overall emotion process, and that much can be learned by studying how the brain processes stimuli and controls objectively measured responses in emotional situations. Since stimulus processing and response control can be studied in animals, but subjective awareness cannot, a focus on processing and responses has facilitated research. These notions are developed further in Chapters 2 and 3.
3. Gazzaniga, Bogen, and Sperry (1962); Gazzaniga, Bogen, and Sperry (1965); Gazzaniga (1970).
4. Bogen and Vogel (1962).
5. D. Wilson, et al (1977).
6. Our research on the Dartmouth patients between 1974 and 1978 is summarized in Gazzaniga and LeDoux (1978).
7. Gazzaniga (1972).

8. Studies of this patient are described in Gazzaniga and LeDoux (1978).
9. See Davidson (1992); Heilman and Satz (1983); Gainotti (1972).
10. A similar point was made by E. Duffy in the 1940s [Duffy (1941)]. However, while Duffy wished to do away with talk about emotion, I want to understand what emotions are. The key is the use of the plural rather than the singular. I don't think that there is anything called "emotion" but I do believe that there are lots of "emotions."

Chapter 2: Souls on Ice

1. This quote was seen on the wall behind the counter at Kim's Underground Video in Greenwich Village in Manhattan.
2. Melville (1930).
3. Bangs (1978).
4. Theories of emotion will be discussed in this chapter and Chapter 3.
5. Fehr and Russell (1984).
6. Plato, *Phaedo,* cited in Flew (1964).
7. Gardner (1987).
8. Watson (1929); Skinner (1938).
9. Actually, psychology, as a field of science, did not exist until the late nineteenth century, when it emerged in Germany as the experimental study of consciousness [see Boring (1950)]. Before that, mental phenomena were the business of philosophers. And following Descartes' proclamation, "I think therefore I am," mind and consciousness came to be equated in Western philosophical discussions, a trend that was inherited by scientific psychology when it emerged. For a translation of some of Descartes' key writings, see Smith (1958). For a summary of the importance of Descartes' views in forcing the modern equation of mind as consciousness, see Rorty (1979). According to Rorty, mind and consciousness were not such interchangeable ideas before Descartes introduced the notion of an all-knowing soul (consciousness), that had no unknowable (unconscious) aspects. If it wasn't knowable (available to conscious) it wasn't mental. In this way, certain things that we consider mental today (like sensations and some aspects of emotions) were demoted to physical states by Descartes.
10. Ryle (1949).
11. The following summary is based on Gardner (1987).
12. Putnam (1960).
13. Rorty (1979).
14. Lashley (1950b).
15. Neisser (1976); Gardner (1987); Kihlstrom (1987).

16. For example, an abacus is a computer made of sticks and stones. It does calculations using an algorithm or program that is built into its design. For some problems, it is as effective as (and in some instances more practical than) an electronic computer.
17. Kihlstrom (1987).
18. Freud (1925). For a reinterpretation of Freudian concepts in terms of cognitive science, see Erdelyi (1985).
19. Erdelyi (1985).
20. These are sometimes called preperceptual or preattentive processes. For example, in visual perception, the determination of the intensity of light reflecting from different parts of a stimulus or the direction of movement of a stimulus occur preconsciously. For a discussion of these processes, see Marr (1982); Ullman (1984).
21. Kosslyn and Koenig (1992); Kosslyn (1983); Kosslyn (1980). For a challenge to Kosslyn's theory, see Pylyshyn (1984).
22. Pinker (1994).
23. Nisbett and Wilson (1977).
24. However, not everyone agrees with the strong claims made by Nisbett and Wilson. For example, following Nisbett and Wilson's study, Ericcson and Simon (1984) attempted to identify whether there might be some kinds of conscious introspections that could be trusted. After an exhaustive study, they concluded that verbal reports about the state of one's mind can be used reliably to indicate the outcome of a decision (whether one thing is bigger than another, whether you like or dislike something, or whether you plan to do something) but that such reports are less reliable about the processes leading up to a decision, especially if there is some delay between the occurrence of the process and the report. They emphasized that information in short-term memory is most accessible, allowing for accurate descriptions of processes as they occur or shortly thereafter, but once information decays from or is displaced from short-term memory, accessibility can decline. Some say that since the events that cause a behavior or mental state typically occur right before the behavior or state, we typically have conscious access to causes because the causal events are still in short-term memory, a view sometimes known as folk psychology [see Goldman (1993); Churchland (1984); Arnold (1960); Johnson-Laird and Oatley (1992); Oatley and Duncan (1994)]. However, in my mind, this view is problematic in at least three respects. First, it assumes that all of the stimuli that have significant mental effects occupy short-term memory and thus are noticed and appreciated. As we will see in the next chapter, some things are not noticed but still influence us, and other things are fully noticed

but their significance is processed implicitly and not consciously appreciated. This latter point means that stimuli that are consciously perceived can have important unconscious effects, influencing our emotions, goals, and attitudes without our being aware that they are being influenced. Second, it assumes that the stimuli that provoke behavior are what cause it, which is not necessarily the case. Innocuous events can set us off if we are in a bad mood. The mood, more than the eliciting stimulus, is the cause in such situations. Third, it assumes that we can correctly identify from the many stimuli that were available the exact ones that actually elicited the response. Obviously, we are often correct, otherwise life would be chaotic and impossible (incidentally, the chaotic and impossible life of persons suffering from mental illness may represent a breakdown in these mechanisms, either the introspective one, the attribution one, or the balance between them). But whether we are correct about causes because we have introspective access to causal events or because we are very good at attributing cause on the basis of noticing correlations is less clear. Regardless, even if we wholly accept the Ericsson and Simon view, that some aspects of cognition can be characterized on the basis of introspective verbal reports, there remains room for much of the cognitive mind to operate below the tip of the iceberg. For a discussion of some additional issues, see Bowers and Meichenbaum (1984); Miller and Gazzaniga (1984); Marcel and Bisiach (1988). Also, see the June 1992 issue of *American Psychologist*, which has numerous articles on this topic.

25. Gazzaniga and LeDoux (1978).
26. For summary, see Gazzaniga (1970).
27. These ideas are elaborated on and expanded in Gazzaniga (1985); Gazzaniga (1988); LeDoux (1985).
28. I am grateful to several people for their discussions with me about unconscious processing, including Daniel Schacter, Matthew Erdelyi, John Bargh, and John Kihlstrom. From their comments, and from my own reading of the literature, several methodological problems that plague studies of unconscious processing of stimuli are apparent. One is that much of the work has involved subliminal perception or masking, both of which involve very brief stimulus exposures. This limits the amount of information that can be presented at one time and also limits the amount of cognitive resources that can be dedicated to the processing task. It is likely that the limits of the unconscious found through such studies reflect, at least to some extent, methodological limitations rather than the real limits of unconscious processing. Another problem with the arguments against the existence of a sophisti-

cated cognitive unconscious is the fact that most of the work has used verbal stimuli (words, sentences) to test processing limits. These are the currency of the systems that are involved in conscious processing, which is an evolutionarily new system. Unconscious processing, on the other hand, is the stock-in-trade of evolutionarily old systems that are likely to be more readily studied with nonverbal measures. Indeed, some of the strongest evidence for unconscious processing comes from studies using pictorial rather than verbal stimuli. These studies will be examined in Chapter 3. Another methodological problem is that of drawing the line between conscious and unconscious processing. Several recent attempts to use more sophisticated analytic techniques to draw the line have been made. Included is work by Merikle, Jacoby, Erdelyi, Bargh, and Kihlstrom (citations below). Each of these concludes that the cognitive unconscious can process significant meanings. Also, as indicated in the text, stimuli that are consciously processed can also be processed by unconscious systems, which may in fact do different things with them; and stimuli that are noticed and attended to can have important unconscious influences because it is their activation of unconscious meanings that is most important, not their physical features. These ideas are elaborated on further in the next chapter. A sampling of citations include: Merikle (1992); Kihlstrom, Barnhardt, and Tataryn (1992); Erdelyi (1992); Bargh (1992); Bargh (1990); Jacoby et al (1992).

29. Bowers (1984); Bowers and Meichenbaum (1984); Bargh (1992); Bargh (1990); Jacoby et al (1992).
30. Posner (1990); Anderson (1990); Kosslyn and Koenig (1992); Gardner (1987).
31. The mind-body problem, the problem of how the mind relates to the brain and the rest of the body, is a deeply troubling philosophical problem. It has always been a thorn in the side of psychology. For a nice summary of the issues involved see Churchland (1984). For a summary of the early impact on psychology, see Boring (1950). The mind-body problem and its relation to cognitive science is discussed in Gardner (1987). Another discussion of the mind-body problem that I like is in Jackendoff (1987).
32. As I was finalizing this book there was a very exciting chess match going on in Philadelphia between grand master Gary Kasparov and a computer. The computer gave Kasparov a run for his money.
33. Gardner (1987).
34. Neisser (1967).
35. Fodor (1975).

36. Von Eckardt (1993).
37. Russell (1905).
38. Fodor (1975).
39. The history of artificial intelligence is nicely summarized in Gardner (1987).
40. The following summary of the work of Johnson-Laird and Kahnemann and Tversky is based on a description in Gardner (1987).
41. Johnson-Laird (1988).
42. Kahneman, Slovic, and Tversky (1982).
43. Frank (1988).
44. Tooby and Cosmides (1990).
45. Goleman (1995).
46. Aristotle (1941); de Sousa (1980); Solomon (1993).
47. Damasio (1994).
48. Dyer (1987); Scherer (1993b); Frijda and Swagerman (1987); Sloman (1987); Grossberg (1982); Armony et al (1995).
49. Johnson-Laird (1988).
50. Simon (1967).
51. Abelson (1963).
52. Miller and Johnson-Laird (1976).
53. Newell, Rosenblum, and Laird (1989).
54. Hilgard (1980).
55. Churchland and Sejnowski (1990).
56. We'll discuss the bodily responses in emotions in Chapters 3–6.
57. James (1884).

Chapter 3: Blood, Sweat, and Tears

1. Crockett (1845).
2. James (1884).
3. Ibid.
4. Cannon (1929).
5. To be fair, though, it is important to note that James proposed that the whole body response in an emotion determines the feedback, not just the ANS response.
6. Modern research on bodily responses suggests that there may be more specificity than proposed by Cannon [see Ekman et al (1983); Levenson (1992); Cacioppo et al (1993)].
7. Watson (1929); Watson and Rayner (1920); Skinner (1938); Duffy (1941); Lindsley (1951).
8. Ryle (1949).

9. Schachter and Singer (1962).
10. Valins (1966).
11. Frijda (1986); Plutchik (1980).
12. Aristotle (1941); Spinoza (1955); Descartes (1958).
13. Arnold (1960).
14. Lazarus (1966).
15. Lazarus (1991).
16. See Frijda (1993); Scherer (1993a); Lazarus (1991); C.A. Smith and P.C. Ellsworth (1985); Ortony and Turner (1990).
17. C.A. Smith and P.C. Ellsworth (1985).
18. See Frijda (1993); Scherer (1988).
19. Zajonc (1980).
20. Bornstein (1992).
21. The New Look in psychology got going as a result of studies by Jerry Bruner [Bruner and Postman (1947)]. It quickly faded, was revived by an article by M. Erdelyi in the mid-1970s [Erdelyi (1974)], slipped away again, but has returned for a third time [Greenwald (1992)].
22. Psychology and psychoanalysis, though both seek to understand mind and behavior, take very different approaches.
23. Erdelyi (1974).
24. The irony is that verbal reports are mainly useful as a means of knowing what a subject is consciously aware of. In adopting verbal reports, behaviorists were befriending the very concept that behaviorism sought to do away with—consciousness.
25. McGinnies (1949); Dixon (1971).
26. Lazarus and McCleary (1951).
27. Loftus and Klinger (1992).
28. Moore (1988).
29. Packard (1957).
30. Eagly and Chaiken (1993).
31. Eriksen (1960).
32. For discussion, see Bowers (1984); Bowers and Meichenbaum (1984).
33. Erdelyi (1974); Erdelyi (1985).
34. Dixon (1971); Dixon (1981); Wolitsky and Wachtel (1973); Erdelyi (1985); Erdelyi (1992); Ionescu and Erdelyi (1992); Greenwald (1992).
35. Merikle (1992); Kihlstrom, Barnhardt, and Tataryn (1992a); Erdelyi (1992); Bargh (1992); Bargh (1990); Jacoby et al (1992); Murphy and Zajonc (1993).
36. Bornstein (1992).
37. Ibid.
38. Murphy and Zajonc (1993).

39. This discussion of the Pöetzl effect is based on Ionescu and Erdelyi (1992).
40. Erdelyi (1985); Erdelyi (1992); Ionescu and Erdelyi (1992).
41. Bowers (1984); Bowers and Meichenbaum (1984).
42. Shevrin et al (1992); also see Shevrin (1992).
43. Bargh (1992); Bargh (1990).
44. Jacoby et al (1992).
45. Bargh (1992).
46. Merikle (1992); Kihlstrom, Barnhardt, and Tataryn (1992b); Erdelyi (1992); Bargh (1992); Bargh (1990); Jacoby et al (1992).
47. Nevertheless, the relevance of unconscious emotional processing for everyday life needs to be carefully considered. I will describe two of the main arguments against the relevance of unconscious processing and attempt to refute both. Both accept that unconscious processing can exist in the laboratory but question whether it is very meaningful in life. The first argument is that life does not work the way unconscious processing tasks (subliminal perception, masking, or shadowing) do since we normally have the opportunity to look at or listen to the stimuli we encounter. However, life may be more like these experiments than it seems. After all, at any one moment there are more stimuli available than can be registered by the limited-capacity serial-processing system that gives rise to conscious content (see Chapter 9). Stimuli that are not noticed can be implicitly perceived and implicitly remembered by unconscious processing systems that chug away outside of awareness (see Chapter 2). At the same time, as we saw in the present chapter, stimuli that are fully perceptible, and even stimuli that are fully perceived and consciously registered, can enter the brain and their implicit meanings can activate goals, attitudes, and emotions without our being aware of these influences. The second argument against the relevance of unconscious processing says that while it may happen during everyday life, it is a fairly unsophisticated form of processing, being limited to physical stimulus features rather than to concepts [for this argument, see: Greenwald (1992); Bruner (1992); Hirst (1994)]. There are three rebuttals to this. The first is that there is in fact evidence for the unconscious processing of conceptual meanings as well as physical features [see Murphy and Zajonc (1993); Öhman (1992); Kihlstrom, Barnhardt, and Tataryn (1992)]. It has been admittedly more difficult to demonstrate unconscious conceptual priming, but it seems to exist. Second, we may not have a very accurate picture of the sophistication of unconscious processing. Most of the work done so far has used verbal stimuli to analyze conceptual processing, but the unconscious mind

may work more fluently in nonverbal modalities (as we'll discuss at the end of this chapter). We will probably not gain an accurate picture of how the more basic unconscious systems work as long as most work on unconscious processing is done using verbal stimuli. Third, some of the best evidence for unconscious processing comes from studies of emotional processing, which is particularly resistant to verbalization. Finally, even if it were true that unconscious perception is limited to stimulus features, as opposed to higher order concepts, there would be important implications for the unconscious emotional processing. As we have seen, primitive physical features of a person, like the color of their skin or the intonation of their voice, are sufficient to unconsciously activate emotions, attitudes, goals, and intentions.

48. Damasio (1994).
49. Arnold (1960); Johnson-Laird and Oatley (1992); Oatley and Duncan (1994); Goldman (1993).
50. S.J. Gould (1977).
51. Hebb (1946).
52. Rorty (1980).
53. Frijda (1993).
54. Scherer (1943a).
55. Bowers (1984).
56. Lazarus (1991).
57. Scherer (1943a).
58. Ericsson and Simon (1984); Nisbett and Wilson (1977); Bowers and Meichenbaum (1984); Miller and Gazzaniga (1984). Also, see Chapter 2 and footnote 24 in Chapter 2 for some discussion of these issues.
59. Erdelyi (1992).
60. Zajonc (1984); Lazarus (1984); Kleinginna and Kleinginna (1985); Leventhal and Scherer (1987).
61. Dyer (1987); K.R. Scherer (1993b); Frijda and Swagerman (1987); Sloman (1987); Grossberg (1982); Armony et al (1995).
62. Some AI proponents assume that emotional feelings and other states of consicousness could be programnmed into a computer if we could just get the right algorithm. For example, see: Newell (1980); Minsky (1985); Sloman and Croucher (1981). For a rebuttal to the notion that computers might have feelings or other conscious states, see Searle (1984).
63. Messick (1963), p. 317.
64. Kelley (1963), p. 222.
65. Ibid.

CHAPTER 4: THE HOLY GRAIL

1. Paraphrased from my memory of a line from a Woody Allen film.
2. MacLean (1949); MacLean (1952).
3. Boring (1950).
4. It is widely believed that Gall was the father of phrenology, but it turns out, according to Mark Rosenzweig [Rosenzweig (1996)], that Gall called his theory "organology" and it was his followers who popularized this view and turned it into "phrenology."
5. See Boring (1950).
6. However, there is still a modern antilocalizationist movement. John (1972) and Freeman (1994) are proponents.
7. Blindness due to cortical damage is called "cortical blindness." However, people who are cortically blind still have some rudimentary visual perception that operates outside of consciousness. Thus, they can reach toward a stimulus in front of them, but claim not to see it. For more information, see Weiskrantz (1988).
8. Van Essen (1985); Ungerleider and Mishkin (1982).
9. Although it is possible for the functions of a cortical area to be taken over by neighboring tissue if that area is damaged in early life, once the function has developed, such replacements do not normally occur.
10. Darwin (1859).
11. For a history of early studies using stimulation and ablation, see Boring (1950).
12. For a history of early neurological studies of humans, see Plum and Volpe (1987).
13. Animal experiments can yield very detailed information and can be used to elucidate the basic neural systems underlying psychological functions. Human findings have traditionally been based on accidents of nature rather than careful experiments. However, new imaging techniques, which allow the visualization of brain activity from outside the brain, are offering new ways of performing detailed studies of human brain function (some applications of imaging are described in Chapter 9). Nevertheless, the findings from such studies represent correlations between brain activity and psychological states and do not show that the brain activity is responsible for the state. Imaging studies often depend on a solid understanding of basic brain processes from studies of animals to help interpret the findings.
14. Summarized in LeDoux (1987).
15. Kaada (1960); Kaada (1967).

16. Head (1921) proposed that the cortex inhibits subcortical areas.
17. Bard (1929); Cannon (1929).
18. See Cannon (1929).
19. Cannon (1929).
20. Papez (1937).
21. Peffiefer (1955).
22. Herrick (1933).
23. Broca (1978).
24. Papez (1937).
25. Descartes (1958).
26. Papez (1937).
27. Klüver and Bucy (1937); Klüver and Bucy (1939).
28. Klüver and Bucy (1937).
29. Weiskrantz (1956); Downer (1961); Horel, Keating, and Misantone (1975); Jones and Mishkin (1972); Aggleton and Mishkin (1986); Rolls (1992b); Ono and Nishijo (1992); Gaffan (1992).
30. MacLean (1949); MacLean (1952).
31. MacLean (1949).
32. MacLean (1949).
33. Ibid.
34. Ibid.
35. Ibid.
36. MacLean (1952).
37. MacLean (1970); MacLean (1990).
38. MacLean (1970).
39. Other contemporary theories with evolutionary perspectives include: Plutchik (1993); Ekman (1992a); Izard (1992a).
40. Historical facts can be found in: Nauta and Karten (1970); Karten and Shimizu (1991); Northcutt and Kaas (1995).
41. Nauta and Karten (1970); Karten and Shimizu (1991); Northcutt and Kaas (1995).
42. Karten and Shimizu (1991); Northcutt and Kaas (1995); Ebbesson (1980); Swanson (1983).
43. Brodal (1982); Swanson (1983); LeDoux (1991).
44. An important exception is the role of the hippocampus in negative feedback control of stress responses, as described in Chapter 8.
45. Others have also proposed getting rid of the limbic system: Brodal (1982); Kotter and Meyer (1992).

Chapter 5: The Way We Were

1. Leonardo da Vinci (1939).
2. Dawkins (1982).
3. Quoted by Dawkins (1982).
4. Dawkins (1982).
5. Pinker (1995).
6. Fodor (1983); Gazzaniga (1988).
7. Nottebohm, Kasparian, Pandazis (1981); Krebs (1990); Sherry, Jacobs, and Gaulin (1992); Sengelaub (1989); Purves, White, and Andrews (1994); Geschwind and Levitsky (1968); Galaburda et al (1987). These references were taken from Finlay and Darlington (1995).
8. Finlay and Darlington (1995).
9. Pinker (1994).
10. This point is nicely made by Plutchik (1980) and will be considered in more detail later.
11. Plato, *Phaedo,* cited in Flew (1964).
12. Darwin (1859).
13. S.J. Gould (1977).
14. Cited by S.J. Gould (1977).
15. Simpson (1953); J.M. Smith (1958); Ayala and Valentine (1979); J.L. Gould (1982).
16. J.L. Gould (1982).
17. Darwin (1872). All references to Darwin below are from this source, unless otherwise indicated.
18. Summarized by Plutchik (1980).
19. Tomkins (1962).
20. Izard (1977); Izard (1992a).
21. Ekman (1984).
22. Plutchik (1980).
23. Frijda (1986).
24. Johnson-Laird and Oatley (1992).
25. Panksepp (1982).
26. Arnold (1960); Fehr and Russell (1984); J.A. Gray (1982).
27. C.A. Smith and R.S. Lazarus (1990).
28. Harré (1986).
29. Averill (1980). Averill's account of the Gururumba is based on research conducted by Newman [P.L. Newman (1960)].
30. Morsbach and Tyler (1986).
31. Ibid.
32. Doi (1973).

33. Heelas (1986); Davitz (1969); Geertz (1959).
34. Wierzbicka (1994).
35. Ekman (1980).
36. Ibid.
37. Twain (1962).
38. Ekman (1980).
39. Ortony and Turner (1990).
40. Gallistel (1980).
41. Ekman (1992a); Izard (1992a).
42. Plutchik (1980).
43. Ibid.
44. Bowlby (1969).
45. Shepherd (1983).
46. Nauta and Karten (1970).
47. S.J. Gould (1977); Pinker (1994).
48. Preuss (1995); Reep (1984); Uylings and van Eden (1990).
49. Nauta and Karten (1970); Karten and Shimizu (1991); Northcutt and Kaas (1995).
50. Preuss (1995); Geschwind (1965).
51. For example, protection from danger, finding food and shelter and suitable mates, and the like.
52. For example, see Jaynes (1976).
53. Ekman (1992a).
54. Johnson-Laird and Oatley (1992).
55. Tooby and Cosmedies (1990).
56. Natural triggers are what ethologists call sign stimuli. These elicit behavioral and/or physiological responses innately. They are also similar to unconditioned stimuli, which elicit responses innately as well (see Chapter 6).
57. Examples include sexual partner recognition or food detection networks.
58. James (1890).
59. Marks (1987).
60. Eibl-Eibesfeldt and Sutterlin (1990).
61. Kierkegaard (1844); Sartre (1943); Heidegger (1927).
62. Lazarus (1991).
63. Marks (1987).
64. D.C. Blanchard and R.J. Blanchard (1989).
65. D.C. Blanchard and R.J. Blanchard (1988); R.J. Blanchard and D.C. Blanchard (1989).

66. Bolles and Fanselow (1980); Watkins and Mayer (1982); Helmstetter (1992).
67. Bolles and Fanselow (1980); Watkins and Mayer (1982); Helmstetter (1992).
68. Among vertebrates, the organizational plan of autonomic nervous system function is similar from amphibians through mammals, including man [Shepherd (1983)].
69. Jacobson and Sapolsky (1991).
70. This will be discussed in some detail in Chapter 8.
71. The neuroendocrine systems, like most other neural systems, is similarly organized in different species. For examples, see: Shepherd (1983); J.A. Gray (1987); McEwen and Sapolsky (1995).
72. Dawkins (1982).
73. This is similar to the approach of ethologists, who look at the invariant, evolutionarily prescribed aspects of behavior, and of evolutionary psychologists, who tend to emphasize evolution's effects on the mind. For a summary of ethological approaches, see J.L. Gould (1982). For an example of the evolutionary psychology approach, see Tooby and Cosmedies (1990).
74. This work is summarized in: J.A. Gray (1987).
75. Wilcock and Broadhurst (1967).
76. See J.A. Gray (1987); Marks (1987).
77. Marks (1987); Kagan and Snidman (1991).
78. J.L. Gould (1982).
79. Tully (1991).
80. Sibley and Ahlquist (1984).
81. Dawkins (1982).

Chapter 6: A Few Degrees of Separation

1. Dickinson (1955).
2. This title is based on the popular play and film *Six Degrees of Separation* by John Guare (1990).
3. Pavlov (1927).
4. D.C. Blanchard and R.J. Blanchard (1972).
5. R.J. Blanchard et al (1993); D.C. Blanchard and R.J. Blanchard (1988).
6. Campeau, Liang, and Davis (1990); Gleitman and Holmes (1967).
7. Pavlov (1927).
8. Bouton (1994); Bouton and D. Swartzentruber (1991).
9. Campbell and Jaynes (1966).

10. Jacobs and Nadel (1985); Marks (1987).
11. Hodes, Cook, and Lang (1985); Hugdahl (1995); Öhman (1992).
12. However, whether fear is experienced is not crucial, since the system is an implicit or unconscious learning system, the processing of which may or may not reach consciousness. Explicit and implicit learning systems will be discussed in Chapter 7.
13. McAllister and McAllister (1971); Brown, Kalish, and Farber (1951); Davis, Hitchcock, and Rosen (1987).
14. For some examples see Carew, Hawkins, and Kandel (1983); Tully (1991); D. H. Cohen (1980); Schneiderman et al (1974); Bolles and Fanselow (1980); Smith et al (1980); Öhman (1992).
15. This is called Occam's razor, or the law of parsimony, and it dictates that we not call upon a complex explanation or process when a simpler one will do. The recent revival of anthropocentric thinking about animal minds [see McDonald (1995); Masson and McCarthy (1995)] intentionally violates this law, which I think is a mistake. If you can't prove consciousness in a particular creature, then you shouldn't use consciousness as an explanation of its behavior.
16. This has always been controversial. However, recent studies by Öhman have shown that conditioning can occur in the absence of conscious awareness of the conditioned stimulus and its relation to the unconditioned stimulus. He uses "backwards masking," which allows the CS to enter the brain but not to enter consciousness [Öhman (1992)].
17. Other ways to study fear include the use of electical brain stimulation to directly elicit fear responses or the use of avoidance conditioning procedures [see LeDoux (1995)].
18. It is possible that some aspects of fear, like fear of failing or fear of being afraid, are not easily modeled by fear conditioning.
19. Freezing occurs in many species in response to sudden danger [Marks (1987)] but has been studied as a conditioned response mostly in rats.
20. Von Uexkull (1934).
21. Archer (1979).
22. Cannon (1929); Hilton (1979); Mancia and Zanchetti (1981).
23. Mason (1968); van de Kar et al (1991).
24. Bolles and Fanselow (1980); Watkins and Mayer (1982); Helmstetter (1992).
25. Brown, Kalish, and Farber (1951); Davis (1992b); Weisz, Harden, and Xiang (1992).
26. D.H. Cohen (1980).
27. Kapp et al (1992).
28. McCabe et al (1992).

29. Powell and Levine-Bryce (1989).
30. Fanselow (1994).
31. Davis (1992).
32. O.A. Smith et al (1980).
33. LeDoux (1994), (1995).
34. Most of the older data was based on the placement of a lesion in a brain area. The area then degenerated and the nerve fibers that originated there would in turn degenerate. By using special stains, the degenerating fibers could then be seen. However, because fibers passing through but not originating in the lesioned area were also damaged, it was possible to get false results. Although newer techniques using tracer chemicals also have this problem to some extent (since the tracer can be taken up by passing axons in some cases), it is much less of a problem than with the older techniques.
35. Studies by Cohen in the pigeon and Kapp in rabbits, however, had used fear conditioning to identify brain mechanisms of fear learning and were important stimuli to me when I was starting to design my early experiments [D.H. Cohen (1980); B.S. Kapp et al (1979)].
36. For a review of studies of the role of limbic areas in fear behavior and other emotional and memory functions, see Isaacson (1982).
37. Many of the lesion and tracing studies described were conducted in the Neurobiology Laboratory at Cornell University Medical College in Manhattan. Don Reis was the director of the lab, and a collaborator. The main collaborator on the anatomical work was David Ruggerio and Claudia Farb. A number of researchers worked on the behavioral studies, including Akira Sakaguchi, Jiro Iwata, and Piera Cicchetti.
38. LeDoux, Sakaguchi, and Reis (1984).
39. Ibid.
40. LeDoux et al (1986).
41. Kapp's work was seminal in starting this field. In 1979 he published the first study showing the lesions of the central nucleus of the amygdala disrupt fear conditioning. Later studies used stimulation, tracing, and unit recording techniques to show without a doubt that the central amygdala was an important structure in fear conditioning [summarized in Kapp, Pascoe, and Bixler (1984)].
42. The effects of central nucleus of the amygdala lesions are summarized in Kapp et al (1990); Davis (1992); LeDoux (1993); LeDoux (1995).
43. Kapp et al (1990); Davis (1992); LeDoux (1993); LeDoux (1995).
44. LeDoux et al (1988).
45. T.S. Gray et al (1993).
46. LeDoux, Farb, and Ruggiero (1990).

47. Ibid.
48. LeDoux et al (1990).
49. Price, Russchen, and Amaral (1987); Amaral et al (1992); Savander et al (1995); Pitkänen et al (1995).
50. Jarrell et al (1987).
51. For discussion of processing differences between auditory thalamus and cortex in fear conditioning, see Weinberger (1995); Bordi and LeDoux (1994a); Bordi and LeDoux (1994b).
52. Nauta and Karten (1970); Northcutt and Kaas (1995).
53. Kapp et al (1992); Davis et al (1992); Fanselow (1994) Weinberger (1995). For an alternative interpretation of the role of the thalamic pathway, see Campeau and Davis (1995). For a rebuttal of their interpretation see Corodimas and LeDoux (1995).
54. O'Keefe and Nadel (1978); Nadel and Willner (1980); Eichenbaum and Otto (1992); Sutherland and Rudy (1989).
55. Amaral (1987); Van Hoesen (1982).
56. O'Keefe and Nadel (1978); Nadel and Willner (1980); Eichenbaum and Otto (1992); Sutherland and Rudy (1989).
57. Phillips and LeDoux (1992); Kim and Fanselow (1992); Maren and Fanselow (1996).
58. Other researchers who have studied contextual fear conditioning include R.J. Blanchard, D.C. Blanchard, and R.A. Fial (1970) and Selden et al (1991).
59. LeDoux (1987); Bandler, Carrive, and Zhang (1991); Kaada (1967).
60. For review see LeDoux (1987). Although there are some situational and species differences in defense response expression, the amygdala is still involved in defense response control.
61. Greenberg, Scott, and Crews (1984); Tarr (1977).
62. Gloor, Olivier, and Quesney (1981); Halgren (1992).
63. LaBar et al (1995).
64. Bechara et al (1995); Adolphs et al (1995); Hamann et al (1995).
65. Aggleton (1992).
66. This brash statement applies to the bodily reaction to danger and not to the cognitive representation of danger or the conscious experience of fear in dangerous situations.
67. Darwin (1872).
68. D.C. Blanchard and R.J. Blanchard (1988).
69. Fuster (1989); Goldman-Rakic (1992).
70. Preuss (1995); Povinelli and Preuss (1995).
71. Luria (1966); Fuster (1989); Nauta (1971); Damasio (1994); Stuss (1991); Milner (1964).

72. Everitt and Robbins (1992); Hiroi and White (1991).
73. Lazarus (1966); Lazarus (1991).

Chapter 7: Remembrance of Emotions Past

1. Dostoyevsky (1864), quoted in Erdelyi (1985).
2. Claparede (1911).
3. Declarative memory and explicit memory are both terms that are used to distinguish conscious recollection from memories that are based on unconscious processes. The two terms, however, come from somewhat different kinds of research. Declarative memory came out of research aimed at understanding the function of the temporal lobe memory system, which we'll have much to say about. In contrast, explicit memory came out of research on the psychology of memory more than the neural basis of memory. Here, the two terms will be used interchangeably to refer to conscious memory and to distinguish memory that involves conscious recollection from memory that is based on unconscious processes, as conscious memory is now clearly established to be a function of the temporal lobe memory system.
4. Lashley (1950a). In this book, Lashley concluded that memory was not localized to any one system of the brain. This conclusion has turned out to be completely wrong. How did one of the most careful researchers in the history of brain science make such a big mistake? Lashley, like most researchers of his day, assumed that any task that measured a change in behavior at some point in time as a result of some earlier experience was as good as any other task in measuring memory. He chose to use various maze learning tasks in his quest to find memory in the brain. We now know that these mazes can be solved in many different ways—a blind animal, for example, can use touch or smell cues. The fact that the maze problems had multiple solutions meant that multiple memory systems were engaged in the learning. As a result, no one brain lesion would interfere with performance. Lashley was thus led to the false conclusion that memory is widely distributed because he used behavioral tasks that called into play multiple memory systems located in different brain regions. We now interpret this in terms of the existence of multiple memory systems in the brain.
5. Scoville and Milner (1957).
6. It is usual in studies of patients to refer to them by their initials in order to protect their identity. It is fairly commonly known, though, that H.M.'s first name was Henry.
7. N.J. Cohen and H. Eichenbaum (1993).

8. Squire (1987).
9. Scoville and Milner (1957).
10. This section is based on descriptions of H.M. found in several publications: Scoville and Milner (1957); Squire (1987); N.J. Cohen and H. Eichenbaum (1993).
11. There is also an intermediate store that has been discovered through studies in which drugs are used to interfere with storage and through studies of animals lacking certain chemicals in their nervous systems.
12. There are patients in whom aspects of short-term memory (STM) are interfered with (they perform poorly on the digit span test, a measure of STM) but they can form long-term memories of other things. However, STM is itself modular and it is unlikely that you could have a long-term memory of some stimulus that you failed to have an STM of.
13. James (1890). James distinguished primary versus secondary memory, which roughly correspond to what we have in mind when we talk about short- and long-term memory today, although there are some subtle differences in the concepts.
14. As we will see in Chapter 9, short-term memory is now often thought of as a working memory system and is believed to involve the prefrontal cortex. For a discussion of the role of the prefrontal cortex in temporary memory processes, see Fuster (1989); Goldman-Rakic (1993).
15. Squire, Knowlton, and Musen (1993); Teyler and DiScenna (1986); McClelland et al (1995).
16. Gaffan (1974).
17. Zola-Morgan and Squire (1993); Murray (1992); Mishkin (1982).
18. Iversen (1976); N.J. Cohen and H. Eichenbaum (1993).
19. N.J. Cohen and H. Eichenbaum (1993).
20. Olton, Becker, and Handleman (1979).
21. Morris (1984); Morris et al (1982).
22. Mishkin (1978).
23. Zola-Morgan, Squire, and Amaral (1986).
24. Zola-Morgan, Squire, and Amaral (1989).
25. Zola-Morgan et al (1991).
26. Meunier et al (1993); Murray (1992).
27. Squire, Knowlton, and Musen (1993); Zola-Morgan and Squire (1993); Eichenbaum, Otto, and Cohen (1994); N.J. Cohen and H. Eichenbaum (1993).
28. Eichenbaum, Otto, and Cohen (1994).
29. Zola-Morgan and Squire (1993); N.J. Cohen and H. Eichenbaum (1993); McClelland, McNaughton, and O'Reilly (1995); Murray (1992).

30. DeLeon et al (1989); Parasuramna and Martin (1994).
31. Milner (1962).
32. Milner (1965).
33. Corkin (1968).
34. N.J. Cohen (1980); N.J. Cohen and L. Squire (1980); N.J. Cohen and S. Corkin (1981).
35. Warrington and Weiskrantz (1973).
36. Weiskrantz and Warrington (1979).
37. Steinmetz and Thompson (1991).
38. N.J. Cohen and L. Squire (1980); Squire and Cohen (1984); Squire, Cohen, and Nadel (1984).
39. Schacter and Graf (1986).
40. Tulving (1983); O'Keefe and Nadel (1978); Olton, Becker, and Handleman (1979); Mishkin, Malamut, and Bachevalier (1984).
41. Graff, Squire, and Mandler (1984).
42. Cohen and Eichenbaum (1993).
43. Amaral (1987).
44. O'Keefe (1976).
45. O'Keefe and Nadel (1978).
46. Olton, Becker, and Handleman (1979).
47. Morris et al (1982).
48. O'Keefe (1993).
49. McNaughton and Barnes (1990); Barnes et al (1995); Wilson and McNaughton (1994).
50. Olton, Becker, and Handleman (1979).
51. Morris (1984); Morris et al (1982).
52. Kubie, Muller, and Bostock (1990); Kubie and Ranck (1983); Muller, Ranck, and Taube (1996).
53. Eichenbaum, Otto, and Cohen (1994).
54. Rudy and Sutherland (1992).
55. MacLean (1949; 1952).
56. McClelland, McNaughton, and O'Reilly (1995); Gluck and Myers (1995).
57. Eysenck (1979).
58. Jacobs and Nadel (1985).
59. Freud (1966).
60. Jacobs and Nadel (1985).
61. Rudy and Morledge (1994).
62. I became aware of this Gary Larson cartoon by attending lectures on memory by J. McGaugh, who often shows a slide of the drawing.
63. R. Brown and J. Kulik (1977); Christianson (1989).

64. McGaugh et al (1995); Cahill et al (1994); McGaugh et al (1993); McGaugh (1990).
65. Adrenaline doesn't actually get into the brain directly under normal circumstances and McGaugh believes that it has its effects by way of the vagus nerve, which then influences several brain systems, including the amygdala and hippocampus indirectly. Although McGaugh emphasizes action in the amygdala, it would seem that the hippocampus may also be affected, since explicit memory strength is altered. This could be a parallel effect on the hippocampus and amygdala. Also, it is possible that the amygdala is more important in his animal learning studies and the hippocampus in the human studies because of differences in the tasks used. Alternatively, it is possible that the effects are on the amygdala, which then influences the hippocampus. Paul Gold (1992) has a somewhat different view of the effects of adrenaline on memory. His work suggests that adrenaline releases glucose into the blood. Blood-borne glucose readily gets into the brain and serves as a source of energy for neurons in areas like the hippocampus. This increase in hippocampal energy resources might then help strengthen memories being created through the temporal lobe memory system.
66. Christianson (1992b).
67. Ibid.
68. Bartlett (1932).
69. Erdelyi (1985).
70. Loftus (1993); Loftus and Hoffman (1989).
71. Christianson (1992a).
72. Neisser and Harsch (1992).
73. Freud (1966).
74. Bower (1992).
75. Bower, 1992; Lang, 1984.
76. Hebb (1949).
77. Brown et al (1989); Cotman, Monaghan, and Ganong (1988).
78. Bliss and Lømo (1973).
79. Cotman et al (1988); Nicoll and Malenka (1995); Madison et al (1991); Lynch (1986); Staübli (1995); McNaughton and Barnes (1990).
80. Cotman et al (1988); Nicoll and Malenka (1995); Madison et al (1991); Staübli (1995); McNaughton and Barnes (1990).
81. Cotman et al (1988); Nicoll and Malenka (1995); Madison et al (1991); Lynch (1986); Staübli (1995); McNaughton and Barnes (1990).
82. AMPA and NMDA are the two major classes of glutamate receptors [Collingridge and Lester (1989); Cotman et al (1988)].
83. Collingridge and Lester (1989); Cotman et al (1988).

84. Bliss and Collingridge (1993); Brown et al (1988); Cotman et al (1988); Staübli (1995); Lynch (1986); McNaughton and Barnes (1990).
85. Morris et al (1986); but see Saucier and Cain (1996); Bannerman et al (1996).
86. Skelton et al (1987); Berger (1984); Laroche et al (1995); Barnes (1995); Staübli (1995); Rogan and LeDoux (1995); Barnes et al (1995); Dudai (1995).
87. Clugnet and LeDoux (1990); Rogan and LeDoux (1995); Chapman et al (1990).
88. Miserendino et al (1990); Fanselow and Kim (1994).
89. Nicoll and Malenka (1995); Staübli (1995).
90. Squire and Davis (1975); Rose (1995); Rosenzweig (1996).
91. Kandel (1989); Lisman (1995).
92. Kandel and Schwartz (1982).
93. Frey, Huang, and Kandel (1993).
94. Yin et al (1994).
95. Mayford, Abel, and Kandel (1995); Bourtchouladze et al (1994).
96. Eichenbaum and Otto (1992).

Chapter 8: Where the Wild Things Are

1. Phillips (1993).
2. Wilson (1968).
3. Shattuck (1980).
4. Shakespeare, quoted in Grey Walter (1953).
5. Manderscheid and Sonnenschein (1994).
6. This paragraph is based on Kramer (1993).
7. Klein (1981).
8. *Diagnostic and statistical manual of mental disorders* (1994).
9. Manderscheid and Sonnenschein (1994).
10. Öhman (1992); Epstein (1972).
11. Öhman (1992); Lader and Marks (1973).
12. Zuckerman (1991).
13. Ibid.
14. Freud included dysthymia and somatoform disorders under anxiety. DSM IV includes dysthymia with depressive illness under mood disorders and has a separate classification for somatoform disorders.
15. The following brief descriptions of anxiety disorders are taken from the longer DSM IV descriptions.
16. Öhman (1992).

17. Breuer and Freud, quoted in Erdelyi (1985).
18. Erdelyi (1985).
19. It's not necessary for a fear conditioning interpretation of anxiety to be correct in order to make the point I want to make here—that anxiety disorders reflect the operation of the fear system of the brain. However, since the most thorough understanding of the fear system has come from studies of fear conditioning, my job is made much easier if I can piggyback on fear conditioning explanations of anxiety. As the following discussion will, I hope, show, the conditioning theory is in fact plausible.
20. This history is summarized in Chapter 2.
21. Watson and Rayner (1920).
22. Waton's position is summarized by Eysenck (1979).
23. Thorndike (1913); Skinner (1938); Hull (1943); Tolman (1932).
24. Mowrer (1939).
25. Ibid.
26. Ibid.
27. N.E. Miller (1948).
28. This experiment was described by Hall and Lindzey (1957).
29. Dollard and Miller (1950).
30. This point was made by Hall and Lindzey (1957).
31. Dollard and Miller (1950).
32. Freud (1909).
33. Wolpe and Rachman (1960).
34. Eysenck and Rachman (1965).
35. Seligman (1971).
36. Reviewed by Seligman (1971).
37. Mineka et al (1984).
38. Bandura (1969).
39. Öhman (1992).
40. Jacobs and Nadel (1985).
41. Ibid.
42. For a summary of the adrenal steroid response to stress, see: J. A. Gray (1987); McEwen and Sapolsky (1995).
43. Jacobson and Sapolsky (1991).
44. Diamond and Rose (1994); Diamond and Rose (1993); Diamond et al (1994); Luine (1994).
45. Shors et al (1990); Pavlides, Watanabe, and McEwen (1993); Diamond et al (1994); Diamond and Rose (1994).
46. McNally et al (1995); Bremner et al (1993); Newcomer et al (1994); Wolkowitz, Reuss, and Weingartner (1990); McEwen and Sapolsky (1995).

47. McEwen (1992).
48. Bekkers and Stevens (1989); Coss and Perkel (1985); Koch, Zador, and Brown (1992).
49. Sapolsky (1990); Uno et al (1989).
50. McKittrick et al (1995); Blanchard et al (1995).
51. Bremner et al (1995).
52. McEwen and Sapolsky (1995).
53. Diamond and Rose (1994); Diamond et al (1993); Diamond et al (1994); Luine (1994).
54. It is important to point out that damage to the hippocampus can result in a retrograde amnesia as well as an anterograde one. This is important since it takes time for the steroids to build up and have their effect. So even though the hippocampus may participate in the initial phases of memory formation while the trauma is just beginning, and it may take a while for the steroids to build up. Once the hippocampus is interfered with, the effects can act to prevent the solidification of the memories that happened at the beginning of the trauma as well.
55. Loftus and Hoffman (1989); Loftus et al (1989); Loftus (1993).
56. Erdelyi (1984).
57. Dali (1948).
58. Makino, Gold, and Schulkin (1994); Swanson and Simmons (1989).
59. Corodimas et al (1994).
60. Servatius and Shors (1994).
61. Jacobs and Nadel (1985).
62. LeDoux, Romanski, and Xagoraris (1989).
63. Amaral et al (1992).
64. Morgan, Romanski, and LeDoux (1993).
65. Luria (1966); Fuster (1989); Nauta (1971); Damasio (1994); Stuss (1991); Petrides (1994); Stuss (1991); Shimamura (1995); Milner (1964).
66. Morgan, Romanski, and LeDoux (1993).
67. Morgan and LeDoux (1995).
68. Thorpe, Rolls, and Maddison (1983); Rolls (1985); Rolls (1992b).
69. Damasio (1994); Stuss (1991); Luria (1966); Fuster (1989); Nauta (1971).
70. Diorio, Viau, and Meaney (1993).
71. LeDoux, Romanski, and Xagoraris (1989).
72. Bouton and Peck (1989); Bouton and Swartzentruber (1991); Bouton (1994).
73. Jacobs and Nadel (1985).
74. Quirk, Repa, and LeDoux (1995).

75. Hebb (1949).
76. Shalev, Rogel-Fuchs, and Pitman (1992).
77. Kramer (1993).
78. I will have relatively little to say about generalized anxiety and obsessive compulsive disorder. For a theory of general anxiety, see J.A. Gray (1982). And for a critique of his theory, especially of the fact that the theory does not include a major role for the amygdala in anxiety, see LeDoux (1993). For the record, though, it should be noted that Neil McNaughton and Gray are currently working on a revision of *The Neuropsychology of Anxiety,* based on the large body of work that has emerged since 1982, which will give the amygdala a more prominent role.
79. Blanchard et al (1991).
80. Bordi and LeDoux (1992).
81. Of course, genetic preparation to respond to certain stimuli and past learning about stimuli that are important to the species probably both contribute.
82. Rolls (1992a); Allman and Brothers (1994).
83. Öhman (1992).
84. Charney et al (1993); Kolb (1987).
85. Charney et al (1993).
86. Kolb (1987).
87. Charney et al (1993); Shalev, Rogel-Fuchs, and Pitman (1992).
88. *Diagnostic and statistical manual of mental disorders* (1987).
89. *Diagnostic and statistical manual of mental disorders* (1987); Öhman (1992).
90. *Diagnostic and statistical manual of mental disorders* (1987).
91. Margraf, Ehlers, and Roth (1986b); Margraf, Ehlers, and Roth (1986a); Klein (1993).
92. Ehlers and Margraf (1987).
93. Ackerman and Sachar (1974); Margraf, Ehlers, and Roth (1986b); Wolpe (1988).
94. Ackerman and Sachar (1974); Margraf, Ehlers, and Roth (1986b); Wolpe (1988).
95. Margraf, Ehlers, and Roth (1986a).
96. Ibid.
97. Klein (1993).
98. Wolpe (1988).
99. Benarroch et al (1986); Ruggiero et al (1991).
100. Ruggiero et al (1991).
101. Cechetto and Calaresu (1984).

102. J.A. Gray (1982); J.A. Gray (1987); Sarter and Markowitsch (1985); LeDoux (1993); Isaacson (1982).
103. J.A. Gray (1982); Nagy, Zambo, and Decsi (1979).
104. J.A. Gray (1982).
105. Cognitive therapy attempts to eliminate pathological emotions by changing appraisals and thoughts. Some representative cognitive approaches to phobias and other anxiety disorders are: Lang (1979); Lang (1993); Koa and Kozak (1986); Beck and Emery (1985).
106. Reid (1989).
107. Summarized by Erdelyi (1985).
108. Erdelyi (1985).
109. Falls, Miserendino, and Davis (1992).
110. Amaral et al (1992).

Chapter 9: Once More, with Feelings

1. Spinoza (1955).
2. Nabokov (1966).
3. Feelings constitute the subjective experiences we know our emotions by and are the hallmark of an emotion from the point of view of the person experiencing the feeling. Not all feelings are emotions, but all conscious emotional experiences are feelings. This point is nicely made by Damasio (1994).
4. There are some scientists who have studied aspects of emotions other than their conscious properties, but these have been in the minority and often they too have in the end been mostly concerned with the conscious aspects. Some theorists who have included unconscious processes include: Izard (1992b); Zajonc (1980); Ekman (1980); Mandler (1975); Mandler (1992).
5. Churchland (1984); Boring (1950); Gardner (1987); Jackendoff (1987); Rorty (1979); Searle (1992); Eccles (1990); Picton and Stuss (1994); Chalmers (1996); Humphrey (1992).
6. There are of course exceptions. See note 4 above.
7. See Chapter 2.
8. See Chapters 2 and 3.
9. This statement raises as many questions as it answers, and it certainly does not solve the problem of how conscious emotional experiences occur in the brain. However, it does two other things that are important. First, it provides a way of conceptualizing what an emotional experience is. Second, it shows that we are no worse off in under-

standing where conscious emotions come from than we are in understanding where conscious perceptions or memories come from. The latter is a crucial realization, for it puts the emotion on a par with other aspects of the mind as a scientific topic for the first time since the dawn of behaviorism.

10. Some recent discussions of consciousness include: Dennett (1991); Johnson-Laird (1988); Minsky (1985); Penrose (1989); Humphrey (1992); Gazzaniga (1992); Shallice (1988); Kinsbourne (1988); Churchland (1988); Posner and Snyder (1975); Shiffrin and Schneider (1977); Baars (1988); Kosslyn and Koenig (1992); Mandler (1988); Norman and Shallice (1980); Churchland (1984); Jackendoff (1987); Rorty (1979); Searle (1992); Eccles (1990); Picton and Stuss (1994); Harnad (1982); Hirst (1994); Chalmers (1996); Velams (1991); Dennett and Kinsbourne (1992); Crick (1995); Sperry (1969); Maccel and Bisiach (1988); Crick and Koch (1992); Edelman (1989).
11. A number of contemporary theories assume that the contents of working memory are what we are conscious of (some of these are described below in the text). The working memory theories are varied but all assume the existence of some executive, or supervisory, mechanism involved in the focusing of attention, such that the thing you are attending to is what working memory works on. I'm going to embrace the working memory theory of consciousness, not because I believe that it completely explains consciousness, but because I believe that it gives us a framework for illustrating how feelings are created. I'm going to describe a feeling as the representation of the activity of an emotional processing system in working memory. This framework can be easily transported to some other framework if the working memory theory proves to be inadequate.
12. Baddeley (1982).
13. Baddeley and Hitch (1974); Baddeley (1992).
14. Miller (1956).
15. Baars (1988).
16. Kosslyn and Koenig (1992).
17. Jacobsen and Nissen (1937).
18. Preuss (1995).
19. Fuster (1989); Goldman-Rakic (1987); Goldman-Rakic (1993); Wilson, Scalaidhe, and Goldman-Rakic (1993).
20. Petrides (1994); Fuster (1989).
21. Petrides et al (1993); Jonides et al (1993); Grasby et al (1993); Schwartz et al (1995).
22. D'Esposito et al (1995).

23. Fuster (1989); Goldman-Rakic (1987); Reep (1984); Uylings and van Eden (1990).
24. Fuster (1989); Goldman-Rakic (1987).
25. Van Essen (1985).
26. Ungerleider and Mishkin (1982); Ungerleider and Haxby (1994).
27. Goldman-Rakic (1988).
28. Desimone et al (1995).
29. This simple story involving specialized short-term buffers in the sensory systems and a general-purpose working memory mechanism in the prefrontal cortex is somewhat more complicated than the way I have presented it. The prefrontal cortex itself seems to have regions that are specialized, at least to some degree, for specific kinds of working memory functions. Such findings, however, do not discredit the notion that the prefrontal cortex is involved in the general-purpose or executive aspects of working memory since only some cells in these areas play specialized roles. Interactions between the general-purpose cells in different areas may coordinate the overall activity of working memory. It is thus possible that the executive functions of the prefrontal cortex might be mediated by cells that are distributed across the different prefrontal subsystems rather than by cells that are collected together in one region.
30. F.A.W. Wilson, S.P.O Scalaidhe, and P.S. Goldman-Rakic (1993).
31. The studies have examined one kind of memory at a time rather than putting different kinds of memory in competition.
32. Petrides (1994).
33. D'Esposito et al (1995); Corbetta et al (1991); Posner and Petersen (1990).
34. Goldman-Rakic (1988); J.M. Fuster (1989).
35. Posner (1992).
36. Gaffan, Murray, and Fabre-Thorpe (1993).
37. Thorpe, Rolls, and Maddison (1983); Rolls (1992b); Ono and Nishijo (1992).
38. Damasio (1994).
39. Kosslyn and Koenig (1992); Shimamura (1995).
40. Williams (1964).
41. Kosslyn and Koenig (1992).
42. Kihlstrom (1987).
43. Johnson-Laird (1988).
44. Baars (1988).
45. Shallice (1988); Posner and Snyder (1975); Shiffrin and Schneider (1977); Norman and Shallice (1980).

46. However, under some conditions, consciousness can be divided: Hirst et al (1980); Kihlstrom (1984).
47. Summarized by Dennett (1991).
48. Newell, Rosenbloom, and Laird (1989); Newell and Simon (1972).
49. Johnson-Laird (1988).
50. Smolensky (1990); Rumelhart et al (1988).
51. Johnson-Laird (1988).
52. That symbolic architecture is the basis of consciousness is somewhat counterintuitive since cognitive science was founded on the symbol manipulation approach, but has mostly been about unconscious processes. However, we are not aware of the processes by which symbols are manipulated, but only of the outcome of the manipulations. Symbol manipulation may be the architecture of consciousness, but there is still something missing between symbolic representation and conscious awareness. And that is the big question about consciousness.
53. Johnson-Laird (1988).
54. Tom Nagel's discussion of what it is like to be a bat is relevant here: Nagel (1974).
55. For a discussion of the difference between phenomenal consciousness and access consciousness, see Jackendoff (1987); N. Block (1995).
56. Right now, we are concerned with what happens if a snake is a potent emotional stimulus for you rather than with the manner in which emotional potency is created (emotional learning in neutral and evolutionarily prepared situations was discussed in earlier chapters).
57. But if you have, in your past, experienced rabbits in association with some trauma or stress, then the rabbit too could serve as a trigger stimulus that would turn on the amygdala and its outputs.
58. Amaral et al (1992).
59. Moruzzi and Magoun (1949).
60. Hobson and Steriade (1986); McCormick and Bal (1994).
61. This applies to arousal occurring during waking states. Arousal also occurs during sleep, especially dream or REM sleep. In this case the cortex becomes insensitive in external inputs and is focused instead on internal stimuli [Hobson and Steriade (1986); McCormick and Bal (1994)].
62. This is generally known as the Yerkes-Dodson law in psychology.
63. Kapp et al (1992); Weinberger (1995).
64. Amygdala interactions with brainstem arousal systems is described in LeDoux (1995); Gallagher and Holland (1994).
65. Kapp et al (1992).

66. It appears in fact that the cortex arouses itself since sensory stimuli first go to the cortex and are then sent back to the brain stem and these inputs trigger the arousal system, which then arouses the cortex [Lindsley (1951)].
67. Ekman, Levenson, and Friesen (1983); R.W. Levenson (1992).
68. Tomkins (1962).
69. Izard (1971); Izard (1992a).
70. Damasio (1994).
71. Hohmann (1966).
72. B. Bermond, B. Nieuwenhuyse, L. Fasotti, and J. Schuerman (1991).
73. Also, the patients were not tested during emotional experiences but were asked to recall past emotions. This approach, as we saw in Chapters 2 and 3, is plagued with problems.
74. James (1890).
75. Ekman (1992b); Ekman (1993); Adelman and Zajonc (1989).
76. There are now several patients with amygdala damage. [Adolphs et al (1995); Bechara et al (1995); Young et al (1995).] However, these people have a congenital disorder. Whenever the brain is damaged in early life, there are numerous compensatory mechanisms. For example, if the visual cortex is damaged, the auditory cortex can take on some visual functions. We have to be very cautious in using negative findings in patients with developmental disorders to infer what normally goes on in the brain.
77. Scherer (1993a); Leventhal and Scherer (1987); Scherer (1984).
78. Pinker (1994).
79. Jerison (1973).
80. Preuss (1995); Reep (1984); Uylings and van Eden (1990).
81. Preuss (1995); Povinelli and Preuss (1995).
82. Gallup (1991).
83. Pinker (1994).
84. The relation between language and consciousness is complex and controversial. Some propose that all thought (and our consciousness of our thoughts) takes place in a propositional mode, a language of thought, whereas others argue that thought can occur in nonpropositional, say pictorial or visual, terms. My view is that while language is not a necessary precursor to consciousness, the presence of language (or at least the cognitive capacities that make language possible) allows a unique kind of awareness in humans. This does not mean that one must be able to speak or understand speech in order to be conscious. Deaf and dumb individuals, for example, are no less conscious than the rest of us. They have the cognitive capacities that make language, and linguistically

based thought, possible. They simply cannot use those capacities to understand speech or produce it themselves.

85. Kihlstrom (1987); LeDoux (1989).
86. Dawkins (1982).
87. Wilde (1909).

BIBLIOGRAPHY

Abelson, R. P. (1963). Computer simulation of "hot" cognition. In *Computer simulation of personality,* S. S. Tomkins and S. Messick, eds. (New York: Wiley).

Ackerman, S., and Sachar, E. (1974). The lactate theory of anxiety: A review and reevaluation. *Psychosomatic Medicine 36,* 69–81.

Adelman, P. K., and Zajonc, R. B. (1989). Facial efference and the experience of emotion. *Annual Review of Psychology* 40, 249-80.

Adolphs, R., Tranel, D., Damasio, H., and Damasio, A. R. (1995). Fear and the human amygdala. *Journal of Neuroscience 15,* 5879–91

Aggleton, J. P. (1992). *The amygdala: Neurobiological aspects of emotion, memory, and mental dysfunction* (New York: Wiley-Liss).

Aggleton, J. P., and Mishkin, M. (1986). The amygdala: Sensory gateway to the emotions. In *Emotion: Theory, research and experience* (Vol. 3), R. Plutchik and H. Kellerman, eds. (Orlando: Academic Press), pp. 281–99.

Allman, J., and Brothers, L. (1994). Faces, fear and the amygdala. *Nature 372,* 613–14.

Amaral, D. G. (1987). Memory: Anatomical organization of candidate brain regions. In *Handbook of Physiology. Section 1: The Nervous System. Vol. 5: Higher Functions of the Brain,* F. Plum, ed. (Bethesda, MD: American Physiological Society), pp. 211–94.

Amaral, D. G., Price, J. L., Pitkänen, A., and Carmichael, S. T. (1992). Anatomical organization of the primate amygdaloid complex. In *The amygdala: Neurobiological aspects of emotion, memory, and mental dysfunction,* J. P. Aggleton, ed. (New York: Wiley-Liss), pp. 1–66.

Anderson, J. R. (1990). *Cognitive psychology and its implications,* 3rd edition (New York: Freeman).

Archer, J. (1979). Behavioral aspects of fear. In *Fear in animals and man,* W. Sluckin, ed. (New York: Van Nostrand Reinhold).

Aristotle (1941). In *The basic works of Aristotle,* R. McKeon, ed. (New York: Random House).

Armony, J.L., Servan-Schreiber, D., Cohen, J.D., and LeDoux, J.E. (1995). An anatomically constrained neural network model of fear conditioning. *Behavioral Neuroscience 109,* 246–57.

Arnold, M. B. (1960). *Emotion and personality* (New York: Columbia University Press).

Averill, J. (1980). Emotion and anxiety: Sociocultural, biological, and psychological determinants. In *Explaining emotions,* A. O. Rorty, ed. (Berkeley: University of California Press).

Ayala, E. J., and Valentine, J. W. (1979). *Evolving.* Benjamin Cummings.

Baars, B. J. (1988). *A cognitive theory of consciousness* (New York: Cambridge University Press).

Baddeley, A. (1982). *Your memory: A user's guide* (New York: Macmillan).

Baddeley, A. (1992). Working memory. *Science* 255, 556–59.

Baddeley, A., and Hitch, G. J. (1974). Working memory. In *The psychology of learning and motivation,* vol. 8, G. Bower, ed. (New York: Academic Press).

Bandler, R., Carrive, P., and Zhang, S. P. (1991). Integration of somatic and autonomic reactions within the midbrain periaqueductal grey: Viscerotopic, somatotopic and functional organization. *Progress in Brain Research* 87, 269–305.

Bandura, A. (1969). *Principles of behavior modification* (New York: Holt).

Bangs, L. (1978). *Gig* (New York: Gig Enterprises).

Bannerman D. M., Good, M. A., Butcher, S. P., Ramsay, M., and Morris, R. G. M. (1995). Distinct components of spatial learning revealed by prior training and NMDA receptor blockade. *Nature* 378, 182–86.

Bard, P. (1929). The central representation of the sympathetic system: As indicated by certain physiological observations. *Archives of Neurology and Psychiatry* 22, 230–46.

Bargh, J. A. (1990). Auto-motives: Preconscious determinants of social interaction. In *Handbook of motivation and cognition,* T. Higgins and R. M. Sorrentino, eds., pp. 93–130 (New York: Guilford).

Bargh, J. A. (1992). Being unaware of the stimulus vs. unaware of its interpretation: Why subliminality per se does matter to social psychology. In *Perception without awareness,* R. Bornstein and T. Pittman, eds. (New York: Guilford).

Barnes, C. A. (1995). Involvement of LTP in memory: Are we "searching under the streetlight?" *Neuron 15,* 751–54.

Barnes, C. A., Erickson, C. A., Davis, S., and McNaughton, B. L. (1995). Hippocampal synaptic enhancement as a basis for learning and memory: A selected review of current evidence from behaving animals. In *Brain and memory: Modulation and mediation of neuroplasticity,* J. L. McGaugh, N. M. Weinberger, and G. Lynch, eds. (New York: Oxford University Press), pp. 259–76.

Bartlett, F. C. (1932). *Remembering* (Cambridge: Cambridge University Press).

Bechara, A., Tranel, D., Damasio, H., Adolphs, R., Rockland, C., and Damasio, A. R. (1995). Double dissociation of conditioning and declarative knowledge relative to the amygdala and hippocampus in humans. *Science 269,* 1115–18.

Beck, A. T., and Emery, G. (1985). *Anxiety disorders and phobias: A cognitive perspective* (New York: Basic Books).

Bekkers, J. M., and Stevens, C. F. (1989). NMDA and non-NMDA receptors are co-localized at individual excitatory synapses in cultured rat hippocampus. *Nature 341,* 230–33.

Benarroch, E. E., Granata, A. R., Ruggiero, D. A., Park, D. H., and Reis, D. J. (1986). Neurons of C1 area mediate cardiovascular responses initiated from ventral medullary surface. *American Physiological Society,* R932–R945.

Berger, T. W. (1984). Long-term potentiation of hippocampal synaptic transmission affects rate of behavioral learning. *Science* 224, 627–29.

Bermond, B., and Nieuwenhuyse, B., Fasotti, L. and Schuerman, J. (1995). Spinal cord lesions, peripheral feedback, and intensities of emotional feelings. *Cognition and Emotion* 5, 201–20.

Blanchard, C., Spencer, R. L., Weiss, S. M., Blanchard, R., McEwen, B. S., and Sakai, R. (1995). Visible burrow system as a model of chronic social stress. *Behavioral Neuroendocrinology 20,* 117–39.

Blanchard, D. C., and Blanchard, R. J. (1972). Innate and conditioned reactions to threat in rats with amygdaloid lesions. *Journal of Comparative Physiological Psychology 81,* 281–90.

Blanchard, D. C., and Blanchard, R. J. (1988). Ethoexperimental approaches to the biology of emotion. *Annual Review of Psychology* 39, 43–68

Blanchard, D. C., and Blanchard, R. J. (1989). Experimental animal models of aggression: what do they say about human behaviour? In *Human aggression: Naturalistic approaches,* J. Archer and K. Browne, eds. (New York: Routledge), pp. 94–121.

Blanchard, R. J., and Blanchard, D. C. (1989). Antipredator defensive behaviors in a visible burrow system. *Journal of Comparative Psychology 103,* 70–82.

Blanchard, R. J., Blanchard, D. C., and Fial, R. A. (1970). Hippocampal lesions in rats and their effect on activity, avoidance, and aggression. *Journal of Comparative Physiological Psychology 71(1),* 92–102.

Blanchard, R. J., Weiss, S., Agullana, R., Flores, T., and Blanchard, D. C. (1991). Antipredator ultrasounds: Sex differences and drug effects. *Neuroscience Abstracts 17.*

Blanchard, R. J., Yudko, E. B., Rodgers, R. J., and Blanchard, D. C. (1993). Defense system psychopharmacology: An ethological approach to the pharmacology of fear and anxiety. *Behavioural Brain Research* 58, 155–66.

Bliss, T. V. P., and Collingridge, G. L. (1993). A synaptic model of memory: Long-term potentiation in the hippocampus. *Nature 361,* 31–39.

Bliss, T. V. P., and Lomo, T. (1973). Long-lasting potentiation of synaptic transmission in the dentate area of the anaesthetized rabbit following stimulation of the perforant path. *Journal of Physiology* 232, 331–56.

Block, N. (1995). On a confusion about a function of consciousness. *Behavioral and Brain Sciences 18,* 227–87.

Bogen, J. E., and Vogel, P. J. (1962). Cerebral commissurotomy: A case report. *Bulletin of the Los Angeles Neurological Society* 27, 169.

Bolles, R. C., and Fanselow, M. S. (1980). A perceptual-defensive-recuperative model of fear and pain. *Behavioral and Brain Sciences* 3, 291–323.

Bordi, F., and LeDoux, J. (1992). Sensory tuning beyond the sensory system: An initial analysis of auditory properties of neurons in the lateral amygdaloid nucleus and overlying areas of the striatum. *Journal of Neuroscience 12 (7),* 2493–2503.

Bordi, F., and LeDoux, J. E. (1994a). Response properties of single units in areas of rat auditory thalamus that project to the amygdala. I: Acoustic discharge patterns and frequency receptive fields. *Experimental Brain Research* 98, 261–74.

Bordi, F., and LeDoux, J. E. (1994b). Response properties of single units in areas of rat auditory thalamus that project to the amygdala. II: Cells receiving convergent auditory and somatosensory inputs and cells antidromically activated by amygdala stimulation. *Experimental Brain Research* 98, 275–86.

Boring, E. G. (1950). *A history of experimental psychology* (New York: Appleton-Century-Crofts).

Bornstein, R. F. (1992). Subliminal mere exposure effects. In *Perception without awareness: Cognitive, clinical, and social perspectives,* R. F. Bornstein and T. S. Pittman, eds. (New York: Guilford), pp. 191–210.

Bourtchouladze, R., Frengeulli, B., Blendy, J., Cioffi, D., Schutz, G., and Silva, A. J. (1994). Deficient long-term memory in mice with a targeted mutation of the cAMP-responsive element binding protein. *Cell 79,* 59–68.

Bouton, M. E. (1994). Conditioning, remembering, and forgetting. *Journal of Experimental Psychology: Animal Behavior Processes 20,* 219–31.

Bouton, M. E., and Peck, C. A. (1989). Context effects on conditioning, extinction, and reinstatement in an appetitive conditioning preparation. *Animal Learning and Behavior 17,* 188–98.

Bouton, M. E., and Swartzentruber, D. (1991). Sources of relapse after extinction in Pavlovian and instrumental learning. *Clinical Psychology Review 11,* 123–40.

Bower, G. (1992). How might emotions affect learning? In *Handbook of emotion and memory: Research and theory,* S.-A. Christianson, ed. (Hillsdale, NJ: Erlbaum).

Bowers, K. S. (1984). On being unconsciously influenced and informed. In *The unconscious reconsidered,* K. S. Bowers and D. Meichenbaum, eds. (New York: Wiley) 227–72.

Bowers, K. S., and Meichenbaum, D. (1984). *The unconscious reconsidered* (New York: Wiley).

Bowlby, J. (1969). *Attachment and Loss: Vol. 1, Attachment* (New York: Basic Books).

Bremner, J. D., Randall, T., Scott, T. M., Brunen, R. A., Seibyl, J. P., Southwick, S. M., Delaney, R. C., McCarthy, G., Charney, D. S., and Innis, R. B. (1995). MRI-based measurement of hippocampal volume in patients with combat-related PTSD. *American Journal of Psychiatry 152,* 973–81.

Bremner, J. D., Scott, T. M., Delaney, R. C., Southwick, S. M., Mason, J. W., Johnson, C. R., Innis, R. B., McCarthy, G., and Charney, D. S. (1993). Deficits in short-term memory in posttraumatic stress disorder. *American Journal of Psychiatry 150,* 1015–19.

Broca, P. (1978). Anatomie comparée des circonvolutions cérébrales. Le grand lobe limbique et la scissure limbique dans le série des mammifères. *Revue Anthropologique, Ser. 21 21,* 385–498.

Brodal, A. (1982). *Neurological anatomy* (New York: Oxford University Press).

Brown, J. S., Kalish, H. I., and Farber, I. E. (1951). Conditioned fear as revealed by magnitude of startle response to an auditory stimulus. *Journal of Experimental Psychology 41,* 317–28.

Brown, R., and Kulik, J. (1977). Flashbulb memories. *Cognition* 5, 73–99.

Brown, T. H., Chapman, P. F., Kairiss, E. W., and Keenan, C. L. (1988). Long-term synaptic potentiation. *Science* 242, 724–28.

Brown, T. H., Ganong, A. H., Kairiss, E. W., Keenan, C. L., and Kelso, S. R. (1989). Long-term potentiation in two synaptic systems of the hippocampal brain slice. In *Neural models of plasticity,* J. H. Byrne and W. O. Berry, eds. (San Diego: Academic Press), pp. 266–306.

Bruner, J. (1992). Another look at New Look 1. *American Psychologist* 47, 780–83.

Bruner, J. S., and Postman, L. (1947). Emotional selectivity in perception and reaction. *Journal of Personality* 16, 60–77.

Cacioppo, J. T., Klein, D. J., Berntson, G. G., and Hatfield, E. (1993). The psychophysiology of emotion. In *Handbook of emotions,* M. Lewis and J. M. Haviland, eds. (New York: Guilford), pp. 119–42.

Cahill, L., Prins, B., Weber, M., and McGaugh, J. L. (1994). Beta-adrenergic activation and memory for emotional events. *Nature* 371, 702–4.

Campbell, B. A., and Jaynes, J. (1966). Reinstatement. *Psychological Review* 73, 478–80.

Campeau, S. and Davis, M. (1995). Involvement of subcortical and cortical afferents to the lateral nucleus of the amygdala in fear conditioning measured with fear-potentiated startle in rats trained concurrently with auditory and visual conditioned stimuli. *Journal of Neuroscience* 15, 2312–27.

Campeau, S., Liang, K. C., and Davis, M. (1990). Long-term retention of fear-potentiated startle following a short training session. *Animal Learning and Behavior* 18(4), 462–68.

Cannon, W. B. (1929). *Bodily changes in pain, hunger, fear, and rage,* vol. 2 (New York: Appleton).

Carew, T. J., Hawkins, R. D., and Kandel, E. R. (1983). Differential classical conditioning of a defensive withdrawal reflex in Aplysia californica. *Science* 219, 397–400.

Cechetto, D. F., and Calaresu, F. R. (1984). Units in the amygdala responding to activation of carotid baro- and chemoreceptors. *American Journal of Physiology* 246, R832–R836.

Chalmers, D. (1996). The Conscious Mind (New York: Oxford).

Chapman, P. F., Kairiss, E. W., Keenan, C. L., and Brown, T. H. (1990). Long-term synaptic potentiation in the amygdala. *Synapse* 6, 271–278.

Charney, D. S., Deutch, A. V., Krystal, J. H., Southwick, A. M., and Davis, M. (1993). Psychobiologic mechanisms of posttraumatic stress disorder. *Archives of General Psychiatry* 50, 295–305.

Christianson, S.-A. (1989). Flashbulb memories: Special, but not so special. *Memory and Cognition 17,* 435–43

Christianson, S.-A. (1992a). Eyewitness memory for stressful events: Methodological quandaries and ethical dilemmas. In *Handbook of emotion and memory: Research and theory,* S.-A. Christianson, ed. (Hillsdale, NJ: Erlbaum).

Christianson, S.-A. (1992b). Remembering emotional events: Potential mechanisms. In *Handbook of emotion and memory: Research and theory,* S.-A. Christianson, ed. (Hillsdale, NJ: Erlbaum).

Churchland, P. (1984). *Matter and consciousness* (Cambridge: MIT Press).

Churchland, P. (1988). Reduction and the neurobiological basis of consciousness. In *Consciousness in contemporary science,* A. Marcel and E. Bisiach, eds. (Oxford: Clarendon Press).

Churchland, P. S., and Sejnowski, T. J. (1990). In *Neural connections, mental computation,* L. Nadel, L. Cooper, P. Culicover, and M. Harnish, eds. (Cambridge: MIT Press).

Claparede, E. (1911). Recognition and "me-ness." In *Organization and pathology of thought* (1951), D. Rapaport, ed. (New York: Columbia University Press), pp. 58–75.

Clugnet, M. C., and LeDoux, J. E. (1990). Synaptic plasticity in fear conditioning circuits: Induction of LTP in the lateral nucleus of the amygdala by stimulation of the medial geniculate body. *Journal of Neuroscience 10,* 2818–24.

Cohen, D. H. (1980). The functional neuroanatomy of a conditioned response. In *Neural mechanisms of goal-directed behavior and learning,* R. F. Thompson, L. H. Hicks, and B. Shvyrkov, eds. (New York: Academic Press), pp. 283–302.

Cohen, N. J. (1980). *Neuropsychological evidence for a distinction between procedural and declarative knowledge in human memory and amnesia* (San Diego: University of California Press).

Cohen, N. J., and Corkin, S. (1981). The amnestic patient H.M.: Learning and retention of cognitive skills. *Society for Neuroscience Abstracts* 7, 517–18.

Cohen, N. J., and Eichenbaum, H. (1993). *Memory, amnesia, and the hippocampal system* (Cambridge: MIT Press).

Cohen, N. J., and Squire, L. (1980). Preserved learning and retention of pattern-analyzing skill in amnesia: Dissociation of knowing how and knowing that. *Science 210,* 207–9.

Collingridge, G. L., and Lester, R. A. J. (1989). Excitatory amino acid receptors in the vertebrate central nervous system. *Pharmacological Reviews 40,* 143–210.

Corbetta, M., Miezin, F. M., Dobmeyer, S., Shulman, G. L., and Petersen, S. E. (1991). Selective and divided attention during visual discriminations of shape, color, and speed: Functional anatomy by positron emission tomography. *Journal of Neuroscience 11*, 2383–2402.

Corkin, S. (1968). Acquisition of motor skill after bilateral medial temporal lobe excision. *Neuropsychologia 6*, 255–65.

Corodimas, K. P. and LeDoux, J. E. (1995) Disruptive effects of posttraining perihinal cortex lesions on conditioned fear: Contributions of contextual cues. *Behavioral Neuroscience 109*, 613–19.

Corodimas, K. P., LeDoux J. E., Gold, P. W., and Schulkin, J. (1994). Corticosterone potentiation of learned fear. *Annals of the New York Academy of Sciences* 746, 392–93.

Coss, R. G., and Perkel, D. H. (1985). The function of dendritic spines: A review of theoretical issues. *Behavioral and Neural Biology 44*, 151–85.

Cotman, C. W., Monaghan, D. T., and Ganong, A. H. (1988). Excitatory amino acid neurotransmission: NMDA receptors and Hebb-type synaptic plasticity. *Annual Review of Neuroscience 11*, 61–80.

Crick, F. (1994). *The Astonishing Hypothesis* (New York: Scribners).

Crick, F. and Koch, C. (1990). Toward a neurobiological theory of consciousness. *The Neurosciences* 2, 263–75.

Crockett, D. (1845). *A narrative of the life of David Crockett* (New York: Nafis & Cornish).

Dali, S. (1976). *The secret life of Salvador Dali* (London: Vision Press).

Damasio, A. (1994). *Descarte's error: Emotion, reason, and the human brain* (New York: Grosset/Putnam).

Darwin, C. (1859). *The origin of species by means of natural selection; Or, the preservation of favored races in the struggle for life* (New York: Collier).

Darwin, C. (1872). *The expression of the emotions in man and animals* (Chicago: University of Chicago Press, 1965).

Davidson, R. (1992). Emotion and affective style: Hemispheric substrates. *Psychological Science* 3, 39–43.

Davis, M. (1992a). The role of amygdala in conditioned fear. In *The amygdala: Neurobiological aspects of emotion, memory, and mental dysfunction,* J. P. Aggleton, ed. (New York: Wiley-Liss), pp. 255–306.

Davis, M. (1992b). The role of the amygdala in fear-potentiated startle: Implications for animal models of anxiety. *Trends in Pharmacological Science 13*, 35–41.

Davis, M., Hitchcock, J. M., and Rosen, J. B. (1987). Anxiety and the amygdala: pharmacological and anatomical analysis of the fear-potentiated startle paradigm. In *The psychology of learning and motivation,* Vol. 21, G. H. Bower, ed. (San Diego: Academic Press), pp. 263–305.

Davitz, H. J. (1969). *The language of emotion* (London: Academic Press).

Dawkins, R. (1982). *The extended phenotype: The gene as the unit of selection* (San Francisco: Freeman).

DeLeon, M. J., George, A. E., Stylopoulos, L. A., Smith, G., and Miller, D. C. (1989). Early marker for Alzheimer's disease: The atrophic hippocampus. *Lancet* September 16, 672–73.

Dennett, D. C. (1991). *Consciousness explained* (Boston: Little, Brown).

Dennett, D. C. and Kinsbourne, M. (1992). Time and the observer: The where and when of consciousness in the brain. *Behavioral and Brain Sciences, 15,* 183–247.

Descartes, R. (1958). *Philosophical writings,* N. K. Smith, ed. (New York: Modern Library).

Desimone, R., Miller, E. K., Chelazzi, L., and Lueschow, A. (1995). Multiple memory systems in the visual cortex. In *The cognitive neurosciences,* M. S. Gazzaniga, ed. (Cambridge: MIT Press), pp. 475–86.

de Sousa, R. (1980). The rationality of emotions. In *Explaining emotions,* A. O. Rorty, ed. (Berkeley: University of California Press).

D'Esposito, M., Detre, J., Alsop, D., Shin, R., Atlas, S., and Grossman, M. (1995). The neural basis of the central executive system of working memory. *Nature* 378, 279–81.

Diagnostic and statistical manual of mental disorders (1994), 4th edition (Washington, D.C.: American Psychiatric Association).

Diamond, D. M. and Rose, G. M. (1993). Psychological stress interferes with working, but not reference, spatial memory. *Society for Neuroscience Abstracts 19,* 366.

Diamond, D. M., Fleshner, M. and Rose, G. M. (1994). Psychological stress repeatedly blocks hippocampal primed burst potentiation in behaving rats. *Behavioural Brain Research* 62, 1–9.

Diamond, D. M., Branch, B. J., Rose, G. M., and Tocco, G. (1994). Stress effects on memory and AMPA receptors are abolished by adrenalectomy. *Society for Neuroscience Abstracts 20,* 1215.

Diamond, D. M., and Rose, G. (1994). Stress impairs LTP and hippocampal-dependent memory. *Annals of the New York Academy of Sciences* 746, 411–14.

Dickinson, E. (1955). The brain (#632). In T. H. Johnson (ed.) *The Poems of Emily Dickinson* (Cambridge, MA: Belknap).

Diorio, D., Viau, V., and Meaney, M. J. (1993). The role of the medial prefrontal cortex (cingulate gyrus) in the regulation of hypothalamic-pituitary-adrenal responses to stress. *Journal of Neuroscience 13,* 3839–47.

Dixon, N. F. (1971). *Subliminal perception: The nature of controversy* (London: McGraw-Hill).

Dixon, N. F. (1981). *Preconscious processing* (New York: Wiley).

Doi, T. (1973). *The anatomy of dependence* (Tokyo: Kodansha International).

Dollard, J. C., and Miller, N. E.,(1950). *Personality and psychotherapy* (New York: McGraw-Hill).

Dostoyevsky, F. (1864). *Notes from the underground* (New York: Dell).

Downer, J. D. C. (1961). Changes in visual gnostic function and emotional behavior following unilateral temporal lobe damage in the "split-brain" monkey. *Nature 191,* 50–51.

Dreiser, T. (1900). *Sister Carrie* (New York: Doubleday).

Dudai, Y. (1995). On the relevance of long-term potentiation to learning and memory. In J. L. McGaugh, N. M. Weinberger, and G. Lynch, ed. *Brain and memory: Modulation and mediation of neuroplasticity.* (New York: Oxford University Press).

Duffy, E. (1941). An explanation of "emotional" phenomena without the use of the concept "emotion." *Journal of General Psychology* 25, 283–93.

Dyer, M. G. (1987). Emotions and their computations: Three computer models. *Cognition and Emotion 1,* 323–47.

Eagly, A., and Chaiken, S. (1993). *The psychology of attitudes* (Fort Worth: Harcourt Brace Jovanovich).

Ebbesson, S. O. E. (1980). The parcellation theory and its relation to interspecific variability in brain organization, evolutionary and ontogenetic development, and neural plasticity. *Cell and Tissue Research* 213, 179–212.

Eccles, J. C. (1990). A unitary hypothesis of mind-brain interaction in the cerebral cortex. *Proceedings of the Royal Society of London* 240, 433–51.

Edelman, G. (1989). *The Remembered Present: A Biological Theory of Consciousness* (New York: Basic Books).

Edmunds, M. (1974). *Defence in animals: A survey of anti-predator defences* (New York: Longman).

Ehlers, A., and Margraf, J. (1987). Anxiety induced by false heart rate feedback in patients with panic disorder. *Behaviour Research and Therapy 26,* 1–11.

Eibl-Eibesfeldt, I., and Sutterlin, C. (1990). Fear, defence and aggression in animals and man: Some ethological perspectives. In *Fear and defense,* P. F. Brain, S. Parmigiani, R. Blanchard, and D. Mainardi, eds. (London: Harwood), pp. 381–408.

Eichenbaum, H., and Otto, T. (1992). The hippocampus: What does it do? *Behavioral and Neural Biology* 57, 2–36.

Eichenbaum, H., Otto, T., and Cohen, N. J. (1994). Two functional components of the hippocampal memory system. *Behavioral and Brain Sciences* 17, 449–518.

Ekman, P. (1980). Biological and cultural contributions to body and facial movement in the expression of emotions. In *Explaining emotions,* A. O. Rorty, ed. (Berkeley: University of California Press).

Ekman, P. (1984). Expression and nature of emotion. In *Approaches to emotion,* K. Scherer and P. Ekman, eds. (Hillsdale, NJ: Erlbaum), pp. 319–43.

Ekman, P. (1992a). An argument for basic emotions. *Cognition and Emotion* 6, 169–200.

Ekman, P. (1992b). Facial expressions of emotion: New findings, new questions. *Psychological Science* 3, 34–38.

Ekman, P. (1993). Facial expression and emotion. *American Psychologist* 48.

Ekman, P., Levenson, R. W., and Friesen, W. V. (1983). Autonomic nervous system activity distinguishes among emotions. *Science* 221, 1208–10.

Epstein, S. (1972). The nature of anxiety with emphasis upon its relationship to expectancy. In *Anxiety: Current trends in theory and research,* C. D. Speilberger, ed. (New York: Academic Press).

Erdelyi, M. H. (1974). A new look at the new look: Perceptual defense and vigilance. *Psychological Review* 81, 1–25.

Erdelyi, M. H. (1984). The recovery of unconscious (inaccessible) memories: Laboratory studies of hypermnesia. In *The psychology of learning and motivation: Advances in research and theory,* G. Bower, ed. (New York: Academic Press), pp. 95–127.

Erdelyi, M. (1985). *Psychoanalysis: Freud's cognitive psychology* (New York: Freeman).

Erdelyi, M. H. (1992). Psychodynamics and the unconscious. *American Psychologist* 47, 784–87.

Ericcson, K. A., and Simon, H. (1984). *Protocol analysis: Verbal reports as data* (Cambridge: MIT Press).

Eriksen, C. W. (1960). Discrimination and learning without awareness: A methodological survey and evaluation. *Psychological Review* 67, 279–300.

Everitt, B. J. and Robbins, T. W. (1942). Amygdala—ventral striatal interactions and reward related processes. In J. Aggleton (ed.) *The Amygdala: Neurobiological Aspects of Emotion, Memory and Mental Dysfunction* (New York: Wiley-Liss).

Eysenck, H. J. (1979). The conditioning model of neurosis. *Behavioral and Brain Sciences* 2, 155–99.

Eysenck, H. J., and Rachman, S. (1965). *The causes and cures of neuroses* (San Diego: Knapp).

Falls, W. A., Miserendino, M. J. D., and Davis, M. (1992). Extinction of fear-potentiated startle: Blockade by infusion of an NMDA antagonist into the amygdala. *Journal of Neuroscience 12(3),* 854–63.

Fanselow, M. S. (1994). Neural organization of the defensive behavior system responsible for fear. *Psychonomic Bulletin and Review 1,* 429–38.

Fanselow, M. S., and Kim, J. J. (1994). Acquisition of contextual Pavlovian fear conditioning is blocked by application of an NMDA receptor antagonist DL-2-amino-5-phosphonovaleric acid to the basolateral amygdala. *Behavioral Neuroscience 108,* 210–12.

Fehr, F. S., and Russell, J. A. (1984). Concept of emotion viewed from a prototype perspective. *Journal of Experimental Psychology, General 113,* 464–86.

Finlay, B., and Darlington, R. (1995). Linked regularities in the development and evolution of mammalian brains. *Science,* 1578–84.

Flew, A. (1964). *Body, mind and death* (New York: Macmillan).

Fodor, J. (1975). *The language of thought* (Cambridge: Harvard University Press).

Fodor, J. (1983). *The modularity of mind* (Cambridge: MIT Press).

Frank, R. H. (1988). *Passions within reason: The strategic role of the emotions* (New York: Norton).

Freeman, W. J. (1994). Role of chaotic dynamics in neural plasticity. *Progress in Brain Research 102,* 319–33.

Freud, S. (1909). The analysis of a phobia in a five-year-old boy. In *Collected papers* (London: Hogarth).

Freud, S. (1925). The unconscious. In *Collected papers* (London: Hogarth).

Freud, S. (1966). *Introductory lectures on psychoanalysis,* Standard Edition, J. Strachey, ed. (New York: Norton).

Frey, U., Huang, Y.-Y., and Kandel, E. R. (1993). Effects of cAMP simulate a late stage of LTP in hippocampal CA1 neurons. *Science 260,* 1661–64.

Frijda, N. (1986). *The emotions* (Cambridge: Cambridge University Press).

Frijda, N. H. (1993). The place of appraisal in emotion. *Cognition and Emotion 7,* 357–88.

Frijda, N., and Swagerman, J. (1987). Can computers feel? Theory and design of an emotional system. *Cognition and Emotion 1,* 235–57.

Fuster, J. M. (1989). *The prefrontal cortex* (New York: Raven).

Gaffan, D. (1974). Recognition impaired and association intact in the memory of monkeys after transection of the fornix. *Journal of Comparative and Physiological Psychology 86,* 1100–1109.

Gaffan, D. (1992). Amygdala and the memory of reward. In *The amygdala: Neurobiological aspects of emotion, memory, and mental dysfunction,* J. P. Aggleton, ed. (New York: Wiley-Liss), pp. 471–83.

Gaffan, D., Murray, E. A., and Fabre-Thorpe, M. (1993). Interaction of the amygdala with the frontal lobe in reward memory. *European Journal of Neuroscience* 5, 968–75.

Gainotti, G. (1972). Emotional behavior and hemispheric side of the lesion. *Cortex* 8, 41–55.

Galaburda, A. M., Corsiglia, J., Rosen, G. D., and Sherman, G. F. (1987). Planum temporale asymmetry, reappraisal since Geschwind and Levitsky. *Neuropsychologia* 25, 853–68.

Gallagher, M., and Holland, P. (1994). The amygdala complex. Proceedings of the National Academy of Sciences, U.S.A. 91, 11, 771–76.

Gallistel, R. (1980). *The organization of action: A new synthesis* (Hillsdale, NJ: Erlbaum).

Gallup, G. (1991). Toward a comparative psychology of self-awareness: Species limitations and cognitive consequences. In *The self: Interdisciplinary approaches,* J. Strauss and G. R. Goethals, eds. (New York: Springer).

Gardner, H. (1987). *The mind's new science: A history of the cognitive revolution* (New York: Basic Books).

Gazzaniga, M. S. (1970). *The bisected brain* (New York: Appleton-Century-Crofts).

Gazzaniga, M. S. (1972). One brain—two minds. *American Scientist* 60, 311–17.

Gazzaniga, M. S. (1985). *The social brain* (New York: Basic Books).

Gazzaniga, M. S. (1988). Brain modularity: Towards a philosophy of conscious experience. In *Consciousness in contemporary science,* A. J. Marcel and E. Bisiach, eds. (Oxford: Clarendon Press).

Gazzaniga, M. S. (1992). *Nature's mind* (New York: Basic Books).

Gazzaniga, M. S., Bogen, J. E., and Sperry, R. W. (1962). Some functional effects of sectioning the cerebral commissures in man. *Proceedings of the National Academy of Sciences USA* 48, 1765–69.

Gazzaniga, M. S., Bogen, J. E., and Sperry, R. W. (1965). Cerebral commissurotomy in man: Minor hemisphere dominance for certain visuo-spatial functions. *Journal of Neurosurgery* 23, 394–99.

Gazzaniga, M. S., and LeDoux, J. E. (1978). *The Integrated Mind* (New York: Plenum).

Geertz, H. (1959). The vocabulary of emotion. *Psychiatry* 22, 225–37.

Geschwind, N. (1965). The disconnexion syndromes in animals and man. I. *Brain* 88, 237–94.

Geschwind, N., and Levitsky, W. (1968). Human brain: Left-right asymmetries in temporal speech region. *Science 161,* 186–87.

Gleitman, H., and Holmes, P. A. (1967). Retention of incompletely learned CER in rats. *Psychonomic Science* 7, 19–20.

Gloor, P., Olivier, A., and Quesney, L. F. (1981). The role of the amygdala in the expression of psychic phenomena in temporal lobe seizures. In *The amygdaloid complex,* Y. Ben-Ari, ed. (New York: Elsevier/North-Holland Biomedical Press), pp. 489–98.

Gluck, M. A., and Myers, C. E. (1995). Representation and association in memory: A neurocomputational view of hippocampal function. *Current Directions in Psychological Science 4,* 23–29.

Gold, P. E. (1992). Modulation of memory processing: enhancement of memory in rodents and humans. In L. R. Squire and N. Butters *Neuropsychology of Memory* (New York: Guilford), 402–14.

Goldman, A. I. (1993). The psychology of folk psychology. *Behavioral and Brain Sciences 16,* 15–28.

Goldman-Rakic, P. S. (1988). Topography of cognition: Parallel distributed networks in primate association cortex. *Annual Review of Neuroscience 11,* 137–56.

Goldman-Rakic, P. S. (1987). Circuitry of primate prefrontal cortex and regulation of behavior by representational memory. In *Handbook of physiology. Section 1: The nervous system. Vol. 5: Higher Functions of the Brain,* F. Plum, ed. (Bethesda, MD: American Physiological Society, pp. 373–417.

Goldman-Rakic, P. S. (1993). Working memory and the mind. In *Mind and brain: Readings from Scientific American magazine,* W. H. Freeman, ed. (New York: Freeman), pp. 66–77.

Goleman, D. (1995). *Emotional intelligence* (New York: Bantam).

Gould, J. L. (1982). *Ethology: The mechanisms and evolution of behavior* (New York: Norton).

Gould, S. J. (1977). *Ever since Darwin: Reflections in natural history* (New York: Norton).

Graff, P., Squire, L. R., and Mandler, G. (1984). The information that amnesic patients do not forget. *Journal of Experimental Psychology: Learning, Memory and Cognition 10,* 16–178.

Grasby, P. M., Firth, C. D., Friston, K. J., Bench, C., Frackowiak, R. S. J., and Dolan, R. J. (1993). Functional mapping of brain areas implicated in auditory-verbal memory function. *Brain 116,* 1–20.

Gray, J. A. (1982). *The neuropsychology of anxiety* (New York: Oxford University Press).

Gray, J. A. (1987). *The psychology of fear and stress,* Vol. 2 (New York: Cambridge University Press).

Gray, T. S., Piechowski, R. A., Yracheta, J. M., Rittenhouse, P. A., Betha, C. L., and van der Kar, L. D. (1993). Ibotenic acid lesions in the bed nucleus of the stria terminalis attenuate conditioned stress induced increases in prolactin, ACTH, and corticosterone. *Neuroendocrinology* 57, 517–24.

Greenberg, N., Scott, M., and Crews, D. (1984). Role of the amygdala in the reproductive and aggressive behavior of the lizard. *Physiology and Behavior* 32, 147–51.

Greenwald, A. G. (1992). New look 3: Unconscious cognition reclaimed. *American Psychologist* 47, 766–79.

Grey Walter, W. (1953). *The living brain* (New York: Norton).

Grossberg, S. (1982). A psychophysiological theory of reinforcement, drive, motivation and attention. *Journal of Theoretical Biology 1,* 286–369.

Guare, J. (1990). *Six degrees of separation* (New York: Random House).

Halgren, E. (1992). Emotional neurophysiology of the amygdala within the context of human cognition. In *The amygdala: Neurobiological aspects of emotion, memory, and mental dysfunction,* J. Aggleton, ed. (New York: Wiley-Liss), pp. 191–228.

Hall, C. S., and Lindzey, G. (1957). *Theories of personality* (New York: Wiley).

Hamann, S. B., Stefanacci, L., Squire, L., Adolphs, R., Tranel, D., Damasio, H., Damasio, A. (1996). Recognizing facial emotion. *Nature 379,* 497.

Harnad, S. (1982). Consciousness: An afterthought. *Cognition and Brain Theory* 5, 29–47.

Harré, R. (1986). *The social construction of emotions* (New York: Blackwell).

Head, H. (1921). Release function in the nervous system. *Proceedings of the Royal Society of London: Biology 92B,* 184–87.

Hebb, D. O. (1946). Emotion in man and animal: An analysis of the intuitive processes of recognition. *Psychological Review* 53, 88–106.

Hebb, D. O. (1949). *The organization of behavior* (New York: Wiley).

Heelas, P. (1986). Emotion talk across cultures. In *The social construction of emotions,* R. Harré, ed. (New York: Blackwell).

Heidegger, M. (1927). *Being and time* (New York: SUNY Press).

Heilman, K. and Satz, P., eds. (1983). *Neuropsychology of Human Emotion.* (New York: Guilford Press).

Helmstetter, F. (1992). The amygdala is essential for the expression of conditioned hypoalgesia. *Behavioral Neuroscience 106,* 518–28.

Herrick, C. J. (1933). The functions of the olfactory parts of the cerebral cortex. *Proceedings of the National Academy of Sciences USA 19,* 7–14.

Hilgard, E. R. (1980). The trilogy of mind: Cognition, affection, and conation. *Journal of the History of the Behavioral Sciences 16,* 107–17.

Hilton, S. M. (1979). The defense reaction as a paradigm for cardiovascular control. In *Integrative functions of the autonomnic nervous system,* C. M. Brooks, K. Koizuni, and A. Sato, eds. (Tokyo: University of Tokyo Press), pp. 443–49.

Hirst, W. (1994). Cognitive aspects of consciousness. In *The cognitive neurosciences,* M. S. Gazzaniga, ed. (Cambridge: MIT Press).

Hirst, W., Spelke, E. S., Reaves, C. C., Charack, G., and Neisser, U. (1980). Dividing attention without alternation or automaticity. *Journal of Experimental Psychology, General 109,* 98–117.

Hobson, J. A., and Steriade, M. (1986). Neuronal basis of behavioral state control. In Handbook of Physiology. Section 1: The Nervous System. Vol. 4: *Intrinsic Regulatory Systems of the Brain.* V. B. Mountcastle, ed. (Bethesda, MD: American Physiological Society), pp. 701–823.

Hodes, R. L., Cook, E. W., and Lang, P. J. (1985). Individual differences in autonomic response: Conditioned association or conditioned fear? *Psychophysiology* 22, 545–60.

Hohmann, G. W. (1966). Some effects of spinal cord lesions on experienced emotional feelings. *Psychophysiology* 3.

Horel, J. A., Keating, E. G., and Misantone, L. J. (1975). Partial Kluver-Bucy syndrome produced by destroying temporal neocortex or amygdala. *Brain Research 94,* 347–59.

Hugdahl, K., (1995). Psychophysiology: The Mind-Body Perspective (Cambridge: Harvard University Press).

Hull, C. L. (1943). *Principles of behavior* (New York: Appleton-Century-Crofts).

Humphrey, N. (1992). *A history of the mind* (New York: Simon & Schuster).

Ionescu, M. D., and Erdelyi, M. H. (1992). The direct recovery of subliminal stimuli. In *Perception without awareness: Cognitive, clinical, and social perspectives,* R. F. Bornstein and T. S. Pittman, eds. (New York: Guilford), pp. 143–69.

Isaacson, R. L. (1982). The limbic system (New York: Plenum).

Iversen, S. (1976). Do temporal lobe lesions produce amnesia in animals? *International Review of Neurobiology 19,* 1–49.

Izard, C. E. (1971). *The face of emotion* (New York: Appleton-Century-Crofts).

Izard, C. E. (1977). *Human emotions* (New York: Plenum).

Izard, C. E. (1992a). Basic emotions, relations among emotions, and emotion-cognition relations. *Psychological Review 99,* 561–65.

Izard, C. E. (1992b). Four systems for emotion activation: Cognitive and noncognitive. *Psychological Review 100,* 68–90.

Jackendoff, R. (1987). *Consciousness and the computational mind* (Cambridge: Bradford Books, MIT Press).

Jacobs, W. J., and Nadel, L. (1985). Stress-induced recovery of fears and phobias. *Psychological Review* 92, 512–31.

Jacobsen, C. F., and Nissen, H. W. (1937). Studies of cerebral function in primates: IV. The effects of frontal lobe lesions on the delayed alternation habit in monkeys. *Journal of Comparative and Physiological Psychology* 23, 101–12.

Jacobson, L., and Sapolsky, R. (1991). The role of the hippocampus in feedback regulation of the hypothalamic-pituitary-adrenocortical axis. *Endocrine Reviews 12(2),* 118–34.

Jacoby, L. L., Toth, J. P., Lindsay, D. S., and Debner, J. A. (1992). Lectures for a layperson: Methods for revealing unconscious processes. In *Perception without awareness: Cognitive, clinical, and social perspectives,* R. F. Bornstein and T. S. Pittman, eds. (New York: Guilford), pp. 81–120.

James, W. (1884). What is an emotion? *Mind 9,* 188–205.

James, W. (1890). *Principles of psychology* (New York: Holt).

Jarrell, T. W., Gentile, C. G., Romanski, L. M., McCabe, P. M., and Schneiderman, N. (1987). Involvement of cortical and thalamic auditory regions in retention of differential bradycardia conditioning to acoustic conditioned stimuli in rabbits. *Brain Research 412,* 285–94.

Jaynes, J. (1976). *The origin of consciousness in the breakdown of the bicameral mind* (Boston: Houghton Mifflin).

Jerison, H. (1973). *Evolution of brain and intelligence* (New York: Academic Press).

John, E. R. (1972). Switchboard versus statistical theories of learning. *Science 177,* 850–64.

Johnson-Laird, P. N. (1988). *The computer and the mind: An introduction to cognitive science* (Cambridge: Harvard University Press).

Johnson-Laird, P. N., and Oatley, K. (1992). Basic emotions, rationality, and folk theory. *Cognition and Emotion 6,* 201–23.

Jones, B., and Mishkin, M. (1972). Limbic lesions and the problem of stimulus-reinforcement associations. *Experimental Neurology 36,* 362–77.

Jonides, J., Smith, E. E., Keoppe, R. A., Awh, E., Minoshima, S., and Mintun, M. A. (1993). Spatial working memory humans as revealed by PET. *Nature 363,* 623–25.

Kaada, B. R. (1960). Cingulate, posterior orbital, anterior insular and temporal pole cortex. In *Handbook of physiology. Section 1, Vol. 2: Neuro-*

physiology, J. Field, H. J. Magoun and V. E. Hall, eds. (Washington, D.C.: American Physiological Society), pp. 1345–72.

Kaada, B. R. (1967). Brain mechanisms related to aggressive behavior. In *Aggression and defense—Neural mechanisms and social patterns,* C. Clemente and D. B. Lindsley, eds. (Berkeley: University of California Press), pp. 95–133.

Kagan, J. and Snidman, N. (1991). Infant predictors of inhibited and uninhibited profiles. *Psychological Science* 2, 40–43.

Kahneman, D., Slovic, P., and Tversky, A. (1982). *Judgement under uncertainty: Heuristics and biases* (Cambridge: Cambridge University Press).

Kandel, E., and Schwartz, J. (1982). Molecular biology of an elementary form of learning: Modulation of transmitter release by cAMP. *Science 218,* 433–43.

Kandel, E. R. (1989). Genes, nerve cells, and the remembrance of things past. *Journal of Neuropsychiatry,* 103–25.

Kapp, B. S., Whalen, P. J., Supple, W. F., and Pascoe, J. P. (1992). Amygdaloid contributions to conditioned arousal and sensory information processing. In *The amygdala: Neurobiological aspects of emotion, memory, and mental dysfunction,* J. P. Aggleton, ed. (New York: Wiley-Liss).

Kapp, B. S., Frysinger, R. C., Gallagher, M., and Haselton, J. (1979). Amygdala central nucleus lesions: Effect on heart rate conditioning in the rabbit. *Physiology and Behavior* 23, 1109–17.

Kapp, B. S., Pascoe, J. P., and Bixler, M. A. (1984). The amygdala: A neuroanatomical systems approach to its contributions to aversive conditioning. In *Neuropsychology of memory,* N. Buttlers and L. R. Squire, eds. (New York: Guilford), pp. 473–88.

Kapp, B. S., Wilson, A., Pascoe, J., Supple, W., and Whalen, P. J. (1990). A neuroanatomical systems analysis of conditioned bradycardia in the rabbit. In *Learning and computational neuroscience: Foundations of adaptive networks.,* M. Gabriel and J. Moore, eds. (Cambridge: MIT Press), pp. 53–90.

Karten, H. J., and Shimizu, T. (1991). Are visual hierarchies in the brains of the beholders? Constancy and variability in the visual system of birds and mammals. In *The changing visual system,* P. Bagnoli and W. Hodos, eds. (New York: Plenum), pp. 51–59.

Keating, G. E., Kormann, L. A., and Horel, J. A. (1970). The behavioral effects of stimulating and ablating the reptilian amygdala (Caiman sklerops). *Physiology and Behavior* 5, 55–59.

Kelley, G. A. (1963). Discussion: Aldous, the personable computer. In *Computer simulation of personality: Frontier of psychological theory,* S. S. Tomkins and S. Messick, eds. (New York: Wiley).

Kierkegaard, S. (1844). *The concept of dread* (Princeton: Princeton University Press).

Kihlstrom, J. F. (1984). Conscious, subconscious, unconscious: A cognitive perspective. In *The unconscious reconsidered,* K. S. Bowers and D. Meichenbaum, eds. (New York: Wiley), pp. 149–211.

Kihlstrom, J. F. (1987). The cognitive unconscious. *Science* 237, 1445–52.

Kihlstrom, J. F., Barnhardt, T. M., and Tataryn, D. J. (1992a). Implicit perception. In *Perception without awareness: Cognitive, clinical, and social perspectives,* R. F. Bornstein and T. S. Pittman, eds. (New York: Guilford), pp. 17–54.

Kihlstrom, J. F., Barnhardt, T. M., and Tatryn, D. J. (1992b); The psychological unconscious: Found, lost, regained. *American Psychologist* 47, 788–91.

Kim, J. J., and Fanselow, M. S. (1992). Modality-specific retrograde amnesia of fear. *Science* 256, 675–77.

Kinsbourne, M. (1988). Integrated field theory of consciousness. In *Consciousness in contemporary science,* A. Marcel and E. Bisiach, eds. (Oxford: Oxford University Press).

Klein, D. (1981). Anxiety reconceptualized. In *New research and changing concepts,* D. Klein and J. Rabkin, eds. (New York: Raven).

Klein, D. F. (1993). False suffocation alarms, spontaneous panics, and related conditions: An integrative hypothesis. *Archives of General Psychiatry 50,* 306–17.

Kleinginna, P. R., and Kleinginna, A. M. (1985). Cognition and affect: A reply to Lazarus and Zajonc. *American Psychologist 40,* 470–71.

Klüver, H., and Bucy, P. C. (1939). Preliminary analysis of functions of the temporal lobes in monkeys. *Archives of Neurology and Psychiatry* 42, 979–1000.

Klüver, H., and Bucy, P. C. (1937). "Psychic blindness" and other symptoms following bilateral temporal lobectomy in rhesus monkeys. *American Journal of Physiology 119,* 352–53.

Koa, E. B., and Kozak, E. J. (1986). Emotional processing of fear: Exposure to corrective information. *Psychological Bulletin 99,* 20–35.

Koch, C., Zador, A., and Brown, T. H. (1992). Dendritic spines: Convergence of theory and experiment. *Science 256,* 973–74.

Kolb, L. C. (1987). A neuropsychological hypothesis explaining post-traumatic stress disorders. *American Journal of Psychiatry 144,* 989–995.

Kosslyn, S. M. (1980). *Image and mind* (Cambridge: Harvard University Press).

Kosslyn, S. M. (1983). *Ghosts in the mind's machine* (New York: Norton).

Kosslyn, S. M., and Koenig, O. (1992). *Wet mind: The new cognitive neuroscience* (New York: Macmillan).

Kotter, R., and Meyer, N. (1992). The limbic system: a review of its empirical foundation. *Behavioural Brain Research* 52, 105–27.

Kramer, P. (1993). *Listening to Prozac* (New York: Viking).

Krebs, J. R. (1990). Food-storage birds: Adaptive specialization in brain and behavior? *Philosophical Transactions of the Royal Society. London. Series B: Biological Sciences 329,* 153–60.

Kubie, J., and Ranck, J. (1983). Sensory-behavioral correlates of individual hippocampal neurons in three situations: Space and context. In *The neurobiology of the hippocampus,* W. Seifert, ed. (New York: Academic Press).

Kubie, J. L., Muller, R. U., and Bostock, E. (1990). Spatial firing properties of hippocampal theta cells. *Journal of Neuroscience 10(4),* 1110–23.

LaBar, K. S., LeDoux, J. E., Spencer, D. D., and Phelps, E. A. (1995). Impaired fear conditioning following unilateral temporal lobectomy in humans. *Journal of Neuroscience 15,* 6846–55.

Lader, M., and Marks, I. (1973). *Clinical anxiety* (London: Heinemann).

Lang, P. (1979). A bioinformational theory of emotional imagery. *Psychophysiology 16,* 495–512.

Lang, P. (1993). The network model of emotion: Motivational concerns. In *Advances in social cognition,* R. S. Wyer and T. K. Srull, eds. (Hillsdale, NJ: Erlbaum), pp. 109–33.

Laroche, S., Doyere, V., Redini-Del Negro, C., and Burette, F. (1995). Neural mechanisms of associative memory: Role of long-term potentiation. In *Brain and memory: Modulation and mediation of neuroplasticity,* J. L. McGaugh, N. M. Weinberger, and G. Lynch, eds. (New York: Oxford University Press), pp. 277–302.

Lashley, K. S. (1950a). In search of the engram. *Symposia of the Society for Experimental Biology IV,* 454–82.

Lashley, K. (1950b). The problem of serial order in behavior. In *Cerebral mechanisms in behavior,* L. A. Jeffers, ed. (New York: Wiley).

Lazarus, R. S. (1966). Psychological stress and the coping process (New York: McGraw Hill).

Lazarus, R. S. (1984). On the primacy of cognition. *American Psychologist,* 39, 124–29.

Lazarus, R. S. (1991). Cognition and motivation in emotion. *American Psychologist 46(4),* 352–67.

Lazarus, R., and McCleary, R. (1951). Autonomic discrimination without awareness: A study of subception. *Psychological Review* 58, 113–22.

LeDoux, J. E. (1985). Brain, mind, and language. In *Brain and mind,* D. A. Oakley, ed. (London: Methuen).

LeDoux, J. E. (1987). Emotion. In *Handbook of Physiology. Section 1: The Nervous System. Vol. 5; Higher Functions of the Brain,* F. Plum, ed. (Bethesda, MD: American Physiological Society), pp. 419–60.

LeDoux, J. E. (1989). Cognitive-emotional interactions in the brain. *Cognition and Emotion* 3, 267–289.

LeDoux, J. E. (1991). Emotion and the limbic system concept. *Concepts in Neuroscience* 2, 169–99.

LeDoux, J. E. (1993). Emotional memory systems in the brain. *Behavioural Brain Research* 58, 69–79.

LeDoux, J. E. (1994). Emotion, memory and the brain. *Scientific American 270,* 32–39.

LeDoux, J. E. (1995). Emotion: Clues from the brain. *Annual Review of Psychology 46,* 209–35.

LeDoux, J. E., Cicchetti, P., Xagoraris, A., and Romanski, L. M. (1990). The lateral amygdaloid nucleus: Sensory interface of the amygdala in fear conditioning. *Journal of Neuroscience 10,* 1062–69.

LeDoux, J. E., Farb, C. F., and Ruggiero, D. A. (1990). Topographic organization of neurons in the acoustic thalamus that project to the amygdala. *Journal of Neuroscience 10,* 1043–54.

LeDoux, J. E., Iwata, J., Cicchetti, P., and Reis, D. J. (1988). Different projections of the central amygdaloid nucleus mediate autonomic and behavioral correlates of conditioned fear. *Journal of Neuroscience* 8, 2517–29.

LeDoux, J. E., Romanski, L. M., and Xagoraris, A. E. (1989). Indelibility of subcortical emotional memories. *Journal of Cognitive Neuroscience l,* 238–43.

LeDoux, J. E., Sakaguchi, A., Iwata, J., and Reis, D. J. (1986). Interruption of projections from the medial geniculate body to an archi-neostriatal field disrupts the classical conditioning of emotional responses to acoustic stimuli in the rat. *Neuroscience 17,* 615–27.

LeDoux, J. E., Sakaguchi, A., and Reis, D. J. (1984). Subcortical efferent projections of the medial geniculate nucleus mediate emotional responses conditioned by acoustic stimuli. *Journal of Neuroscience* 4(3), 683–98.

Leonardo da Vinci (1939). *The notebooks of Leonardo da Vinci.* (New York: Reynal & Hitchcock).

Levenson, R. W. (1992). Autonomic nervous system differences among emotions. *Psychological Science* 3, 23–27.

Leventhal, H., and Scherer, K. (1987). The relationship of emotion to cognition: A functional approach to a semantic controversy. *Cognition and Emotion 1*, 3–28.

Lindsley, D. B. (1951). Emotions. In *Handbook of Experimental Psychology*, S. S. Stevens, ed. (New York: Wiley), pp. 473–516.

Lisman, J. (1995). What does the nucleus know about memories? *Journal of NIH Research* 7, 43–46.

Loftus, E. (1993). The reality of repressed memories. *American Psychologist* 48, 518–37.

Loftus, E. F., Donders, K., Hoffman, H. G., and Schooler, J. W. (1989). Creating new memories that are quickly accessed and confidently held. *Memory and Cognition* 17, 607–16.

Loftus, E. F., and Hoffman, H. G. (1989). Misinformation and memory: The creation of new memories. *Journal of Experimental Psychology: General 118*, 100–104.

Loftus, E. F., and Klinger, M. R. (1992). Is the unconscious smart or dumb? *American Psychologist* 47, 761–65.

Luine, V. N. (1994). Steroid hormone influences on spatial memory. *Annals of the New York Academy of Sciences* 743, 201–11.

Luria, A. (1966). Higher cortical functions in man (New York: Basic Books).

Lynch, G. (1986). *Synapses, circuits, and the beginnings of memory* (Cambridge: MIT Press).

MacLean, P. D. (1949). Psychosomatic disease and the "visceral brain": recent developments bearing on the Papez theory of emotion. *Psychosomatic Medicine 11*, 338–53.

MacLean, P. D. (1952). Some psychiatric implications of physiological studies on frontotemporal portion of limbic system (visceral brain). *Electroencephalography and Clinical Neurophysiology* 4, 407–18.

MacLean, P. D. (1970). The triune brain, emotion and scientific bias. In *The neurosciences: Second study program*, F. O. Schmitt, ed. (New York: Rockefeller University Press), pp. 336–49.

MacLean, P. D. (1990). *The triune brain in evolution: Role in paleocerebral functions* (New York: Plenum).

Madison, D. V., Malenka, R. C., and Nicoll, R. A. (1991). Mechanisms underlying long-term potentiation of synaptic transmission. *Annual Review of Neuroscience 14*, 379–97.

Makino, S., Gold, P. W., and Schulkin, J. (1994). Corticosterone effects on corticotropin-releasing hormone mRNA in the central nucleus of the amygdala and the parvocellular region of the paraventricular nucleus of the hypothalamus. *Brain Research 640*, 105–12.

Mancia, G., and Zanchetti, A. (1981). Hypothalamic control of autonomic functions. In *Handbook of the hypothalamus Vol. 3: Behavioral studies of the hypothalamus,* P. J. Morgane and J. Panksepp, eds. (New York: Marcel Dekker), pp. 147–202.

Manderscheid, R. W., and Sonnenschein, M. A. (1994). *Mental health, United States 1994* (Rockville, MD: U.S. Department of Public Health and Human Services).

Mandler, G. (1975). *Mind and emotion* (New York: Wiley).

Mandler, G. (1988). Memory: Conscious and unconscious. In *Memory: Interdisciplinary approaches,* P. R. Solomon, G. R. Goethals, C. M. Kelly, and B. R. Stephens, eds. (New York: Springer).

Mandler, G. (1992). Memory, arousal, and mood. In *Handbook of emotion and memory: Research and theory,* S.-A. Christianson, ed. (Hillsdale, NJ: Erlbaum).

Marcel, A. J., and Bisiach, E. (1988). *Consciousness in contemporary science* (Oxford: Clarendon Press).

Margraf, J., Ehlers, A., and Roth, W. T. (1986a). Biological models of panic disorder and agoraphobia—a review. *Behaviour Research and Therapy 24,* 553–67.

Margraf, J., Ehlers, A., and Roth, W. T. (1986b). Sodium lactate infusions and panic attacks: A review and critique. *Psychosomatic Medicine* 48, 23– 51.

Marks, I. (1987). *Fears, phobias, and rituals: Panic, anxiety and their disorders* (New York: Oxford University Press).

Marr, D. (1982). *Vision: A computational investigation into the human representation and processing of visual information* (San Francisco: Freeman).

Mason, J. W. (1968). A review of psychoendocrine research on the sympathetic-adrenal medullary system. *Psychosomatic Medicine 30,* 631–53.

Masson, J. M., and McCarthy, S. (1995). *When elephants weep: The emotional lives of animals* (New York: Delacorte).

Mayford, M., Abel, T., and Kandel, E. R. (1995). Transgenic approaches to cognition. *Current Opinions in Neurobiology* 5, 141–48.

McAllister, W. R., and McAllister, D. E. (1971). Behavioral measurement of conditioned fear. In *Aversive conditioning and learning,* F. R. Brush, ed. (New York: Academic Press), pp. 105–79.

McCabe, P. M., Schneiderman, N., Jarrell, T. W., Gentile, C. G., Teich, A. H., Winters, R. W., and Liskowsky, D. R. (1992). Central pathways involved in differential classical conditioning of heart rate responses. In *Learning and memory: The behavioral and biological substrates,* I. Gormenzano, E.A., ed. (Hillsdale, NJ: Erlbaum), pp. 321–46.

McClelland, J. L., McNaughton, B. L., and O'Reilly, R. C. (1995). Why there are complementary learning systems in the hippocampus and neocortex: Insights from the successes and failures of connectionist models of learning and memory. *Psychological Review 102,* 419–57.

McCormick, D. A., and Bal, T. (1994). Sensory gating mechanisms of the thalamus. *Current Opinion in Neurobiology 4,* 550–56.

McDonald, K. A. (1995). Scientists rethink anthropomorphism. *The Chronicle of Higher Education,* February 24, 1995.

McEwen, B. S. (1992). Paradoxical effects of adrenal steroids on the brain: Protection versus degeneration. *Biological Psychiatry 31,* 177–99.

McEwen, B., and Sapolsky, R. (1995). Stress and cognitive functioning. *Current Opinion in Neurobiology* 5, 205–16.

McGaugh, J. L. (1990). Significance and remembrance: The role of neuromodulatory systems. *Psychological Science 1,* 15–25.

McGaugh, J. L., Cahill, L., Parent, M. B., Mesches, M. H., Coleman-Mesches, K., and Salinas, J. A. (1995). Involvement of the amygdala in the regulation of memory storage. In *Plasticity in the central nervous system: Learning and memory,* J. L. McGaugh, F. Bermudez-Rattoni, and R. A. Prado-Alcala, eds. (Hillsdale, NJ: Erlbaum).

McGaugh, J. L., Introini-Collison, I. B., Cahill, L. F., Castellano, C., Dalmaz, C., Parent, M. B., and Williams, C. L. (1993). Neuromodulatory systems and memory storage: Role of the amygdala. *Behavioural Brain Research* 58, 81–90.

McGinnies, E. (1949). Emotionality and perceptual defense. *Psychological Review* 56, 244–51.

McKittrick, C., Blanchard, C., Blanchard, R., McEwen, B. S., and Sakai, R. (1995). Serotonin receptor binding in a colony model of chronic social stress. *Biological Psychiatry* 37, 383–93.

McNally, R. J., Lasko, N. B., Macklin, M. L., and Pitman, R. K. (1995). Autobiographical memory disturbance in combat-related posttraumatic stress disorder. *Behavior Research and Therapy* 33, 619–30.

McNaughton, B. L., and Barnes, C. A. (1990). From cooperative synaptic enhancement to associative memory: Bridging the abyss. *Seminars in the Neurosciences* 2, 403–16.

Melville, H. (1930). *Moby-Dick* (New York: Penguin).

Merikle, P. M. (1992). Perception without awareness. *American Psychologist* 47, 792–95.

Messick, S. (1963). Computer models and personality theory. In *Computer simulation of personality: Frontier of psychological theory,* S. S. Tomkins and S. Mesnick, eds. (New York: Wiley), 305–17.

Meunier, M., Bachevalier, J., Mishkin, M., and Murray, E. A. (1993). Effects on visual recognition of combined and separate ablations of the entorhinal and perirhinal cortex in rhesus monkeys. *Journal of Neuroscience 13,* 5418–32.

Miller, G. (1956). The magical number seven, plus or minus two: Some limits on our capacity for processing information. *Psychological Review* 63, 81–97.

Miller, G. A., and Gazzaniga, M. S. (1984). The cognitive sciences. In *Handbook of cognitive neuroscience,* M. S. Gazzaniga, ed. (New York: Plenum).

Miller, G. A., and Johnson-Laird, P. (1976). *Language and perception* (Cambridge: Cambridge University Press).

Miller, N. E. (1948). Studies of fear as an acquirable drive: I. Fear as motivation and fear reduction as reinforcement in the learning of new responses. *Journal of Experimental Psychology* 38, 89–101.

Milner, B. (1962). Les troubles de la mémoire accompagnant des lésions hippocampiques bilaterales. In *Physiologie de l'hippocampe,* P. Plassouant, ed. (Paris: Centre de la Recherche Scientifique).

Milner, B. (1964). Some effects of frontal lobectomy in man. In J. M. Warren and K. Akert, eds. *The Frontal Granular Cortex and Behavior.* (New York: McGraw-Hill), pp. 313–34.

Milner, B. (1965). Memory disturbances after bilateral hippocampal lesions in man. In *Cognitive processes and brain,* P. M. Milner and S. E. Glickman, eds. (Princeton: Van Nostrand).

Mineka, S., Davidson, M., Cook, M. and Keir, R. (1984). Observational conditioning of snake fear in rhesus monkeys. *Journal of Abnormal Psychology* 93, 355–72.

Minsky, M. (1985). *The society of mind* (New York: Touchstone Books/Simon & Schuster).

Miserendino, M. J. D., Sananes, C. B., Melia, K. R., and Davis, M. (1990). Blocking of acquisition but not expression of conditioned fear-potentiated startle by NMDA antagonists in the amygdala. *Nature* 345, 716–18.

Mishkin, M. (1978). Memory in monkeys severely impaired by combined but not separate removal amygdala and hippocampus. *Nature* 273, 297–98.

Mishkin, M. (1982). A memory system in the monkey. *Philosophical Transactions of the Royal Society, London, Series B: Biological Sciences.* 298, 85–95.

Mishkin, M., Malamut, B., and Bachevalier, J. (1984). Memories and habits: Two neural systems. In *The neurobiology of learning and memory,* J. L. McGaugh, G. Lynch, and N. M. Weinberger, eds. (New York: Guilford).

Moore, T. E. (1988). The case against subliminal manipulation. *Psychology and Marketing* 5, 297–316.

Morgan, M., and LeDoux, J. E. (1995). Differential contribution of dorsal and ventral medial prefrontal cortex to the acquisition and extinction of conditioned fear. *Behavioral Neuroscience 109,* 681–88.

Morgan, M. A., Romanski, L. M., and LeDoux, J. E. (1993). Extinction of emotional learning: Contribution of medial prefrontal cortex. *Neuroscience Letters 163,* 109–13.

Morris, R. G. M. (1984). Development of a water-maze procedure for studying spatial learning in the rat. *Journal of Neuroscience Methods 11,* 47–60.

Morris, R. G. M., Anderson, E., Lynch, G. S., and Baudry, M. (1986). Selective impairment of learning and blockade of long-term potentiation by and N-methyl-D-asparate receptor antagonist, AP5. *Nature 319,* 774–76.

Morris, R. G. M., Garrard, P., Rawlins, J. N. P., and O'Keefe, J. (1982). Place navigation impaired in rats with hippocampal lesions. *Nature* 273, 297–98.

Morsbach, H., and Tyler, W. J. (1986). A Japanese emotion: Amae. In *The social construction of emotions,* R. Harré, ed. (New York: Blackwell).

Moruzzi, G., and Magoun, H. W. (1949). Brain stem reticular formation and activation of the EEG. *Electroencephalography and Clinical Neurophysiology 1,* 455–73.

Mowrer, O. H. (1939). A stimulus-response analysis of anxiety and its role as a reinforcing agent. *Psychological Review 46,* 553–65.

Muller, R., Ranck, J., and Taube, J. (1996). Head direction cells: Properties and functional significance. *Current Opinion in Neurobiology* (in press).

Murphy, S., and Zajonc, R. (1993). Affect, cognition, and awareness: Affective priming with suboptimal and optimal stimuli. *Journal of Personality and Social Psychology 64,* 723–39.

Murray, E. A. (1992). Medial temporal lobe structures contributing to recognition memory: The amygdaloid complex versus the rhinal cortex. In J. P. Aggleton, ed. *The Amygdala: Neurobiological Aspects of Emotion, Memory, and Mental Dysfunction.* (New York: Wiley-Liss, Inc.).

Nabokov, V. (1966). Speak, memory: An autobiography revisited (New York: Putnam).

Nadel, L., and Willner, J. (1980). Context and conditioning: A place for space. *Physiological Psychology* 8, 218–28.

Nagel, T. (1974). What is it like to be a bat? *Philosophical Review* 83, 4435–50.

Nagy, J., Zambo, K., and Decsi, L. (1979). Anti-anxiety action of diazepam after intraamygdaloid application in the rat. *Neuropharmacology 18,* 573–76.

Nauta, W. J. H. (1971). The problem of the frontal lobe: A reinterpretation. *Journal of Psychiatric Research* 8, 167–87.

Nauta, W. J. H., and Karten, H. J. (1970). A general profile of the vertebrate brain, with sidelights on the ancestry of cerebral cortex. In *The neurosciences: Second study program,* F. O. Schmitt, ed. (New York: Rockefeller University Press), pp. 7–26.

Neisser, U. (1976). *Cognition and reality* (San Francisco: Freeman).

Neisser, U. (1967). *Cognitive psychology* (New York: Appleton-Century-Crofts).

Neisser, U., and Harsch, N. (1992). Phantom flashbulbs: False recollections of hearing the news about *Challenger.* In *Affect and accuracy in recall: Studies of "flashbulb" memories,* E. Winograd and U. Neisser, eds. (New York: Cambridge University Press).

Newcomer, J. W., Craft, S., Hershey, T., Askins, K., and Bardgett, M. E. (1994). Glucocorticoid-induced impairment in declarative memory performance in adult humans. *Journal of Neuroscience 14,* 2047–53.

Newell, A. (1980). Physical symbol systems. *Cognition 4,* 135–43.

Newell, A., Rosenbloom, P. S., and Laird, J. E. (1989). Symbolic architecture for cognition. In *Foundations of cognitive science,* M. Posner, ed. (Cambridge: MIT Press).

Newell, A., and Simon, H. (1972). *Human problem solving* (Boston: Little, Brown).

Newman, P. L. (1960). "Wild man" behavior in a New Guinea highlands community. *American Anthropologist 66,* 1–19.

Nicoll, R. A., and Malenka, R. C. (1995). Contrasting properties of two forms of long-term potentiation in the hippocampus. *Nature* 377, 115–18.

Nisbett, R. E., and Wilson, T. D. (1977). Telling more than we can know: Verbal reports on mental processes. *Psychological Review 84,* 231–59.

Norman, D. A., and Shallice, T. (1980). Attention to action: Willed and automatic control of behavior. In *Consciousness and self-regulation,* R. J. Davidson, G. E. Schwartz, and D. Shapiro, eds. (New York: Plenum).

Northcutt, R. G., and Kaas, J. H. (1995). The emergence and evolution of mammalian neocortex. *Trends in Neuroscience 18,* 373–79.

Nottebohm, F., Kasparian, S., and Pandazis, C. (1981). Brain space for a learned task. *Brain Research 213,* 99–109.

Oatley, K., and Duncan, E. (1994). The experience of emotions in everyday life. *Cognition and Emotion* 8, 369–81.

Öhman, A. (1992). Fear and anxiety as emotional phenomena: Clinical, phenomenological, evolutionary perspectives, and information-processing mechanisms. In *Handbook of the emotions,* M. Lewis and J. M. Haviland, eds. (New York: Guilford), pp. 511–36.

O'Keefe, J. (1976). Place units in the hippocampus of the freely moving rat. *Experimental Neurology* 51, 78–109.

O'Keefe, J. (1993). Hippocampus, theta, and spatial memory. *Current Opinion in Neurobiology* 3, 917–24.

O'Keefe, J., and Nadel, L. (1978). *The hippocampus as a cognitive map* (Oxford: Clarendon Press).

Olton, D., Becker, J. T., and Handleman, G. E. (1979). Hippocampus, space and memory. *Behavioral and Brain Sciences* 2, 313–65.

Ono, T., and Nishijo, H. (1992). Neurophysiological basis of the Klüver-Bucy syndrome: Responses of monkey amygdaloid neurons to biologically significant objects. In *The amygdala: Neurobiological aspects of emotion, memory, and mental dysfunction,* J. P. Aggleton, ed. (New York: Wiley-Liss), pp. 167–90.

Ortony, A., and Turner, T. J. (1990). What's basic about basic emotions? *Psychological Review* 97, 315–31.

Packard, V. (1957). *The hidden persuaders* (New York: D. M. McKay).

Panksepp, J. (1982). Toward a general psychobiological theory of emotions. *Behavioral and Brain Sciences* 5, 407–67.

Papez, J. W. (1937). A proposed mechanism of emotion. *Archives of Neurology and Psychiatry* 79, 217–24.

Parasuramna, R., and Martin, A. (1994). Cognition in Alzheimer's disease. *Current Opinion in Neurobiology* 4, 237–44.

Pavlides, C., Watanabe, Y., and McEwen, B. S. (1993). Effects of glucocorticoids on hippocampal long-term potentiation. *Hippocampus* 3, 183–192.

Pavlov, I. P. (1927). *Conditioned reflexes* (New York: Dover).

Peffiefer, J. (1955). *The human brain* (New York: Harper & Row).

Penrose, R. (1989). *The emperor's new mind: Concerning computers, minds, and the laws of physics* (New York: Penguin).

Peterson E. (1980). Behavioral studies of telencephalic function in reptiles. In: Ebbesson S. O. E., ed. *Comparative Neurology of the Telencephalon* (New York: Plenum Press), pp. 343–88.

Petrides, M. (1994). Frontal lobes and behaviour. *Current Opinion in Neurobiology* 4, 207–11.

Petrides, M., Alivsatos, B., Meyer, E., and Evans, A. C. (1993). Functional activation of the human frontal cortex during the performance of verbal working memory tasks. *Proceedings of the National Academy of Sciences USA 90*, 878–82.

Phillips, A. (1993). *On kissing, tickling, and being bored: Psychoanalytic essays on the unexamined life* (Cambridge: Harvard University Press).

Phillips, R. G., and LeDoux, J. E. (1992). Differential contribution of amygdala and hippocampus to cued and contextual fear conditioning. *Behavioral Neuroscience 106*, 274–85.

Picton, T. W., and Stuss, D. T. (1994). Neurobiology of conscious experience. *Current Opinion in Neurobiology 4*, 256–65.

Pinker, S. (1994). *The language instinct: How the mind creates language* (New York: Morrow).

Pinker, S. (1995). Language is a human instinct. In *The third culture*, J. Brockman, ed. (New York: Simon & Schuster).

Pitkänen, A., Stefanacci, L., Farb, C. R., Go, C.-G., LeDoux, J. E., and Amaral, D. G. (1995). Intrinsic connections of the rat amygdaloid complex: Projections originating in the lateral nucleus. *Journal of Comparative Neurology 356*, 288–310.

Plum, F., and Volpe, B. T. (1987). Neuroscience and higher brain function: From myth to public responsibility. In *Handbook of physiology. Section 1: The nervous system, Vol. 5: Higher Functions of the Brain*, F. Plum, ed. (Bethesda, MD: American Physiological Society).

Plutchik, R. (1980). Emotion: A psychoevolutionary synthesis (New York: Harper & Row).

Plutchik, R. (1993). Emotions and their vicissitudes: Emotions and psychopathology. In *Handbook of emotions*, M. Lewis and J. M. Haviland, eds. (New York: Guilford), pp. 53–65.

Posner, M. I. (1990). *Foundations of cognitive science* (Cambridge: MIT Press).

Posner, M. (1992). Attention as a cognitive and neural system. *Current Directions in Psychological Science 1*, 11–14.

Posner, M., and Petersen, S. (1990). The attention system of the human brain. *Annual Review of Neuroscience 13*, 25–42.

Posner, M., and Snyder, C. (1975). Facilitation and inhibition in the processing of signals. In *Attention and performance V*, P. Rabbitt and S. Domic, eds. (London: Academic Press).

Povinelli, D. J., and Preuss, T. M. (1995). Theory of mind: Evolutionary history of a cognitive specialization. *Trends in Neuroscience 18*, 418–24.

Powell, D. A., and Levine-Bryce, D. (1989). A comparison of two model systems of associative learning: Heart rate and eyeblink conditioning in the rabbit. *Psychophysiology 25*, 672–82.

Preuss, T. M. (1995). Do rats have prefrontal cortex? The Rose-Woolsey-Akert program reconsidered. *Journal of Cognitive Neuroscience* 7, 1–24.

Price, J. L., Russchen, F. T., and Amaral, D. G. (1987). The limbic region. II: The amygdaloid complex. In *Handbook of Chemical Neuroanatomy. Vol. 5: Integrated Systems of the CNS, Part 1,* A. Bjorklund, T. Hokfelt, and L. W. Swanson, eds. (Amsterdam: Elsevier), pp. 279–388.

Purves, D., White, L. E., and Andrews, T. J. (1994). Manual asymmetry and handedness. *Proceedings of the National Academy of Sciences USA 91,* 5030–32.

Putnam, H. (1960). Minds and machines. In *Dimensions of mind,* S. Hook, ed. (New York: Collier).

Pylyshyn, Z. (1984). *Computation and cognition: Toward a foundation for cognitive science* (Cambridge, MA: Bradford Books, MIT Press).

Quirk, G. J., Repa, J. C., and LeDoux, J. E. (1995). Fear conditioning enhances auditory short-latency responses of single units in the lateral nucleus of the amygdala: Simultaneous multichannel recordings in freely behaving rats. *Neuron 15,* 1029–39.

Reep, R. (1984). Relationship between prefrontal and limbic cortex: A comparative anatomical review. *Brain, Behavior and Evolution* 25, 5–80.

Reid, W. H. (1989). *The treatment of psychiatric disorders: Revised for the DSM-III-R* (New York: Brunner/Mazel).

Rogan, M. T., and LeDoux, J. E. (1995). LTP is accompanied by commensurate enhancement of auditory-evoked responses in a fear conditioning circuit. *Neuron 15,* 127–36.

Rolls, E. T. (1985). Connections, functions and dysfunctions of limbic structures, the prefrontal cortex, and hypothalamus. In *The scientific basis of clinical neurology,* M. Swash and C. Kennard, eds. (London: Churchill Livingstone), pp. 201–13.

Rolls, E. T. (1992a). Neurophysiological mechanisms underlying face processing within and beyond the temporal cortical visual areas. *Philosophical Transactions of the Royal Society, London, Series B, Biological Sciences* 335, 11–21.

Rolls, E. T. (1992b). Neurophysiology and functions of the primate amygdala. In *The amygdala: Neurobiological aspects of emotion, memory, and mental dysfunction,* J. P. Aggleton, ed. (New York: Wiley-Liss), pp. 143–65.

Rose, S. P. R. (1995). Glycoproteins and memory formation. *Behavioural Brain Research* 66, 73–78.

Rorty, A. O. (1980). Explaining emotions. In *Explaining emotions,* A. O. Rorty, ed. (Berkeley: University of California Press).

Rorty, R. (1979). *Philosophy and the mirror of nature* (Princeton: Princeton University Press).

Rosenzweig, M. (1996). Aspects of the search for neural mechanisms of memory. *Annual Review of Psychology 47,* 1–32.

Rudy, J. W., and Morledge, P. (1994). Ontogeny of contextual fear conditioning in rats: Implications for consolidation, infantile amnesia, and hippocampal system function. *Behavioral Neuroscience 108,* 227–34.

Rudy, J. W., and Sutherland, R. J. (1992). Configural and elemental associations and the memory coherence problem. *Journal of Cognitive Neuroscience 4(3),* 208–16.

Ruggiero, D. A., Gomez, R. E., Cravo, S. L., Mtui, E., Anwar, M., and Reis, D. J. (1991). The rostral ventrolateral medulla: Anatomical substrates of cardiopulmonary integration. In *Cardiorespiratory and motor coordination,* H.-P. Koepchen and T. Huopaniemi, eds. (New York: Springer), pp. 89–102.

Rumelhart, D. E. and McClelland, J. E. (1988). *Parallel Distributed Processing: Explorations in the Microstructure of Cognition.* (Cambridge: Bradford Books, MIT Press).

Russell, B. (1905). On denoting. *Mind 14,* 479–93.

Ryle, G. (1949). *The concept of mind* (New York: Barnes & Noble).

Sapolsky, R. M. (1990). Stress in the wild. *Scientific American 262,* 116–23.

Sarter, M. F., and Markowitsch, H. J. (1985). Involvement of the amygdala in learning and memory: A critical review, with emphasis on anatomical relations. *Behavioral Neuroscience 99,* 342–80.

Sartre, J.-P. (1943). *Being and nothingness* (New York: Philosophical Library).

Saucier, D. and Cain, D. P. (1995). Spatial learning without NMDA receptor-dependent long-term potentiation. *Nature 378,* 186–89.

Savander, V., Go, C. G., LeDoux, J. E., and Pitkänen, A. (1995). Intrinsic connections of the rat amygdaloid complex: Projections originating in the basal nucleus. *Journal of Comparative Neurology 361,* 345–68.

Schachter, S., and Singer, J. E. (1962). Cognitive, social, and physiological determinants of emotional state. *Psychological Review 69,* 379–99.

Schacter, D. L., and Graf, P. (1986). Effects of elaborative processing on implicit and explicit memory for new associations. *Journal of Experimental Psychology: Learning, Memory, and Cognition 12(3),* 432–44.

Scherer, K. R. (1984). On the nature and function of emotion: A component process approach. In *Approaches to emotion,* K. R. Scherer and P. Ekman, eds. (Hillsdale, NJ: Erlbaum), pp. 293–317.

Scherer, K. R. (1988). Criteria for emotion-antecedent appraisal: A review. In *Cognitive perspectives on emotion and motivation,* V. Hamilton, G. H.

Bower, and N. H. Frijda, eds. (Norwell, MA: Kluwer Academic Publishers), pp. 89–126.

Scherer, K. R. (1993a). Neuroscience projections to current debates in emotion psychology. *Cognition and Emotion* 7, 1–41.

Scherer, K. R. (1993b). Studying the emotion-antecedent appraisal process: An expert system approach. *Cognition and Emotion* 7, 325–55.

Schneiderman, N., Francis, J., Sampson, L. D., and Schwaber, J. S. (1974). CNS integration of learned cardiovascular behavior. In *Limbic and autonomic nervous system research,* L. V. DiCara, ed. (New York: Plenum), pp. 277–309.

Schwartz, B. E., Halgren, E., Fuster, J. M., Simpkins, E., Gee, M., and Mandelkern, M. (1995). Cortical metabolic activation in humans during a visual memory task. *Cerebral Cortex* 5.

Scoville, W. B., and Milner, B. (1957). Loss of recent memory after bilateral hippocampal lesions. *Journal of Neurology and Psychiatry 20,* 11–21.

Searle, J. (1984). *Minds, brains, science* (Cambridge: Harvard University Press).

Searle, J. (1992). *The rediscovery of the mind* (Cambridge: MIT Press).

Selden, N. R. W., Everitt, B. J., Jarrard, L. E., and Robbins, T. W. (1991). Complementary roles for the amygdala and hippocampus in aversive conditioning to explicit and contextual cues. *Neuroscience* 42(2), 335–50.

Seligman, M. E. P. (1971). Phobias and Preparedness. *Behavior Therapy* 2, 307–20.

Sengelaub, D. R. (1989). Cell generation, migration, death and growth in neural systems mediating social behavior. In *Advances in Comparative and Environmental Physiology* 3: *Molecular and Cellular Basis of Social Behavior in Vertebrates,* J. Balthazart, ed. (New York: Springer), pp. 239–67.

Servatius, R. J., and Shors, T. J. (1994). Exposure to inescapable stress persistently facilitates associative and nonassociative learning in rats. *Behavioral Neuroscience 108,* 1101–06.

Shalev, A. Y., Rogel-Fuchs, Y., and Pitman, R. K. (1992). Conditioned fear and psychological trauma. *Biological Psychiatry 31,* 863–65.

Shallice, T. (1988). Information processing models of consciousness. In *Consciousness in contemporary science,* A. Marcel and E. Bisiach, eds. (Oxford: Clarendon Press).

Shattuck, R. (1980). *The forbidden experiment* (New York: Farrar, Straus & Giroux).

Shepherd, G. (1983). *Neurobiology* (New York: Oxford University Press).

Sherry, D. F., Jacobs, L. F., and Gaulin, S. J. C. (1992). Spatial memory and adaptive specialization of the hippocampus. *Trends in Neuroscience 15,* 298–303.

Shevrin, H. (1992). Subliminal perception, memory, and consciousness: Cognitive and dynamic perspectives. In *Perception without awareness: Cognitive, clinical, and social perspectives,* R. F. Bornstein and T. S. Pittman, eds. (New York: Guilford), pp. 123–42.

Shevrin, H., Williams, W. J., Marshall, R. E., Hertel, R. K., Bond, J. A., and Brakel, L. A. (1992). Event-related potential indicators of the dynamic unconscious. *Consciousness and Cognition 1,* 340–66.

Shiffrin, M., and Schneider, W. (1977). Controlled and automatic human information processing: II. Perceptual learning, automatic attending, and a general theory. *Psychological Review 84,* 127–90.

Shimamura, A. (1995). Memory and frontal lobe function. In *The cognitive neurosciences,* M. S. Gazzaniga, ed. (Cambridge: MIT Press).

Shors, T. J., Foy, M. R., Levine, S., and Thompson, R. F. (1990). Unpredictable and uncontrollable stress impairs neuronal plasticity in the rat hippocampus. *Brain Research Bulletin 24,* 663–67.

Sibley, C. G., and Ahlquist, J. E. (1984). The phylogeny of the hominoid primates, as indicated by DNA-DNA hybridization. *Journal of Molecular Evolution 20,* 2–15.

Simon, H. A. (1967). Motivational and emotional controls of cognition. *Psychological Review 74,* 29–39.

Simpson, G. G. (1953). *The major features of evolution* (New York: Columbia University Press).

Simpson, G. G. (1967). *The meaning of evolution,* revised edition (New Haven: Yale University Press).

Skelton, R. W., Scarth, A. S., Wilkie, D. M., Miller, J. J., and Philips, G. (1987). Long-term increases in dentate granule cell responsivity accompany operant conditioning. *Journal of Neuroscience* 7, 3081–3087.

Skinner, B. F. (1938). *The behavior of organisms: An experimental analysis* (New York: Appleton-Century-Crofts).

Sloman, A. (1987). Motives, mechanisms and emotions. *Cognition and Emotion.* 1:217–33.

Sloman, A., and Croucher, M. (1981). Why robots will have emotions. In *Seventh Proceedings of the International Joint Conference on Artificial Intelligence* (Vancouver, British Columbia), pp. 197–202.

Smith, C. A., and Ellsworth, P. C. (1985). Patterns of cognitive appraisal in emotion. *Journal of Personality and Social Psychology* 56, 339–53.

Smith, C. A., and Lazarus, R. S. (1990). Emotion and adaptation. In *Handbook of personality: Theory and research,* L. A. Pervin, ed. (New York: Guilford), pp. 609–37.

Smith, J. M. (1958). *The theory of evolution* (Middlesex, England: Penguin).

Smith, O. A., Astley, C. A., Devito, J. L., Stein, J. M., and Walsh, R. E. (1980). Functional analysis of hypothalamic control of the cardiovascular responses accompanying emotional behavior. *Federation Proceedings 39(8),* 2487–94.

Smolensky, P. (1990). Connectionist modeling: Neural computation/mental connections. In *Neural connections, mental computation,* L. Nadel, L. Cooper, P. Culicover, and M. Harnish, eds. (Cambridge: MIT Press).

Solomon, R. C. (1993). The philosophy of emotions. In *Handbook of emotions,* M. Lewis and J. Haviland, eds. (New York: Guilford).

Sperry, R. W. (1969). A modified concept of consciousness. *Psychological Review* 76, 532–36.

Spinoza, B. (1955). *Works of Spinoza* (New York: Dover).

Squire, L. (1987). *Memory and the brain* (New York: Oxford University Press).

Squire, L. R., and Cohen, N. J. (1984). Human memory and amnesia. In *Neurobiology of learning and memory,* G. Lynch, J. L. McGaugh, and N. M. Weinberger, eds. (New York: Guilford).

Squire, L. R., Cohen, N. J., and Nadel, L. (1984). The medial temporal region and memory consolidation: A new hypothesis. In *Memory consolidation,* H. Eingartner and E. Parker, eds. (Hillsdale, NJ: Erlbaum).

Squire, L. R., and Davis, H. P. (1975). Cerebral protein synthesis inhibition and discrimination training: Effects of extent and duration of inhibition. *Behavioral Biology* 13, 49–57.

Squire, L. R., Knowlton, B., and Musen, G. (1993). The structure and organization of memory. *Annual Review of Psychology,* 44, 453–95.

Staübli, U. V. (1995). Parallel properties of long-term potentiation and memory. In *Brain and memory: Modulation and mediation of neuroplasticity,* J. L. McGaugh, N. M. Weinberger, and G. Lynch, eds. (New York: Oxford University Press), pp. 303–18.

Steinmetz, J. E., and Thompson, R. F. (1991). Brain substrates of aversive classical conditioning. In *Neurobiology of learning, emotion and affect,* J. I. Madden, ed. (New York: Raven), pp. 97–120.

Stuss, D. T. (1991). Self, awareness, and the frontal lobes: A neuropsychological perspective. In *The self: Interdisciplinary approaches,* J. Strauss and G. R. Goethals, eds. (New York: Springer).

Sutherland, R. J., and Rudy, J. W. (1989). Configural association theory: The role of the hippocampal formation in learning, memory, and amnesia. *Psychobiology* 17, 129–44.

Swanson, L. W. (1983). The hippocampus and the concept of the limbic system. In *Neurobiology of the hippocampus,* W. Seifert, ed. (London: Academic Press), pp. 3–19

Swanson, L. W. and Simmons, D. M. (1989). Differential steroid hormone and neural influences on peptide mRNA levels in CRH cells of the paraventricular nucleus: A hybridization histochemical study in the rat. *Journal of Comparative Neurology* 285, 413–35.

Swartz, B. E., Halgren, E., Fuster, J. M., Simpkins, E., Gee, M., and Mandelkern, M. (1995). Cortical metabolic activation in humans during a visual memory task. *Cerebral Cortex* 5, 205–14.

Tarr, R. S. (1977). Role of the amygdala in the intraspecies aggressive behavior of the iguanid lizard. *Physiology and Behavior 18,* 1153–58.

Teyler, T. J., and DiScenna, P. (1986). The hippocampal memory indexing theory. *Behavioral Neuroscience 100,* 147–54.

Thorndike, E. L. (1913). *The psychology of learning* (New York: Teachers College Press).

Thorpe, S. J., Rolls, E. T., and Maddison, S. (1983). The orbitofrontal cortex: Neuronal activity in the behaving monkey. *Experimental Brain Research 49,* 93–115.

Tolman, E. C. (1932). *Purposive behavior* (New York: Appleton-Century-Crofts).

Tomkins, S. S. (1962). Affect, imagery, consciousness (New York: Springer).

Tooby, J., and Cosmides, L. (1990). The past explains the present: Emotional adaptations and the structure of ancestral environments. *Ethological Sociobiology 11,* 375–424.

Tully, T. (1991). Genetic dissection of learning and memory in drosophila melanogaster. In *Neurobiology of learning, emotion and affect,* J. I. Madden, ed. (New York: Raven), pp. 29–66.

Tulving, E. (1983). *Elements of episodic memory* (New York: Oxford University Press).

Twain, M. (1962). *Letters from the earth: "From an unfinished burlesque of books on etiquette." Part 1, "At the funeral,"* B. DeVoto, ed. (New York: Harper and Row).

Ullman, S. (1984). Early processing of visual information. In *Handbook of cognitive neuroscience,* M. S. Gazzaniga, ed. (New York: Plenum).

Ungerleider, L. G., and Haxby, J. (1994). What and where in the human brain. *Current Opinion in Neurobiology* 4, 157–65.

Ungerleider, L. G., and Mishkin, M. (1982). Two cortical visual systems. In *Analysis of visual behavior,* D. J. Ingle, M. A. Goodale, and R. J. W. Mansfield, eds. (Cambridge: MIT Press), pp. 549–86.

Uno, H., Ross, T., Else, J., Suleman, M., and Sapolsky, R. (1989). Hippocampal damage associated with prolonged and fatal stress in primates. *Journal of Neuroscience* 9, 1705–11.

Uylings, H. B. M., and van Eden, C. G. (1990). Qualitative and quantitative comparison of the prefrontal cortex in rat and in primates, including humans. *Progress in Brain Research* 85, 31–62.

Valins, S. (1966). Cognitive effects of false heart-rate feedback. *Journal of Personality and Social Psychology* 4, 400–408.

van de Kar, L. D., Piechowski, R. A., Rittenhouse, P. A., and Gray, T. S. (1991). Amygdaloid lesions: Differential effect on conditioned stress and immobilization-induced increases in corticosterone and renin secretion. *Neuroendocrinology* 54, 89–95.

Van Essen, D. C. (1985). Functional organization of primate visual cortex. In *Cerebral cortex,* A. Peters and E. G. Jones, eds. (New York: Plenum), pp. 259–328.

Van Hoesen, G. W. (1982). The parahippocampal gyrus: New observations regarding its cortical connections in the monkey. *Trends in Neuroscience* 5, 345–50.

Velams, M. (1991). Is human information processing conscious? *Behavioral and Brain Sciences 14,* 651–726.

von Eckardt, B. (1993). *What is cognitive science?* (Cambridge: MIT Press).

von Uexkull, J. (1934). A stroll through the world of animals and man. In *Instinctive behavior: The development of a modern concept,* C. H. Chiller, ed. (London: Methuen).

Walter, W. G. (1953). *The living brain* (New York: Norton).

Warrington, E., and Weiskrantz, L. (1973). The effect of prior learning on subsequent retention in amnesic patients. *Neuropsychologia 20,* 233–48.

Watkins, L. R., and Mayer, D. J. (1982). Organization of endogenous opiate and nonopiate pain control systems. *Science 216,* 1185–92.

Watson, J. B. (1929). *Behaviorism* (New York: Norton).

Watson, J. B., and Rayner, R. (1920). Conditioned emotional reactions. *Journal of Experimental Psychology* 3, 1–14.

Weinberger, N. M. (1995). Retuning the brain by fear conditioning. In *The cognitive neurosciences,* M. S. Gazzaniga, ed. (Cambridge: MIT Press), pp. 1071–90.

Weiskrantz, L. (1956). Behavioral changes associated with ablation of the amygdaloid complex in monkeys. *Journal of Comparative Physiological Psychology* 49, 381–91.

Weiskrantz, L. (1988). Some contributions of neuropsychology of vision and memory to the problem of consciousness. In *Consciousness in contemporary science,* A. Marcel and E. Bisiach, eds. (Oxford: Clarendon Press).

Weiskrantz, L., and Warrington, E. (1979). Conditioning in amnesic patients. *Neuropsychologia 17,* 187–94

Weisz, D. J., Harden, D. G., and Xiang, Z. (1992). Effects of amygdala lesions on reflex facilitation and conditioned response acquisition during nictitating membrane response conditioning in rabbit. *Behavioral Neuroscience 106,* 262–73.

Wierzbicka, A. (1994). Emotion, language and cultural scripts. In S. Kitayama and H. R. Marcus, *Emotion and Culture* (Washington: American Psychological Association).

Wilcock, J., and Broadhurst, P. L. (1967). Strain differences in emotionality: Open-field and conditioned avoidance behavior in the rat. *Journal of Comparative Physiological Psychology* 63, 335–38.

Wilde, O. (1909). Gilbert. *Intentions* (New York: Lamb Publishing).

Williams, T. (1964). *The milk train doesn't stop here anymore* (Norfolk, CT: New Directions).

Wilson, D., Reeves, A., Gazzaniga, M. S., and Culver, C. (1977). Cerebral commissurotomy for the control of intractable epilepsy. *Neurology* 27, 708–15.

Wilson, F. A. W., O Scalaidhe, S. P., and Goldman-Rakic, P. S. (1993). Dissociation of object and spatial processing domains in primate prefrontal cortex. *Science 260,* 1955–58.

Wilson, M. A., and McNaughton, B. L. (1994). Reactivation of hippocampal ensemble memories during sleep. *Science* 265, 676–79.

Wilson, J. R. (1968). *The mind* (New York: Time-Life Books, Life Science Library).

Wolitsky, D. L., and Wachtel, P. L. (1973). *Personality and perception.* In B. B. Wolman (ed.) *Handbook of General Psychology* (Englewood, NJ: Prentice-Hall), pp. 826–57.

Wolkowitz, O., Reuss, V., and Weingartner, H. (1990). Cognitive effects of corticosteroids. *American Journal of Psychiatry* 147, 1297–1303.

Wolpe, J. (1988). Panic disorder: A product of classical conditioning. *Behavior Research and Therapy* 26, 441–50.

Wolpe, J., and Rachman, S. (1960). Psychoanalytic evidence: A critique of Freud's case of Little Hans. *Journal of Nervous and Mental Disease 130,* 198–220.

Yin, J. C. P., Wallach, J. S., Del Vecchio, M., Wilder, E. L., Zhou, H., Quinn, W. G., and Tully, T. (1994). Induction of a dominant-negative CREB transgene specifically blocks long-term memory in drosophila. *Cell 79,* 49–58.

Young, A. W., Aggleton, J. P., Hellawell, D. J., Johnson, M., Broks, P., and Hanley, J. R. (1995). Face processing impairments after amygdalotomy. *Brain 118,* 15–24.

Zajonc, R. (1980). Feeling and thinking: Preferences need no inferences. *American Psychologist* 35, 151–75.

Zajonc, R. B. (1984). On the primacy of affect. *American Psychologist* 39, 117–23.

Zola-Morgan, S., and Squire, L. R. (1993). Neuroanatomy of memory. *Annual Review of Neuroscience 16,* 547–63.

Zola-Morgan, S., Squire, L. R., Alvarez-Royo, P., and Clower, R. P. (1991). Independence of memory functions and emotional behavior: separate contributions of the hippocampal formation and the amygdala. *Hippocampus 1,* 207–20.

Zola-Morgan, S., Squire, L. R., and Amaral, D. G. (1986). Human amnesia and the medial temporal region: Enduring memory impairment following a bilateral lesion limited to field CA1 of the hippocampus. *Journal of Neuroscience 6(10),* 2950–67.

Zola-Morgan, S., Squire, L. R., and Amaral, D. G. (1989). Lesions of the amygdala that spare adjacent cortical regions do not impair memory or exacerbate the impairment following lesions of the hippocampal formation. *Journal of Neuroscience 9,* 1922–36.

Zuckerman, M. (1991). *Psychobiology of personality* (Cambridge: Cambridge University Press).

INDEX